Mathematics and Visualization

For further volumes:
http://www.springer.com/series/4562

David H. Laidlaw • Anna Vilanova
Editors

New Developments in the Visualization and Processing of Tensor Fields

With 133 Figures, 129 in color

Springer

Editors

David H. Laidlaw
Department of Computer Science
Brown University, Providence
Rhode Island, USA

Anna Vilanova
Department of Biomedical Engineering
Einhoven University of Technology
Einhoven, Netherlands

ISBN 978-3-642-27342-1 ISBN 978-3-642-27343-8 (eBook)
DOI 10.1007/978-3-642-27343-8
Springer Heidelberg New York Dordrecht London

Library of Congress Control Number: 2012947651

Mathematical Subject Classification (2010): 68-XX, 70-XX, 15-XX, 76-XX, 92-XX

Printed on acid-free paper

Springer is part of Springer Science+Business Media (www.springer.com)

We dedicate this book to our families.

Preface

In this book, we attempt to capture the excitement and inspiration that has been generated during a series of Dagstuhl Seminars that have looked at visualization and processing of tensor fields. This book includes contributions from attendees of the third meeting held in July 2009. As in the two earlier volumes, the authors report on recent research results as well as opining on future directions for the analysis and visualization of tensor fields. Topics range from applications of the analysis of tensor fields to purer research into their mathematical and analytical properties. One of the goals of this seminar was to bring together researchers from along the axis between pure and applied research, identifying new multidisciplinary research challenges. This book, we hope, will continue to further that goal in a broader context.

The manuscript is organized into seven parts. The parts, to some degree, increase in dimensionality and in mathematical sophistication. One change we implemented in this third in a series of Dagstuhl seminars is that it no longer restricts the underlying data models to second-order tensor fields. Research on representations that go beyond second-order tensors are clearly necessary, as became more evident during the seminars.

Part I, "Structure-Tensor Computation," focuses primarily on the generation of the so-called structure tensor to characterize image structure information. This information is often needed for feature preserving image processing techniques. The chapters present generalizations of the methodology of classic structure tensor estimation and its application beyond scalar fields.

Part II, "Tensor-Field Visualization," presents three results that extend our ability to generate, utilize, and visualize tensor fields. One looks at fabric like visualizations on surfaces. The second uses a Lagrangian metaphor to show the structure of 3D tensor fields. The third presents methods for designing tensor fields that have different applications in visualization such as realizing sampled texture fields.

Part III, "Applications of Tensor-Field Analysis and Visualization," offers contributions from researchers in disciplines where the application of analysis and visualization of second-order tensor fields is relevant. These applied domains include the simulation of combustion and the measurement of material elasticity using magnetic resonance imaging.

Part IV, "Diffusion Weighted MRI Visualization," breaks out one scientific area where second-order tensor fields are often applied. The chapters in this part show techniques for the specific study of diffusion tensor fields produced from magnetic resonance imaging, typically of the brain. These tensor-valued images capture properties of the brain white matter and, therefore, its structural connectivity. Brain connectivity is complex and difficult to understand. These chapters take a step in making that understanding easier. The last chapter especially focuses on uncertainty visualization of features derived from the tensor field, which is a recently emerging research topic.

Part V, "Beyond Second-Order Diffusion Tensor MRI," expands the complexity of the underlying data model used in the earlier chapters. Despite the challenges of working with second-order tensor valued images, the use of second-order tensors has limitations. In diffusion weighted imaging, it implies a model of the underlying anatomy and physics that is not sufficiently accurate in some situations. In this part, models that are more accurate or provide more information are presented. In diffusion weighted imaging, they also require more complex acquisition and processing. As the chapters in this part exemplify, some of the research in this field focuses on limiting the extra complexity.

In Part VI, "Tensor Metrics," some of the challenges of working with sampled tensor fields are addressed. Tensor fields are intrinsically defined over a continuous Euclidean space, and yet they must be represented for computation in some finite way. Typically, this is done by sampling them spatially, keeping only one tensor for each small region of an image. Furthermore, distance measures are needed for several common image processing techniques, e.g., interpolation, segmentation, and registration. However, the distance or metric between tensors is not uniquely defined. The three chapters in this part discuss some of the consequences of various choices for these distances.

Finally, in Part VII, "Tensor Analysis," two mathematical areas are presented that were judged to have relevance to researchers at this seminar, but which were not widely known to them in advance. These two chapters offer moderately deep introductions to the areas with the hope of stimulating cross-disciplinary collaborations.

We hope that these offerings will, overall or individually, inspire in readers some of the enthusiasm and energy that the seminar brought to its participants.

Providence, RI, USA — David H. Laidlaw
Eindhoven, The Netherlands — Anna Vilanova

Contents

Contributors

Carlos Alberola-López Laboratorio de Procesado de Imagen, Universidad de Valladolid, Valladolid, Spain

Mats Andersson Department of Biomedical Engineering, Center for Medical Image Science and Visualization (CMIV), Linköping University, Linköping, Sweden

Samer Barakat Purdue University, Lafayette, IN, USA

P. J. Basser The Eunice Kennedy Shriver National Institute of Child Health and Human Development (NICHD), National Institutes of Health, Bethesda, MD, USA

Steffen Börm Institute of Computer Science, Christian-Albrechts-Universität zu Kiel, Kiel, Germany

J. Braun Department of Radiology, Charité University Medicine, Charitéplatz, Berlin, Germany

Bernhard Burgeth Saarland University, Saarbrücken, Germany

Wei Chen Zhejiang University, Hangzhou, China

Çağatay Demiralp Brown University, Providence, RI, USA

Maxime Descoteaux Computer Science Department, Université de Sherbrooke, Sherbrooke, QC, Canada

Stephan Didas Abteilung Bildverarbeitung, Fraunhofer-Institut für Techno- und Wirtschaftsmathematik ITWM, Kaiserslautern, Germany

Ziáng Ding Zhejiang University, Hangzhou, China

Sebastian Eichelbaum Abteilung für Bild- und Signalverarbeitung, Institut für Informatik, Universität Leipzig, Leipzig, Germany

Miguel Angel Garcia Department of Electronic and Communications Technology, Autonomous University of Madrid, Madrid, Spain

Christoph Garth University of California at Davis, Davis, CA, USA University of Kaiserslautern, Kaiserslautern, Germany

Bart M. ter Haar Romeny Department of Biomedical Engineering, Eindhoven University of Technology, Eindhoven, The Netherlands

Hans Hagen University of Kaiserslautern, Kaiserslautern, Germany

Bernd Hamann Department of Computer Science, Institute for Data Analysis and Visualization (IDAV), University of California, Davis, CA, USA

Abedul Haque Department of Computer Science, University of Pittsburgh, Pittsburgh, PA, USA

Hans-Christian Hege Zuse Institute Berlin, Berlin, Germany

Mario Hlawitschka Institut für Informatik, Universität Leipzig, Leipzig, Germany University of California at Davis, Davis, CA, USA

Radu Jianu Brown University, Providence, RI, USA

D. Klatt Department of Radiology, Charité University Medicine, Chariteplatz, Berlin, Germany

Hans Knutsson Department of Biomedical Engineering, Center for Medical Image Science and Visualization (CMIV), Linköping University, Linköping, Sweden

David H. Laidlaw Brown University, Providence, RI, USA

Rodrigo de Luis-García Laboratory of Mathematics in Imaging, Harvard Medical School, Boston, MA, USA

G. Elisabeta Marai Department of Computer Science, University of Pittsburgh, Pittsburgh, PA, USA

Adrian Maries Department of Computer Science, University of Pittsburgh, Pittsburgh, PA, USA

M. Moakher Laboratory for Mathematical and Numerical Modeling in Engineering Science, National Engineering School at Tunis, University of Tunis El Manar, Tunis-Belvédère, Tunisia

Rodrigo Moreno Department of Medical and Health Sciences (IMH), Center for Medical Image Science and Visualization (CMIV), Linköping University, Linköping, Sweden

Mehdi B. Nik Department of Mechanical Engineering and Materials Science, University of Pittsburgh, Pittsburgh, PA, USA

S. Papazoglou Department of Radiology, Charité University Medicine, Chariteplatz, Berlin, Germany

O. Pasternak Brigham and Women's Hospital, Harvard Medical School, Boston, MA, USA

Luis Pizarro Department of Computing, Imperial College London, London, UK

Kai Pöthkow Zuse Institute Berlin, Berlin, Germany

Cyril Poupon NeuroSpin, CEA Saclay, Gif-sur-Yvette, France

Vesna Prčkovska Department of Biomedical Engineering, Eindhoven University of Technology, Eindhoven, The Netherlands

Domenec Puig Department of Computer Science and Mathematics, Rovira i Virgili University, Intelligent Robotics and Computer Vision Group (IRCV), Tarragona, Spain

I. Sack Department of Radiology, Charité University Medicine, Charitéplatz, Berlin, Germany

Gerik Scheuermann Abteilung für Bild- und Signalverarbeitung, Institut für Informatik, Universität Leipzig, Leipzig, Germany

Thomas Schultz University of Chicago, Computation Institute, Chicago, IL, USA

N. Sochen Department of Applied Mathematics, Tel Aviv University, Tel Aviv, Israel

Xavier Tricoche Purdue University, Lafayette, IN, USA

Anna Vilanova Department of Biomedical Engineering, Eindhoven University of Technology, Eindhoven, The Netherlands

Christian Wagner University of Kaiserslautern, Kaiserslautern, Germany

Joachim Weickert Saarland University, Mathematical Image Analysis Group (MIA), Saarbrücken, Germany

Carl-Fredrik Westin Laboratory of Mathematics in Imaging, Brigham and Women's Hospital, Harvard Medical School, Boston, MA, USA

Department of Biomedical Engineering, Linköping University, Linköping, Sweden

Ruiyi Wu Department of Computer Science and Engineering, Mississippi State University, Starkville, MS, USA

Xiaoyong Yang Department of Computer Science and Engineering, Mississippi State University, Starkville, MS, USA

S. Levent Yilmaz Center for Simulation and Modeling, University of Pittsburgh, Pittsburgh, PA, USA

Eugene Zhang School of Electrical Engineering and Computer Science, Oregon State University, Corvallis, OR, USA

Song Zhang Department of Computer Science and Engineering, Mississippi State University, Starkville, MS, USA

Part I
Structure-Tensor Computation

Structure Tensor Estimation: Introducing Monomial Quadrature Filter Sets

Hans Knutsson, Carl-Fredrik Westin, and Mats Andersson

Abstract Description and estimation of local spatial structure has a long history and numerous analysis tools have been developed. A concept that is widely recognized as fundamental in the analysis is the *structure tensor*. It has, however, a fairly broad and unspecific meaning. This chapter is intended to provide a framework for displaying the differences and similarities of existing structure estimation approaches. A new method for structure tensor estimation, which is a generalization of many of it's predecessors, is presented. The method uses pairs of filter sets having Fourier directional responses in the form of monomials, one odd order set and one even order set. It is shown that such filter sets allow for a particularly simple way of attaining phase invariant, positive semi-definite, local structure tensor estimates. In addition, we show that the chosen filter sets directly links order, scale and the gradient operator. We continue to compare a number of known structure tensor algorithms by formulating them in monomial filter set terms. In conclusion we show how higher order tensors can be estimated using a generalization of the same simple formulation.

H. Knutsson (✉) · M. Andersson
Department of Biomedical Engineering, Linköping University, Linköping, Sweden

Center for Medical Image Science and Visualization (CMIV), Linköping, Sweden
e-mail: knutte@imt.liu.se; matsa@imt.liu.se

C.-F. Westin
Laboratory of Mathematics in Imaging, Brigham and Women's Hospital, Harvard Medical School, Boston, MA, USA

Department of Biomedical Engineering, Linköping University, Linköping, Sweden
e-mail: westin@bwh.harvard.edu

D.H. Laidlaw and A. Vilanova (eds.), *New Developments in the Visualization and Processing of Tensor Fields*, Mathematics and Visualization,
DOI 10.1007/978-3-642-27343-8_1,

1 Introduction

Many of the popular image analysis concepts of today have roots that can be traced to early work in signal processing and optics, e.g. Riesz transforms [64], Zernike moments [9, 78], and Gabor signals [28]. The first steps towards analysis of digital images were taken more than four decades ago [12, 34, 36, 65].

From the very start detecting edges and lines in images was considered a fundamental operation [36, 65]. Since these early days new and more advanced schemes for analysis of local image structure has been suggested in a seemingly never ending stream [1–7, 10, 11, 13–27, 29–33, 37–50, 52–58, 60–63, 66–71, 73–77, 79]. Local image orientation, scale, frequency, phase, motion and locality of estimates are prominent examples of features that have been considered central in the analysis.

Apart from sheer curiosity, the main force driving the research has been the need to analyze data produced by increasingly capable imaging devices. Presently produced data are also often intrinsically more complex. Both the outer and the inner dimensionality can be higher, e.g. volume sequence data and tensor field data respectively.

Regardless of this development the first stages in the analysis remain the same. In most cases the processing starts by performing local linear combinations of image values, e.g. convolution operators. Perhaps somewhat surprising after four decades of research the design of these filters is still debated. In fact the object of this paper is to contribute to this discussion in a way that hopefully will help in bringing it closer to and end.

In particular we will focus on developments relating to the local structure tensor concept [40, 41]. There is potentially an almost unlimited amount of work that could be mentioned and we do by no means claim to present a comprehensive presentation. Some early work on corner and linear symmetry estimation has later been associated with the local structure tensor, although the concept of a tensor was not discussed in the original publications [10, 27]. The earliest use of matrices relating to the local structure tensor is perhaps to be found in the estimation of optical flow [33, 57]. Which contributions that are viewed as relevant in the present context is, in the end, rather subjective. We believe, however, that we present an interesting account of the developments.

1.1 This Chapter Presents

- A new general framework for structure tensor estimation.
  - Introduction of monomial quadrature filter matrices

  - Mathematical derivation displaying the link between scale, order and the gradient operator that exists for lognormal monomial filters.
  - Notation for simple formulation of structure tensor estimation

- Formulations of most of the previously developed structure tensor estimation approaches in terms of the new framework.
- A simple generalization for producing higher order structure tensor estimates.

2 Monomial Filters

All structure tensor estimation approaches are based on the use of a set of filters onto which each local neigborhood is projected. The design of these filters directly determines crucial aspects of the performance of the estimator. The filters should provide an appropriate basis for representing the targeted features of the signal. An important aspect, not further discussed here, is the locality of the estimates, see [44].

In structure tensor estimation the central feature is the distribution of the local signal orientations. A natural requirement is that the estimate directly reflects rotations of the neigborhood, i.e. the estimate should be equivariant with rotation, but be invariant to other transformations, e.g. change of scale. The locality of the estimates is Here we present a class of filters that designed to meet these needs—*monomial filters.* The monomial filters are spherically separable, i.e. defined as a product of one radial and one directional part:

$$F(\boldsymbol{\mu}) = R(\rho)\, D(\hat{\boldsymbol{\mu}}) \tag{1}$$

where $\boldsymbol{\mu}$ is the Fourier domain (FD) coordinate and $\rho = \|\boldsymbol{\mu}\|$.

2.1 *Radial Part*

The radial part is a bandpass (BP) filter ($R(0) = 0$) and the lognormal function is our preferred choice.

$$R(\rho) = \exp\left[-\alpha \ln^2\left(\frac{\rho}{\rho_0}\right)\right] \tag{2}$$

Here ρ_0 is the center frequency and the relative bandwidth is determined by α. The relative bandwidth of the lognormal filter function is constant for varying center frequency. This provides a unique feature relating gradients and scale which will be addressed in Sect. 3.

2.2 Directional Matrix

The directional part consists of monomials i.e. products of non-negative integer powers of the components of $\hat{\boldsymbol{\mu}}$. Performing n repeated outer products of $\hat{\boldsymbol{\mu}}$ will contain all order n component products.

$$\hat{\boldsymbol{\mu}}^{\otimes n} = \underbrace{\hat{\boldsymbol{\mu}} \otimes \hat{\boldsymbol{\mu}} \ldots \otimes \hat{\boldsymbol{\mu}}}_{n \text{ entities}} \tag{3}$$

For convenience, rearrange the terms such that the directional part, $\boldsymbol{D}_n(\hat{\boldsymbol{\mu}})$, becomes a matrix:

$$\boldsymbol{D}_n(\hat{\boldsymbol{\mu}}) = \hat{\boldsymbol{\mu}} \left\lfloor \hat{\boldsymbol{\mu}}^{\otimes(n-1)} \right\rfloor^T \tag{4}$$

Here the " $\lfloor\ \rfloor$ " notation implies a *lineup* operation which arrange the elements of a multi-dimensional array into a lexicographic ordered column vector. The motivation for introducing this notation is that letting $\boldsymbol{D}_n$ contain the elements of $\hat{\boldsymbol{\mu}}^{\otimes n}$ arranged as a matrix greatly simplifies the equations needed in the following analysis.

To handle the special cases of $n = 1$ and $n = 0$ we introduce the following definitions:

$$\hat{\boldsymbol{\mu}} \left\lfloor \hat{\boldsymbol{\mu}}^{\otimes(-1)} \right\rfloor \equiv \hat{\boldsymbol{\mu}}^{\otimes 0} \equiv \boldsymbol{I} \tag{5}$$

where $\boldsymbol{I}$ is the identity matrix. The need to introduce this definitions correspond to the fact that order 0 implies a scalar entity that does not carry orientation information and thus constitutes a special case, i.e. $\boldsymbol{D}_0(\hat{\boldsymbol{\mu}}) = \boldsymbol{I}$. It is also worth noting here that $\boldsymbol{D}_1(\hat{\boldsymbol{\mu}})$ corresponds to the Hilbert transform (i.e. sign($\boldsymbol{\mu}$) in the 1-dimensional case and the Riesz transform for higher dimensions [64].

Using the notation $\hat{\boldsymbol{\mu}} = (\mu, \nu)^T$ Eqs. (6)–(8) give $\boldsymbol{D}_1(\hat{\boldsymbol{\mu}})$, $\boldsymbol{D}_2(\hat{\boldsymbol{\mu}})$ and $\boldsymbol{D}_3(\hat{\boldsymbol{\mu}})$ for the 2D case.

$$\boldsymbol{D}_1(\hat{\boldsymbol{\mu}}) = \begin{pmatrix} \mu \\ \nu \end{pmatrix} \left\lfloor \hat{\boldsymbol{\mu}}^{\otimes 0} \right\rfloor = \begin{pmatrix} \mu \\ \nu \end{pmatrix} \tag{6}$$

ex

$$\begin{aligned} \boldsymbol{D}_2(\hat{\boldsymbol{\mu}}) &= \begin{pmatrix} \mu \\ \nu \end{pmatrix} \left\lfloor \begin{pmatrix} \mu \\ \nu \end{pmatrix} \right\rfloor^T \\ &= \begin{pmatrix} \mu \\ \nu \end{pmatrix} \begin{pmatrix} \mu & \nu \end{pmatrix} \\ &= \begin{pmatrix} \mu^2 & \mu\nu \\ \mu\nu & \nu^2 \end{pmatrix} \end{aligned} \tag{7}$$

Table 1 Table giving the number of different filters needed for an order n monomial filter matrix in dimension d. The cases marked green are shown in Figs. 1–3

Dimension \ Order	0	1	2	3	4	5	6
1	1	1	1	1	1	1	1
2	1	2	3	4	5	6	7
3	1	3	6	10	15	21	28
4	1	4	10	20	35	56	84
5	1	5	15	35	70	126	210

$$
\begin{aligned}
\boldsymbol{D}_3(\hat{\boldsymbol{\mu}}) &= \begin{pmatrix}\mu\\ \nu\end{pmatrix}\left\lfloor \begin{pmatrix}\mu\\ \nu\end{pmatrix}^{\otimes 2}\right\rfloor^T \\
&= \begin{pmatrix}\mu\\ \nu\end{pmatrix}\left\lfloor \begin{matrix}\mu^2 & \mu\nu\\ \mu\nu & \nu^2\end{matrix}\right\rfloor^T \\
&= \begin{pmatrix}\mu\\ \nu\end{pmatrix}\left(\mu^2 \;\; \mu\nu \;\; \mu\nu \;\; \nu^2\right) \\
&= \begin{pmatrix}\mu^3 & \mu^2\nu & \mu^2\nu & \mu\nu^2\\ \mu^2\nu & \mu\nu^2 & \mu\nu^2 & \nu^3\end{pmatrix}
\end{aligned} \tag{8}
$$

For clarity of the presentation most examples in this paper are given in 2D. The proposed concept is, however, valid for any signal dimension. Note that, N_C, the number of different directional matrix components needed grows, in a standard combinatorial fashion, rather quickly with increasing order and dimensionality. The number of components needed for order n and dimensionality d is given by Eq. (9) and shown in Table 1 for a number of cases.

$$
N_C = \binom{n+d-1}{n} = \frac{(n+d-1)!}{n!\,(d-1)!} \tag{9}
$$

2.3 Monomial Filter Matrices

The final step in attaining the filter matrices used in the local structure tensor estimation is simply to multiply the matrix holding the directional responses, $\boldsymbol{D}_n(\hat{\boldsymbol{\mu}})$, with the radial function, $R(\rho)$. For each order $n \geq 0$ a *monomial filter matrix* is defined as:

$$
\boldsymbol{F}_n(\mu) = R(\rho)\;\; \boldsymbol{D}_n(\hat{\boldsymbol{\mu}}) \tag{10}
$$

Figure 1 shows the 2D monomial filter components of order 0 and 1. Figure 2 shows the 2D monomial filter components of order 2. Figure 3 shows 2D monomial

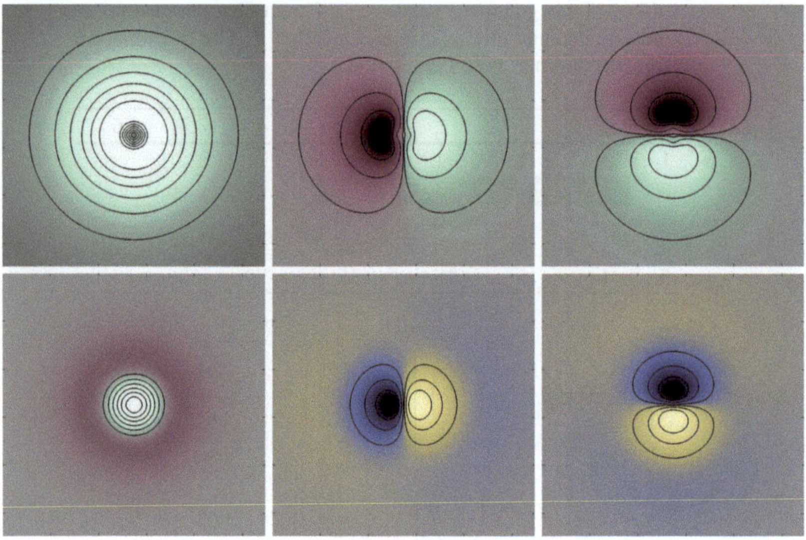

Fig. 1 *Top*: Fourier domain images of $F_0 = R(\rho)$, $F_1(1) = R(\rho)\mu$ and $F_1(2) = R(\rho)\nu$. High values are *bright* and low values are *dark*. *Green* indicates positive real values and *red* indicates negative real values. *Bottom*: Spatial domain images of the corresponding filters in the *top row*. *Yellow* indicates positive imaginary values and *blue* indicates negative imaginary values. The *black* contours are iso-level lines

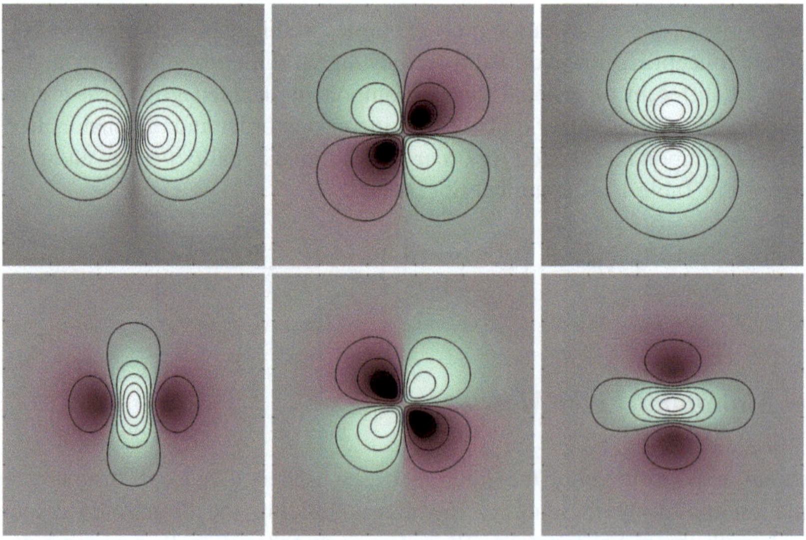

Fig. 2 *Top*: Fourier domain images of $F_2(1,1) = R(\rho)\mu^2$, $F_2(1,2) = R(\rho)\mu\nu$ and $F_2(2,2) = R(\rho)\nu^2$. High values are *bright* and low values are *dark*. *Green* indicates positive real values and *red* indicates negative real values. *Bottom*: Spatial domain images of the corresponding filters in the *top row*. The *black* contours are iso-level lines

Fig. 3 *Top*: Fourier domain images of $F_3(1,1) = R(\rho)\mu^3$, $F_3(1,2) = R(\rho)\mu^2\nu$, $F_3(2,2) = R(\rho)\mu\nu^2$ and $F_3(2,3) = R(\rho)\nu^3$. High values are *bright* and low values are *dark*. *Green* indicates positive real values and *red* indicates negative real values. *Bottom*: Spatial domain images of the corresponding filters in the *top row*. *Yellow* indicates positive imaginary values and *blue* indicates negative imaginary values. The *black* contours are iso-level lines

filter components of order 3. The radial function ($R(\rho)$) used here is a lognormal function having a bandwidth of two octaves, see [30]. Notice that, keeping the radial function fixed, the spatial extent of the filters increase with increasing order thus decreasing spatial locality.

3 Monomials Link Order, Scale and Gradients

Adding an index, j, representing scale we can write Eq. (10) as:

$$\begin{aligned} \boldsymbol{F}_{n,j}(\boldsymbol{u}) &= R_j(\rho)\ \ \boldsymbol{D}_n(\hat{\boldsymbol{u}}) \\ &= R_j(\rho)\ \ \hat{\boldsymbol{u}}\ \left\lfloor \hat{\boldsymbol{u}}^{\otimes(n-1)} \right\rfloor^T \end{aligned} \tag{11}$$

In the Fourier domain the spatial gradient operator corresponds to a multiplication with the frequency coordinate, i.e.

$$\nabla \mathbf{f}_{n,j}(\mathbf{x}) \ \stackrel{\mathscr{F}}{\longleftrightarrow}\ \boldsymbol{u} \otimes \boldsymbol{F}_{n,j}(\boldsymbol{u}) \tag{12}$$

Expanding the right hand side yields:

$$\boldsymbol{u} \otimes \boldsymbol{F}_{n,j}(\boldsymbol{u}) = R_j(\rho)\ \underbrace{\rho\,\hat{\boldsymbol{u}}}_{\boldsymbol{u}}\ \otimes \hat{\boldsymbol{u}}\ \left\lfloor \hat{\boldsymbol{u}}^{\otimes(n-1)} \right\rfloor^T \tag{13}$$

Rearranging terms yields

$$\begin{aligned} \boldsymbol{u} \otimes \boldsymbol{F}_{n,j}(\boldsymbol{u}) &= \rho\, R_j(\rho)\ \hat{\boldsymbol{u}}\ \left\lfloor \hat{\boldsymbol{u}} \otimes \hat{\boldsymbol{u}}^{\otimes(n-1)} \right\rfloor^T \\ &= \rho\, R_j(\rho)\ \hat{\boldsymbol{u}}\ \left\lfloor \hat{\boldsymbol{u}}^{\otimes n} \right\rfloor^T \end{aligned} \tag{14}$$

This shows that the effect of the gradient operator on the directional part, $\boldsymbol{D}$, can be directly identified as an increment in the order of $\boldsymbol{D}$.

The Lognormal Radial Function

In Eqs. (11)–(14) the link between filter order and gradient operator were established. We will now study the effect of the gradient operator when the radial function, $R_j(\rho)$ in Eq. (11), is a lognormal function.

$$R_j(\rho) = \exp\left[-\alpha\ \ln^2\left(\tfrac{\rho}{\rho_j}\right)\right] \tag{15}$$

where ρ_j defines the center frequency and α determines the relative bandwidth of $R_j(\rho)$. We will now show that for this class of functions the effect of the gradient operator on the radial part corresponds to a shift in scale and a change in magnitude. Examining the radial part of Eq. (14) yields:

$$\begin{aligned} \rho\, R_j(\rho) &= \rho\ \exp\left[-\alpha\ \ln^2(\tfrac{\rho}{\rho_j})\right] \\ &= \exp\left[-\alpha\left((\ln\rho - \ln\rho_j)^2 - \tfrac{\ln\rho}{\alpha} \right)\right] \\ &= \exp\left[-\alpha\left(\ln^2\rho + \ln^2\rho_j - 2\ \ln\rho\,(\ln\rho_j + \tfrac{1}{2\alpha})\right)\right] \end{aligned} \tag{16}$$

To simplify the derivation we realize that this is a 2nd order polynomial in $\ln\rho$ and we can directly identify the new center frequency by inspection, i.e. $\ln\rho_{j+1} = \ln\rho_j + \frac{1}{2\alpha}$. Inserting this into Eq. (16) yields:

$$\begin{aligned} \rho R_j(\rho) &= \exp\left[-\alpha\left(\ln^2\rho + (\ln^2\rho_{j+1} - \frac{\ln\rho_{j+1}}{\alpha} + \frac{1}{4\alpha^2}) - 2\ln\rho\,\ln\rho_{j+1}\right)\right] \\ &= \exp\left[-\alpha\left(\ln^2\rho + \ln^2\rho_{j+1} - \frac{\ln\rho_j}{\alpha} - \frac{1}{4\,\alpha^2} - 2\ln\rho\ \ln\rho_{j+1}\right)\right] \\ &= \rho_j\ \exp(\frac{1}{4\,\alpha})\ \exp\left[-\alpha\ \ln^2(\frac{\rho}{\rho_{j+1}})\right] \end{aligned} \tag{17}$$

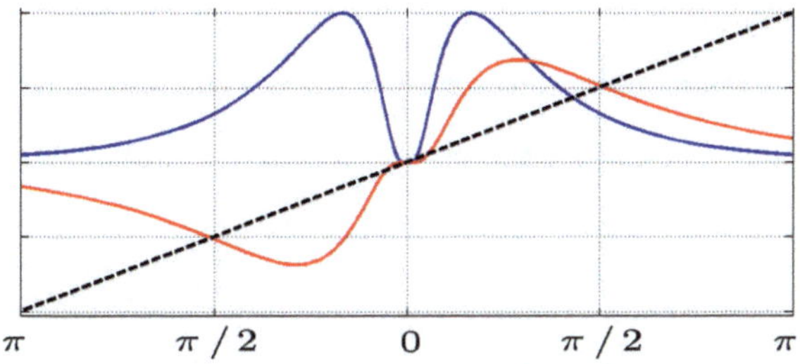

Fig. 4 A 1D example showing the linking of scale, order and gradients for a lognormal filter $F(u)$. Multiplying the lognormal filter (*blue*) with u (*dotted black*) results in a new lognormal filter function (*red*) with the same relative bandwidth at the next scale. Note that the new filter is odd. This corresponds the change in order caused by the gradient operator since, in 1D, only two 'orders' exist, i.e. odd and even

$$= \underbrace{\rho_j \ \exp(\frac{1}{4\alpha})}_{a} \ R_{j+1}(\rho)$$

$$= a \ R_{j+1}(\rho)$$

This shows that a multiplication of the radial filter function $R_j(\rho)$ by ρ results in a new lognormal function, $R_{j+1}(\rho)$, having the same relative bandwidth but scaled in size and amplitude. This unique feature of the lognormal filter function is powerful for scale space applications as implied by Eq. (18).The right hand side of Fig. 4 shows a 1D example of this relation.

3.1 The Gradient Operator Increases Order and Shifts Scale

To summarize, the effect of the gradient operator is, in the frequency domain, given by:

$$\begin{aligned} \boldsymbol{u} \otimes \boldsymbol{F}_{n,j}(\boldsymbol{u}) &= a \ R_{(j+1)}(\rho) \ \ \boldsymbol{D}_{(n+1)}(\hat{\boldsymbol{u}}) \\ &= a \ \boldsymbol{F}_{(n+1),(j+1)}(\boldsymbol{u}) \end{aligned} \tag{18}$$

In the spatial domain this corresponds to:

$$\nabla \mathbf{f}_{n,j}(\mathbf{x}) = a \ \mathbf{f}_{(n+1),(j+1)}(\mathbf{x}) \tag{19}$$

The above equations clearly show that the gradient operator increments the order by one and changes the scale by a constant factor, ρ_∇. The factor is determined by the bandwidth of the lognormal function used, $\rho_\nabla = \exp(1/2\alpha)$, see Eqs. (16) and (17). Figure 5 (and 4) illustrates this relation.

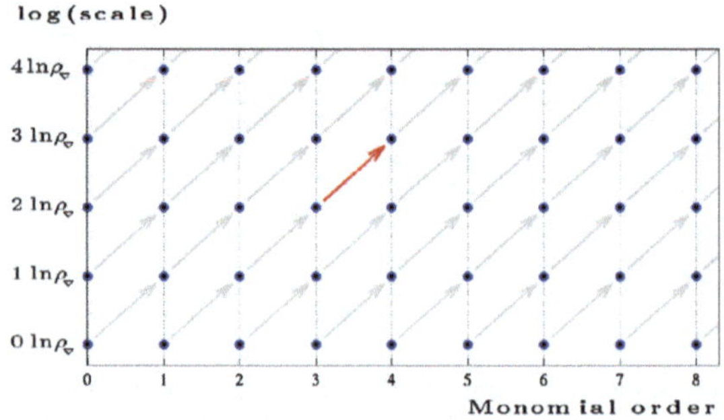

Fig. 5 Figure showing the fundamental linking of scale, order and the gradient operator that is provided by lognormal monomial filter sets. The gradient operator increases the order by one and shifts the center frequency towards higher frequency by a constant factor ρ_∇

4 Monomial Filter Response Matrices

The next step is to apply a monomial filter matrix to a signal, thus obtaining a *monomial filter response matrix*. This can be done by convolving the signal with each of the filters in the filter matrix and storing the results in the corresponding positions. Using an FFT approach the same result can also be obtained multiplying the Fourier transform of the signal by each filter in the Fourier domain.

Let the spatial domain (SD) correspondence of the monomial filter matrix $\boldsymbol{F}_n$ be denoted $\mathbf{F}_n$. Each element of $\mathbf{F}_n$ contains the convolution kernel of the corresponding FD filter function in $\boldsymbol{F}_n$. If the multi-dimensional signal is denoted $\mathsf{s}(\mathbf{x})$ where $\mathbf{x}$ denotes the SD coordinates the monomial filter response matrix, $\mathbf{Q}_n(\mathbf{x})$, is defined as:

$$\mathbf{Q}_n(\mathbf{x}) = \mathbf{F}_n(\mathbf{x}) * \mathsf{s}(\mathbf{x}) \tag{20}$$

where $*$ denotes the convolution operator.

Denoting the Fourier transform of s by $\mathsf{S}_\mathbf{x}$ the same relation is, in the Fourier domain, expressed as:

$$\mathbf{Q}_n(\mathbf{x}) = \mathscr{F}^{-1}\left[\sum_{\Omega_F} \boldsymbol{F}_n(\mu)\ \mathsf{S}_\mathbf{x}(\mu)\right] \tag{21}$$

Here $\mathscr{F}^{-1}$ denotes the inverse Fourier transform and Ω_F defines the Fourier coordinates to sum over.

In this general description each element of $\mathbf{Q}_n(\mathbf{x})$ contains the monomial filter responses for the entire signal. Since all filtering operations in this paper are shift invariant we may, in the interest of clarity, from now on omit to denote the spatial coordinate vector $\mathbf{x}$ and, when doing so, consider each element of $\mathbf{Q}_n$ to contain monomial filter responses for any given spatial coordinate.

4.1 Signal Classes

For clarity it is useful to define different classes of signals. We will here define three different signal classes: sinusoidal, simple and rank p signals.

4.1.1 Sinusoidal Signals

We first present the simplest possible case, a sinusoidal signal with amplitude a, spatial frequency $\boldsymbol{u}$, and zero phase.

$$\mathsf{s}(\mathbf{x}) = \mathsf{a}\cos(\boldsymbol{u}^T\mathbf{x}) \tag{22}$$

For this case the monomial filter response matrix can be described in FD terms as:

$$\mathbf{Q}_n(\boldsymbol{u}) = \ \mathsf{a}\ R(\rho)\ \boldsymbol{D}_n(\hat{\boldsymbol{u}}) \tag{23}$$

Note that in this special case of zero phase, i.e. a symmetric signal, the response will be zero for odd n. For an odd signal, i.e.

$$\mathsf{s}(\mathbf{x}) = \mathsf{a}\sin(\boldsymbol{u}^T\mathbf{x}) \tag{24}$$

even orders will be zero and odd order responses are given by:

$$\mathbf{Q}_n(\boldsymbol{u}) = \ -i\ \mathsf{a}\ R(\rho)\ \boldsymbol{D}_n(\hat{\boldsymbol{u}}) \tag{25}$$

For a general sinusoidal with phase θ, i.e.

$$\mathsf{s}(\mathbf{x}) = \mathsf{a}\cos(\boldsymbol{u}^T\mathbf{x} + \theta) \tag{26}$$

both even and odd order filters will respond and we get:

$$\mathbf{Q}_n = \begin{cases} \mathsf{a}\ \cos(\theta)\ R(\rho)\ \boldsymbol{D}_n(\hat{\boldsymbol{u}}) & \text{for even } n \\ -i\mathsf{a}\ \sin(\theta)\ R(\rho)\ \boldsymbol{D}_n(\hat{\boldsymbol{u}}) & \text{for odd } n \end{cases} \tag{27}$$

4.1.2 Simple Signals

Following [30] we define signals that can be expressed by Eq. (28) to be termed simple signals.

$$\mathsf{s}(\mathbf{x}) = \mathsf{g}(\hat{\boldsymbol{u}}^T\mathbf{x}) \tag{28}$$

Fig. 6 Two simple neighborhoods in 3D. The neighborhoods are constructed using two different signal functions ($\mathsf{g}_1(\hat{\boldsymbol{u}}^T\mathbf{x})$ and $\mathsf{g}_2(\hat{\boldsymbol{u}}^T\mathbf{x})$) but the same signal orienting vector ($\hat{\boldsymbol{u}}$)

Here $\mathsf{g}(\mathrm{x})$ is any real, one variable, function and $\mathrm{x} = \hat{\boldsymbol{u}}^T\mathbf{x}$. $\hat{\boldsymbol{u}}$ is a unit vector giving the orientation of the signal. For this case the monomial filter response matrix can be described in FD terms as:

$$\begin{aligned} \mathsf{Q}_n(\hat{\boldsymbol{u}}) &= \mathsf{A}_n\ \boldsymbol{D}_n(\hat{\boldsymbol{u}}) \\ &= \mathsf{A}_n\ \hat{\boldsymbol{u}}\ \left\lfloor \hat{\boldsymbol{u}}^{\otimes(n-1)} \right\rfloor^T \end{aligned} \tag{29}$$

Here A, the local amplitude of the filter response, is a function of the radial filter function, $R(\rho)$, and the signal generating function, $\mathsf{g}(\mathrm{x})$ The fact that the Fourier transform of a simple signal is non-zero only on a line through the origin makes for a simple solution. Denoting the Fourier transform of $\mathsf{g}(\mathrm{x})$ by $\mathsf{G}(u)$ we find the filter response amplitude as:

$$\begin{cases} \mathsf{A}_n = \mathsf{A}_\mathsf{e} = \int R(|u|)\,\mathsf{G}(u)\,du & \text{for even } n \\ \mathsf{A}_n = \mathsf{A}_\mathsf{o} = \int R(|u|)\,\mathsf{G}(u)\,\mathrm{sign}(u)\,du & \text{for odd } n \end{cases} \tag{30}$$

Unless explicitly mentioned all signals will in the following be regarded simple (Fig. 6).

4.1.3 Rank *p* Signals

It is straight forward to classify more complex signals in a similar manner. Let $\hat{\boldsymbol{U}}$ be a projection operator of rank p and g be a real function of p variables, then

$$\mathsf{s}(\mathbf{x}) = \mathsf{g}(\hat{\boldsymbol{U}}\mathbf{x}) \tag{31}$$

is a rank p signal. Thus, a simple signal is a rank one signal. A full rank signal corresponds to $\hat{\boldsymbol{U}} = \hat{\boldsymbol{I}}$. In the following we will only distinguish between sinusoidal, simple and non-simple signals.

5 Monomial Structure Tensors

The next step towards obtaining a structure tensor is to compute the outer product of the filter matrix.

$$\mathbf{T}_n^2 = \mathbf{Q}_n \, \mathbf{Q}_n^T \tag{32}$$

where " ^{T} " denotes complex conjugate transpose.

5.1 Two Simple Examples

We will present the case where the local neighborhood of the image consists of a simple signal with direction $\hat{\boldsymbol{u}}$. As simple introductory examples we consider 2-dimensional monomial filters of orders 2 and 3. For order 2 we have:

$$\mathbf{Q}_2 = \mathsf{A}_\mathsf{e} \begin{pmatrix} u^2 & uv \\ uv & v^2 \end{pmatrix} \tag{33}$$

Carrying out the sums we get:

$$\begin{aligned} \mathbf{T}_2 &= \mathbf{Q}_2 \, \mathbf{Q}_2^T \\ &= \mathsf{A}_\mathsf{e}^2 \begin{pmatrix} u^2 & uv \\ uv & v^2 \end{pmatrix} \begin{pmatrix} u^2 & uv \\ uv & v^2 \end{pmatrix} \\ &= \mathsf{A}_\mathsf{e}^2 \underbrace{(u^2 + v^2)}_{=1} \begin{pmatrix} u^2 & uv \\ uv & v^2 \end{pmatrix} \\ &= \mathsf{A}_\mathsf{e}^2 \begin{pmatrix} u^2 & uv \\ uv & v^2 \end{pmatrix} \end{aligned} \tag{34}$$

And for order 3 monomial filters we obtain:

$$\mathbf{Q}_3 = \mathsf{A}_\mathsf{o} \begin{pmatrix} u^3 & u^2v & u^2v & uv^2 \\ u^2v & uv^2 & uv^2 & v^3 \end{pmatrix} \tag{35}$$

Carrying out the sums we get:

$$\mathbf{T}_3 = \mathbf{Q}_3 \, \mathbf{Q}_3^T$$

$$
\begin{aligned}
&= \mathsf{A}_o \mathsf{A}_o^* \begin{pmatrix} u^3 & u^2 v & u^2 v & uv^2 \\ u^2 v & uv^2 & uv^2 & v^3 \end{pmatrix} \begin{pmatrix} u^3 & u^2 v \\ u^2 v & uv^2 \\ u^2 v & uv^2 \\ uv^2 & v^3 \end{pmatrix} \\
&= |\mathsf{A}_o|^2 \underbrace{(u^2 + v^2)^2}_{=1} \begin{pmatrix} u^2 & uv \\ uv & v^2 \end{pmatrix} \\
&= |\mathsf{A}_o|^2 \begin{pmatrix} u^2 & uv \\ uv & v^2 \end{pmatrix}
\end{aligned} \tag{36}
$$

5.2 General Structure Tensor Construction

In general the matrix product of Eq. (32) becomes:

$$
\begin{aligned}
\mathbf{T}_n^2 &= \underbrace{\mathsf{A}_n \; \hat{\boldsymbol{u}} \left\lfloor \hat{\boldsymbol{u}}^{\otimes(n-1)} \right\rfloor^T}_{\mathbf{Q}_n} \underbrace{\left\lfloor \hat{\boldsymbol{u}}^{\otimes(n-1)} \right\rfloor \hat{\boldsymbol{u}}^T \; \mathsf{A}_n^*}_{\mathbf{Q}_n^T} \\
&= \mathsf{A}_n \; \hat{\boldsymbol{u}} \underbrace{\left\lfloor \hat{\boldsymbol{u}}^{\otimes(n-1)} \right\rfloor^T \left\lfloor \hat{\boldsymbol{u}}^{\otimes(n-1)} \right\rfloor}_{\text{inner product}} \hat{\boldsymbol{u}}^T \; \mathsf{A}_n^*
\end{aligned} \tag{37}
$$

The inner product of the lined up outer products above can be performed in reversed order. Then, for $n > 0$, Eq. (37) simplifies to:

$$
\mathbf{T}_n^2 = \mathsf{A}_n \; \hat{\boldsymbol{u}} \; \underbrace{(\hat{\boldsymbol{u}}^T \hat{\boldsymbol{u}})^{(n-1)}}_{=1} \; \hat{\boldsymbol{u}}^T \; \mathsf{A}_n^* \tag{38}
$$

By definition $\hat{\boldsymbol{u}}^T \hat{\boldsymbol{u}}$ is equal to one. It follows that the under-braced term also equals one and we have the desired result:

$$
\mathbf{T}_n^2 \; = \; |\mathsf{A}_n|^2 \, \hat{\boldsymbol{u}} \, \hat{\boldsymbol{u}}^T \; ; \qquad n > 0 \tag{39}
$$

For $n = 0$ we have a special case since the filter is isotropic, $\boldsymbol{D}_n = 1$, and $\mathbf{T}_0^2 = \mathbf{Q}_0 \, \mathbf{Q}_0^T = |\mathsf{A}_e|^2 \, \mathbf{I}$.

5.3 Monomial Quadrature

As stated in Eq. (30) even and odd filters will have different local magnitudes.

$$\mathsf{T}_n^2 = \begin{cases} \mathsf{A}_\mathsf{e}^2 \;\; \hat{\boldsymbol{u}}\,\hat{\boldsymbol{u}}^T & \text{for even } n \\ |\,\mathsf{A}_\mathsf{o}|^2\, \hat{\boldsymbol{u}}\,\hat{\boldsymbol{u}}^T & \text{for odd } n \end{cases} \tag{40}$$

A phase invariant monomial quadrature tensor can now be computed as the sum of one even index tensor and odd index tensor. Letting n be even and m odd we obtain:

$$\begin{aligned} \mathsf{T}_{nm}^2 &= \mathsf{T}_n^2 + \mathsf{T}_m^2 \\ &= \mathbf{Q}_n\mathbf{Q}_n^T + \mathbf{Q}_m\mathbf{Q}_m^T \\ &= |\,\mathsf{A}_\mathsf{e} + \mathsf{A}_\mathsf{o}\,|^2\, \hat{\boldsymbol{u}}\,\hat{\boldsymbol{u}}^T \\ &= \mathsf{q}^2\, \hat{\boldsymbol{u}}\,\hat{\boldsymbol{u}}^T \end{aligned} \tag{41}$$

Note that the tensor magnitude, q, will be the same regardless of the order of the filters used.

5.4 Phase Invariance

The term 'phase invariant' stems from the fact that the tensor magnitude will be the same for all phases if the signal is sinusoidal. For a sinusoidal signal with amplitude, a, frequency, ρ and phase, θ, we obtain:

$$\begin{aligned} \mathsf{q}^2 &= |\,\mathsf{A}_\mathsf{e} + \mathsf{A}_\mathsf{o}\,|^2 \\ &= |\,\mathsf{a}\cos(\theta)\; R(\rho) - i\; \mathsf{a}\sin(\theta)\; R(\rho)\,|^2 \\ &= |\,\mathsf{a}\, e^{i\theta}\; R(\rho)\,|^2 \\ &= \mathsf{a}^2\; R^2(\rho) \end{aligned} \tag{42}$$

5.5 Tensor Positivity

A more compact expression of the monomial quadrature tensor can be attained by concatenating the even and the odd filter response matrices to a single matrix. The " , " notation implies concatenation of the arguments left to right.

$$\mathbf{Q}_{nm} = (\mathbf{Q}_n\;,\; \mathbf{Q}_m) \tag{43}$$

and compute the monomial quadrature tensor of origin (m,n) as

$$\mathbf{T}^2_{nm} = \mathbf{Q}_{nm}\,\mathbf{Q}^T_{nm} \tag{44}$$

As the monomial quadrature tensor is computed from products of filter response matrices

$$\mathbf{T}^2_{nm} = \mathbf{Q}_{nm}\mathbf{Q}^T_{nm} = \sum_k \lambda_k\,\hat{\mathbf{e}}_k\hat{\mathbf{e}}_k^T \tag{45}$$

it follows that all $\lambda_k \geq 0$ which allows for robust certainty estimates for the local structure estimation.

6 Structure Tensor Variations

Local structure analysis algorithms are quite complex and involve a lot more than the filters used. This makes comparisons difficult to interpret from a filter point of view. There are, however, a number of interesting similarities between different suggested algorithms. A few previous comparisons can be found in [37] and [60] In the following we point out the relation to the monomial approach for a number of well known approaches to structure tensor estimation. We show that nearly all variants can be formulated as special or modified versions of the monomial approach.

6.1 The Structure Tensor, $\mathbf{T}_{\mathbf{q}}$

The first publications mentioning tensors as a representation for local orientation and structure is due to Knutsson [40, 41]. Similar to the earlier developed vector representation [38, 39], the construction is based on a set of quadrature filters oriented in a number of fixed orientations, q_k, k indicating the orientation. The orientation vector is obtained as:

$$\mathbf{v} = \sum_k \sqrt{\mathsf{q}_k \mathsf{q}_k^*}\;\mathbf{v}_k \tag{46}$$

The structure tensor is obtained in the same fashion:

$$\mathbf{T} = \sum_k \sqrt{\mathsf{q}_k \mathsf{q}_k^*}\;\mathbf{T}_k \tag{47}$$

The vector variant works for two dimensional signals but for three dimension, or more, the tensor formulation is necessary. The loglet based structure tensor estimation suggested in [43] also uses this weighted 'basis tensor' approach but involve a different set of filters allowing higher order orientation ccomponents to be incorporated.

Unlike the methods discussed below this method for structure tensor construction is not possible to describe as a special or modified case of the monomial approach.

6.2 The Gradient Tensor, $\mathbf{T_G}$

The simplest way to obtain a matrix describing local orientation is exemplified by Bigun-Granlund's inertia matrix [10, 11] and Förstners corner detector [27]. This matrix is constructed as the outer product of the local gradient and is, in the notation introduced above and $\mathbf{Q}$ defined by Eq. (20), given by:

$$\begin{aligned} \mathbf{T_G} &= \mathbf{T}_1 \\ &= \mathbf{Q}_1\,\mathbf{Q}_1^T \\ &= \underbrace{\mathsf{A}_{SR}\,\sin(\theta)\,\hat{\boldsymbol{u}}\,\hat{\boldsymbol{u}}^0}_{\mathbf{Q}_1}\;\underbrace{\hat{\boldsymbol{u}}^0\,\hat{\boldsymbol{u}}^T\,\sin(\theta)\,\mathsf{A}_{SR}}_{\mathbf{Q}_1^T} \\ &= \mathsf{A}_{SR}^2\,\sin^2(\theta)\,\hat{\boldsymbol{u}}\,\hat{\boldsymbol{u}}^T \end{aligned} \tag{48}$$

Although the authors never mention tensors in the original work this outer product matrix estimate is often referred to as the gradient tensor or the structure tensor.

Since only a single order, i.e. order 1, is used this tensor is not phase invariant. Another drawback is that the frequency bandwidth of the estimate can become twice that of the original signal which may cause significant aliasing artifacts. Both these shortcomings are in practice, to some extent, remedied by the use of an averaging filter performing a weighted summation of local outer products. On the other hand this decreases the spatial resolution of the estimate [44].

6.3 The Boundary Tensor, $\mathbf{T_B}$

The boundary tensor originally suggested by Köthe [52], uses orders 1 and 2 and constitutes a special case of the monomial quadrature tensor.

$$\begin{aligned} \mathbf{T_B} &= \mathbf{T}_{12} \\ &= \mathbf{Q}_{12}\,\mathbf{Q}_{12}^T \end{aligned} \tag{49}$$

6.4 The Energy Tensor, $\mathbf{T}_{\mathbf{E}}$

The energy tensor, suggested by Felsberg [23], is a variant where filters of different orders are involved in the computed products. The energy tensor uses an isotropically bandpassed filter signal, its gradient and its Hessian. It can in monomial terms be expressed as:

$$\begin{aligned}\mathbf{T}_{\mathbf{E}} &= \mathbf{T}_1 + \mathbf{T}_{(0,2)} \\ &= \mathbf{Q}_1\mathbf{Q}_1^T + q_0\mathbf{Q}_2\end{aligned} \tag{50}$$

Note that the filter response matrices here have different radial frequency response and the mixing of different order terms will not give a positive semi-definite tensor for all image neighborhoods.

6.5 Gradient Energy Tensor, $\mathbf{T}_{\mathbf{GE}}$

The gradient energy tensor (GET) suggested in [24] can be said to use the same formula as the energy tensor with the input signal replaced by it's gradient. In monomial terms the result can be expressed:

$$\mathbf{T}_{\mathbf{GE}} = \mathbf{Q}_2\,\mathbf{Q}_2^T + \frac{1}{2}\left(\mathbf{Q}_1\,\mathbf{Q}_3^T + \mathbf{Q}_3\,\mathbf{Q}_1^T\right) \tag{51}$$

Note that also in this case the filter response matrices here have different radial frequency response and that the mixing of different order terms will not give a positive semi-definite tensor for all image neighborhoods.

6.6 Spatial 2nd Order Polynomial Tensor

Farnebäck [22] The 2nd order polymer tensor suggested by Farnebäck in [22] is a sum of outer products of 1st and 2nd order monomial filters. The difference from the monomial approach is that the filter are designed as windowed 1st and 2nd order polynomials in the spatial domain.

$$\begin{aligned}\mathbf{T}_{\mathbf{SP}} &= \mathbf{T}_1 + \mathbf{T}_2 \\ &= \mathbf{Q}_1\mathbf{Q}_1^T + \mathbf{Q}_2\mathbf{Q}_2^T\end{aligned} \tag{52}$$

This spatial design results in 1st and 2nd order filter that have different radial functions in the frequency domain. For this reason the result is not in general phase invariant i.e. components are not in quadrature. However, since it is a sum of squares, the result is always positive semi-definite.

6.7 *Spherical Harmonics*

A somewhat different way to estimate a local structure tensor is suggested in [44]. This approach is based on sums of products of spherically separable filters. The filter have the same radial function and the directional functions are spherical harmonic functions. The structure tensor carries information about 0th and 2nd order variations in orientation. A product between an order j filter and an order k filter will contain signal components of orders $j-k$ and $j+k$. By an appropriate weighted summation of a number of filter products it is possible to retain only order 0 and order 2 in the correct proportion while canceling out all other orders: i.e:

$$\mathbf{T_{SH}} = \sum_{jk} w_{jk} \mathbf{H}_j \mathbf{H}_k \tag{53}$$

This is a very general approach and, since spherical harmonic filter sets of orders 1 to N span the same function space as monomial filter sets of orders 1 to N, all monomial tensor variations can also be expressed in this way. With proper weights the result can also be made phase-invariant.

6.8 *Sum of Monomial Tensors*

Even more careful weighting of spherical harmonic filter products will give positive semi-definite tensors, in this case the result will be equivalent to a sum of tensor estimates over different order, $n \geq 0$ (even and odd), monomial filter matrices, $\mathbf{Q}_n$.

$$\mathbf{T_{SM}} = \sum_{n} w_n \mathbf{Q}_n \mathbf{Q}_n^T \tag{54}$$

7 Table of Structure Tensor Related Algorithms

The table below is a chronological list of a number of image processing algorithms directly or indirectly related to local structure tensor estimation. The year mentioned refers to the first publication discussing the addressed topic that the authors are aware of.

			Sinusoidal signals		Simple signals		Complex signals	
Algorithm	Dim[a]	Yr	Pos	Quad	Pos	Quad	Pos	Quad
GOP [29]	2	78	*na*	·	*na*	·	*na*	·
Rot Inv Op [17]	2	80	*na*		*na*		*na*	
Quad Vector [38,47]	2	81	*na*	•	*na*	•	*na*	•
Steerable filters [47,48]	2	81	*na*	•	*na*	•	*na*	•
5D Quad Vector [39]	3	85	*na*	•	*na*	•	*na*	•
3D Quad Tensor [40,41]	3	87	•	•	•	•		•
Grad outer prod [10,27]	2	87	•		•		•	
4D Quad Tensor [49]	4	91	•	•	•	•		•
SD Polynomial Tens [21]	n	99	•		•		•	
Loglet Tensor [43]	n	03	•	•	•	•		•
Boundary Tensor [52]	2	03	•	•	•	•	•	
Sp Harmonics Tens [44]	n	05	•	•	•	•		•
Energy Tensor [24]	2	05	•					
Grad Energy Tens [24]	2	05	•					
Monomial Tensors [51]	n	11[b]	•	•	•	•	•	
Sum of M-Tensors	n	11	•	•	•	•	•	•

[a] Dimension discussed in original publication
[b] Presented at Dagstuhl 09

8 Higher Order Structure Tensors

In Eq. (37) the filter matrix is constructed to produce a 2nd order tensor. However, a simple rearrangement of the order n filter matrix components will allow tensors of order $2p$ to be estimated.

$$\begin{aligned} \mathbf{T}^2_{\mathbf{2p},n} &= \underbrace{\mathsf{A}_n \left\lfloor \hat{\boldsymbol{u}}^{\otimes p} \right\rfloor \left\lfloor \hat{\boldsymbol{u}}^{\otimes (n-p)} \right\rfloor^T}_{\mathbf{Q}_{(p)n}} \underbrace{\left\lfloor \hat{\boldsymbol{u}}^{\otimes (n-p)} \right\rfloor \left\lfloor \hat{\boldsymbol{u}}^{\otimes p} \right\rfloor^T \mathsf{A}_n}_{\mathbf{Q}^T_{(p)n}} \\ &= \mathsf{A}_n \left\lfloor \hat{\boldsymbol{u}}^{\otimes p} \right\rfloor \underbrace{\left\lfloor \hat{\boldsymbol{u}}^{\otimes (n-p)} \right\rfloor^T \left\lfloor \hat{\boldsymbol{u}}^{\otimes (n-p)} \right\rfloor}_{\text{inner product}} \left\lfloor \hat{\boldsymbol{u}}^{\otimes p} \right\rfloor^T \mathsf{A}_n \end{aligned} \tag{55}$$

As before the inner product of the *lined up* outer products above can be performed in reversed order. Then, for $n \geq p \geq 0$, Eq. (37) simplifies to:

$$\mathbf{T}^2_{\mathbf{2p},n} = \mathsf{A}_n \left\lfloor \hat{\boldsymbol{u}}^{\otimes p} \right\rfloor \underbrace{(\hat{\boldsymbol{u}}^T \hat{\boldsymbol{u}})^{(n-p)}}_{=1} \left\lfloor \hat{\boldsymbol{u}}^{\otimes p} \right\rfloor^T \mathsf{A}_n \tag{56}$$

By definition $\hat{\boldsymbol{u}}^T\hat{\boldsymbol{u}}$ is equal to one. It follows that the under-braced term also equals one which gives:

$$\mathsf{T}^2_{\mathbf{2p},n} = |\mathsf{A}_n|^2 \left\lfloor \hat{\boldsymbol{u}}^{\otimes p} \right\rfloor \left\lfloor \hat{\boldsymbol{u}}^{\otimes p} \right\rfloor^T \; ; \quad n \geq p \geq 0 \tag{57}$$

The result now holds the components of a tensor of order $2p$. However, due to the use of the $lineup$ operator, the components are stored in matrix form and they need to be re-organized in order to obtain the result as a proper tensor $\boldsymbol{\Upsilon}$ of order $2p$.

$$\boldsymbol{\Upsilon}_{\mathbf{2p},n} = \lceil \mathsf{T}^2_{\mathbf{2p},n} \rceil \tag{58}$$

The " $\lceil \; \rceil$ " notation used here indicates a reshape operation that restores the proper structure of the data, i.e. the result is a tensor having $2p$ indexes.

As 2nd order tensors are naturally represented as matrices the $lineup$ operator greatly simplifies the notation. For higher order tensors, however, the use of standard tensor notation may be preferred by some readers. Equations (57) and (58) can then be jointly expressed as:

$$\Upsilon_n{}^{a_1 \dots a_p}_{b_1 \dots b_p} = |\mathsf{A}_n|^2 \; u^{a_1 \; \dots \; a_p}_{a_{p+1} \dots a_n} \; u^{a_{p+1} \dots a_n}_{b_1 \; \dots \; b_p} \tag{59}$$

According to the Einstein convention a summation is performed over equal indexes and Eq. (59) clearly shows that the difference between tensors of different order is how many indexes are summed over (contracted). Letting $p = 0$ means summing over all indexes and the result is a scalar representing the local energy. For $p = 1$ we obtain the standard structure tensor. For $p > 1$ we obtain higher order tensors having the power to represent more complex local structure.

8.1 Non-simple Signals

For such non-simple signals Eq. (59) is no longer applicable since there is no unique local orientation, u. Directly expressed as a sum of filter products, corresponding to a generalization of the monomial filter response matrix product in Eq. (32), we obtain an order $2p$ structure tensor as:

$$\Upsilon_n{}^{a_1 \dots a_p}_{b_1 \dots b_p} = q^{a_1 \; \dots \; a_p}_{a_{p+1} \dots a_n} \; q^{a_{p+1} \dots a_n}_{b_1 \; \dots \; b_p} \tag{60}$$

Tensors of order 4 have been used to analyze situations with two orientations present, e.g. [8, 59, 72]. Applications where tensors of order higher than four have been used are so far not known to the authors but can be expected to prove a powerful tool when more that two orientation are present.

To produce quadrature type tensors we still need to add a tensor from odd order filter sets and a tensor from even order filter sets. In the most general case we can express the estimation of local structure tensors of order $2p$ as a weighted summation of order $2p$ tensors, here of order $\binom{p}{p}$, from monomial filter sets of different orders, i.e:

$$\Upsilon^{a_1 \dots a_p}_{b_1 \dots b_p} = \sum_n w_n \, q^{a_1 \dots a_p}_{a_{p+1} \dots a_n} \, q^{a_{p+1} \dots a_n}_{b_1 \dots b_p} \tag{61}$$

9 Conclusion

Research concerning 2nd order structure tensor estimation is still continuing after more than two decades. In this chapter we have presented a new general framework for structure tensor estimation. Using this framework enabled a unified presentation of most of the previously suggested estimation approaches. We concluded by presenting a general estimation approach for higher order structure tensors. The higher order tensor estimates produced by Eq. (61) contains a much richer representation of the local structure and we expect that the future will hold considerable effort towards fully understanding these new higher order constructs.

Acknowledgements The Swedish Research Council, Swedish e-Science Research Center, Linköpings Universitet and NIH (grants R01MH074794, R01MH092862, and P41RR013218) are gratefully acknowledged for supporting this work.

References

1. Andersson, M., Knutsson, H.: Orientation estimation in ambiguous neighbourhoods. In: Proceedings of the 7th Scandinavian Conference on Image Analysis, Aalborg. SCIA (1991)
2. Andersson, M., Knutsson, H.: Controllable 3-D filters for low level computer vision. In: Proceedings of the 8th Scandinavian Conference on Image Analysis, Tromsö. SCIA (1993)
3. Andersson, M., Knutsson, H.: Transformation of local spatio-temporal structure tensor fields. In: IEEE International Conference on Acoustics Speech and Signal Processing (ICASSP), Hong Kong (2003). (Presented at ICIP 2003 in Barcelona, Spain, Sept 20)
4. Andersson, K., Andersson, M., Knutsson, H.: A perception based velocity estimator and its use for motion compensated prediction. In: Proceedings of the 12th Scandinavian Conference on Image Analysis, Bergen, pp. 493–499. SCIA (2001)
5. Andersson, K., Andersson, M., Johansson, P., Forcheimer, R., Knutsson, H.: Motion compensation using backward prediction and prediction refinement. Signal Process. Image Commun. **18**, 381 (2003)
6. Andersson, K., Westin, C.-F., Knutsson, H.: Prediction from off-grid samples using continuous normalized convolution. Signal Process. J. **87**(3), 353–365 (2007)
7. Bårman, H., Granlund, G.H., Knutsson, H.: Estimation of curvature in 3-D images using tensor field filtering. In: Faugeras, O. (ed.) Computer Vision – ECCV90. Proceedings, 1990, Antibes, pp. 563–565. Springer, Berlin (1990)
8. Barmpoutis, A., Vemuri, B.C., Forder, J.R.: Registration of high angular resolution diffusion mri images using 4th order tensors. In: Proceedings of the 10th International Conference on Medical Image Computing and Computer-Assisted Intervention – Volume Part I, MICCAI'07, pp. 908–915. Springer, Berlin/Heidelberg (2007)

9. Bhatia, A.B., Wolf, E.: On the circle polynomials of Zernike and related orthogonal sets. In: *Proceedings of the Cambridge Philosophical Society*, vol. 50, pp. 40–48. Cambridge University Press, Cambridge (1954)
10. Bigün, J., Granlund, G.H.: Optimal orientation detection of linear symmetry. In: IEEE First International Conference on Computer Vision, London, pp. 433–438 (1987)
11. Bigün, J., Granlund, G.H.: Optical flow based on the inertia matrix of the frequency domain. In: Proceedings from SSAB Symposium on Picture Processing, Lund University, Sweden (1988). SSAB. Report LiTH–ISY–I–0934, Computer Vision Laboratory, Linköping University, Sweden (1988)
12. Blakemore, C., Campbell, F.W.: On the existence of neurones in the human visual system selectively sensitive to the orientation and size of retinal images. J. Physiol. **203**, 237–260 (1969)
13. Brun, A., Westin, C.-F., Haker, S., Knutsson, H.: A tensor-like representation for averaging, filtering and interpolation of 3-D object orientation data. In: IEEE International Conference on Image Processing (ICIP'05), Genoa (2005)
14. Calway, A.D., Knutsson, H., Wilson, R.: Multiresolution estimation of 2-d disparity using a frequency domain approach. In: Proceedings of British Machine Vision Conference, Leed (1992)
15. Claasen, T.A.C.M., Mecklenbrauker, W.F.G.: The Wigner distribution – a tool for time-frequency signal analysis, part ii: discrete-time signals. Philips J. Res. **35**(4/5), 276–300 (1980)
16. Cyganski, D., Orr, J.A., Pinjo, Z.: A tensor operator method for identifying the affine transformation relating image pairs. In: Proceedings of CVPR'83, IEEE Computer Society Conference on ComputerVision and Pattern Recognition, Washington, pp. 361–363 (1983)
17. Danielsson, P.E.: Rotation invariant operators with directional response. In: Proceedings 5'th International Conference on Pattern Recognition, Miami Beach (1980)
18. Daubechies, I.: The wavelet transform, time-frequency localization and signal analysis. IEEE Trans. Inf. Theory **36**(5), 961–1005 (1990)
19. Daugman,J.G.: Uncertainty relation for resolution in space, spatial frequency, and orientation optimized by two- dimensional visual cortical filters. J. Opt. Soc. Am. **2**, 1160 (1985)
20. Daugman, J.G.: Complete discrete 2-d gabor transform by neural networks for image analysis and compression. IEEE Trans. Acoust. Speech Signal Process. **36**, 1169 (1988)
21. Farnebäck, G.: Fast and accurate motion estimation using orientation tensors and parametric motion models. In: Proceedings of 15th International Conference on Pattern Recognition, Barcelona, vol. 1, pp. 135–139. IAPR (2000)
22. Farnebäck, G.: Polynomial expansion for orientation and motion estimation. Ph.D. thesis, Linköping University, Sweden, SE-581 83 Linköping (2002). Dissertation No 790, ISBN 91-7373-475-6.
23. Felsberg, M., Jonsson, E.: Energy tensors: quadratic, phase invariant image operators. In: Pattern Recognition. Lecture Notes in Computer Science, vol. 3663, pp. 493–500 (2005)
24. Felsberg, M., Köthe,U.: GET: the connection between monogenic scale-space and Gaussian derivatives. In: Kimmel, R., Sochen, N., Weickert, J. (eds.) Scale Space and PDE Methods in Computer Vision. Lecture Notes in Computer Science, vol. 3459, pp. 192–203. Springer, Berlin/London (2005)
25. Felsberg, M., Sommer, G.: The monogenic signal. IEEE Trans. Signal Process, **49**(12), 3136–3144 (2001)
26. Fleet, D.J., Jepson, A.D.: Stability of phase information. IEEE Trans. Pattern Anal. Mach. Intell. **15**(12), 1253–1268 (1993)
27. Forstner, W., Gulch, E.: A fast operator for detection and precise location of distinct points, corners and centres of circular features. In: ISPRS Intercommission Conference on Fast Processing of Photogrammetric Data, Interlaken, pp. 281–305 (1987)
28. Gabor, D.: Theory of communication. J. Inst. Elec. Eng. **93**(26), 429–457 (1946)
29. Granlund, G.H.: In search of a general picture processing operator. Comput. Graph. Image Process. **8**(2), 155–178 (1978)

30. Granlund, G.H., Knutsson, H.: Signal Processing for Computer Vision. Kluwer, Dordrecht/Boston (1995). ISBN 0-7923-9530-1.
31. Haglund, L., Knutsson, H., Granlund, G.H.: On phase representation of image information. In: The 6th Scandinavian Conference on Image Analysis, Oulu, pp. 1082–1089 (1989)
32. Herberthson, M., Brun, A., Knutsson, H.: P-averages of diffusion tensors. In: Proceedings of the SSBA Symposium on Image Analysis, Linköping. SSBA (2007)
33. Horn, B.K.P., Schunck, B.G.: Determining optical flow. Artif. Intell. **17**, 185–204 (1981)
34. Hu, M.K.: Visual pattern recognition by moment invariants. IRE Trans. Inf. Theory **IT-8**(2), 179–187 (1962)
35. Hubel, D.H., Wiesel, T.N.: Sequence regulatiry and geometry of orientation columns in the monkey striate cortex. J. Comp. Neurol. **158**, 267–294 (1974)
36. Hueckel, A.: An operator which locates edges in digital pictures. J. ACM **20**, 113–125 (1971)
37. Johansson, B., Farnebäck, G.: A theoretical comparison of different orientation tensors. In: Proceedings SSAB02 Symposium on Image Analysis, Lund, pp. 69–73. SSAB (2002)
38. Knutsson, H.: Filtering and reconstruction in image processing. Ph.D. thesis, Linköping University, Sweden (1982). Diss. No. 88
39. Knutsson, H.: Producing a continuous and distance preserving 5-D vector representation of 3-D orientation. In: IEEE Computer Society Workshop on Computer Architecture for Pattern Analysis and Image Database Management – CAPAIDM, pages 175–182, Miami Beach. IEEE, Washington (1985). Report LiTH–ISY–I–0843, Linköping University, Sweden (1986)
40. Knutsson, H.: A tensor representation of 3-D structures. In: 5th IEEE-ASSP and EURASIP Workshop on Multidimensional Signal Processing, Noordwijkerhout (1987). Poster presentation
41. Knutsson, H.: Representing local structure using tensors. In: The 6th Scandinavian Conference on Image Analysis, Oulu, pp. 244–251 (1989). Report LiTH–ISY–I–1019, Computer Vision Laboratory, Linköping University, Sweden (1989)
42. Knutsson, H., Andersson, M.: Robust N-dimensional orientation estimation using quadrature filters and tensor whitening. In: Proceedings of IEEE International Conference on Acoustics, Speech, & Signal Processing, Adelaide. IEEE (1994)
43. Knutsson, H., Andersson, M.: Loglets: generalized quadrature and phase for local spatio-temporal structure estimation. In: Proceedings of the Scandinavian Conference on Image Analysis (SCIA), Halmstad (2003)
44. Knutsson, H., Andersson, M.: Implications of invariance and uncertainty for local structure analysis filter sets. Signal Process. Image Commun. **20**(6), 569–581 (2005)
45. Knutsson, H., Granlund, G.H.: Fourier domain design of line and edge detectors. In: Proceedings of the 5th International Conference on Pattern Recognition, Miami (1980)
46. Knutsson, H., Granlund, G.H.: Texture analysis using two-dimensional quadrature filters. In: IEEE Computer Society Workshop on Computer Architecture for Pattern Analysis and Image Database Management – CAPAIDM, Pasadena (1983)
47. Knutsson, H., Wilson, R.G., Granlund, G.H.: Anisotropic filtering operations for image enhancement and their relation to the visual system. In: IEEE Computer Society Conference on Pattern Recognition and Image Processing, Dallas (1981)
48. Knutsson, H., Wilson, R., Granlund, G.H.: Anisotropic non-stationary image estimation and its applications – part I: restoration of noisy images. IEEE Trans. Commun. **31**(3), 388–397 (1983)
49. Knutsson, H., Haglund, L., Bårman, H.: A tensor based approach to structure analysis and enhancement in 2D, 3D and 4D. In: Workshop Program, Seventh Workshop on Multidimentional Signal Processing, Lake Placid (1991). IEEE Signal Processing Society. Poster presentation
50. Knutsson, H., Bårman, H., Haglund, L.: Robust orientation estimation in 2D, 3D and 4D using tensors. In: Proceedings of Second International Conference on Automation, Robotics and Computer Vision, ICARCV'92, Singapore (1992)
51. Knutsson, H., Westin, C.-F., Andersson,M.: Representing local structure using tensors II. In: Proceedings of the 17th Scandinavian conference on image analysis (SCIA 11), Ystad (2011)

52. Köthe, U.: Inegrated edge and junction detection with the boundary tensor. In: Proceedings of ninth IEEE International Conference on Computer Vision (ICCV), Nice (2003)
53. Köthe, U., Felsberg, M.: Riesz-transforms versus derivatives: on the relationship between the boundary tensor and the energy tensor. In: Kimmel, R., Sochen, N., Weickert, J. (eds.) Scale Space and PDE Methods in Computer Vision. Lecture Notes in Computer Science, vol. 3459, pp. 179–191. Springer, Berlin/London (2005)
54. Landelius, T., Haglund, L., Knutsson, H.: Depth and velocity from orientation tensor fields. In: Proceedings of the 8th Scandinavian Conference on Image Analysis, Tromsö. NOBIM, (1993). Report LiTH-ISY-R-1529, Linköping University, Sweden (1993)
55. Lentz, R.: Optimal filters for the detection of linear patterns. Pattern Recognit. Lett. **20**, 163–172 (1987)
56. Lenz, R.: Rotation-invariant operators and scale space filtering. Pattern Recognit. Lett. **6**, 151–154 (1987)
57. Lucas, B., Kanade, T.: An iterative image registration technique with applications to stereo vision. In: Proceedings of Darpa IU Workshop, pp. 121–130 (1981)
58. Marr, D., Hildreth, E.: Theory of edge detection. In: Proceedings of the Royal Society of London, vol. 207, pp. 187–217 (1980)
59. Nordberg, K.: A fourth order tensor for representation of orientation and position of oriented segments. Other academic, Linköping University, Department of Electrical Engineering (2004). diva2:288343
60. Nordberg, K., Farnebäck, G.: Estimation of orientation tensors for simple signals by means of second-order filters. Signal Process. Image Commun. **20**(6), 582–594 (2005)
61. Nordberg, K., Knutsson, H., Granlund, G.: Local curvature from gradients of the orientation tensor field. Report LiTH-ISY-R-1783, Computer Vision Laboratory, SE-581 83 Linköping, Sweden (1995)
62. Oppenheim, A.V., Lim, J.S. The importance of phase in signals. Proc. IEEE **69**, 529 (1981)
63. Pollen, D.A., Ronner, S.F.: Visual cortical neurons as localized spatial frequency filters. IEEE Trans. Syst. Man Cybern. **13**(5), 907–915 (1983)
64. Riesz, M.: Sur les fonctions conjuge'es. Math. Z. **27**, 218–244 (1927)
65. Roberts, L.G.: Machine perception of three-dimensional Solid. In: Tippell, J.T. (ed.) Optical and Electro-Optical Information Processing, pp. 159–197. MIT Press, Cambridge (1965)
66. Svensson, B., Brun, A., Andersson, M., Knutsson, H.: Estimation of non-cartesian local structure tensor fields. In: Proceedings of the 15th Scandinavian conference on image analysis (SCIA'07), Aalborg (2007)
67. Torre, V., Poggio, T.A.: On edge detection. IEEE Trans. Pattern Anal. Mach. Intell. **8**(2), 147–163 (1986)
68. Tretiak, O., Pastor, L.: Velocity estimation from image sequences with second order differential operators. In: Proceedings of the 7th International Conference on Pattern Recognition, Montreal, pp. 20–22 (1984)
69. Wang, Q., Ronneberger, O., Burkhardt, H.: Fourier analysis in polar and spherical coordinates. Technical Report Internal Report 1/08, IIF-LMB, Computer Science Department, University of Freiburg (2008)
70. Westin, C.-F., Knutsson, H.: Extraction of local symmetries using tensor field filtering. In: Proceedings of 2nd Singapore International Conference on Image Processing, pp. 371–375, Singapore. IEEE Singapore Section (1992). LiTH–ISY–R–1515, Linköping University, Sweden.
71. Westin, C.-F., Knutsson, H.: Estimation of motion vector fields using tensor field filtering. In: Proceedings of the IEEE International Conference on Image Processing, Austin, pp. 237–242. IEEE (1994)
72. Westin, C.-F., Knutsson, H.: Visualization and Processing of Tensor Fields: Proceedings of the Dagstuhl Workshop, chapter Representation and Estimation of Tensors-Pairs. 2010. submitted.
73. Westin, C.-F., Martin-Fernandez, M., Alberola-Lopez, C., Ruiz-Alzola, J., Knutsson, H.: Tensor field regularization using normalized convolution and markov random fields in a bayesian framework. In: Weickert, J., Hagen, H. (eds.) Visualization and Image Processing of

Tensor Fields. Series: Mathematics and Visualization, pp. 381–398, 464–467. Springer, Berlin (2006). ISBN:3-540-25032-8
74. Wiklund, J., Nicolas, W., Alface, P.R., Andersson, M., Knutsson, H.: T-flash: tensor visualization in medical studio. In: Aja-Fernandez, S., de Luis Garcia, R., Tao, D., Li, X. (eds.) Tensors in Image Processing and Computer Vision, pp. 455–466. Springer, London (2009). ISBN 978-1-84882-298-6
75. Wilson, R., Calway, A.D. A general multiresolution signal descriptor and its application to image analysis. In: Signal Processing, pages 663–666. EURASIP, 1988.
76. Wilson, R., Knutsson, H.: Uncertainty and inference in the visual system. IEEE Trans. Syst. Man Cybern. **18**(2), 305–312 (1988)
77. Yuille, A.L., Kammen, D.M., Cohen, D.S.: Quadrature and the development of orientation selective cortical cells by hebb rules. Biol. Cybern. **61**, 183–194 (1989)
78. Zernike, F.: Diffraction theory of the cut procedure and its improved form, the phase contrast method. Physica, **1**, 689–704 (1934)
79. Zucker, S.W., Hummel, R.M.: A three-dimensional edge operator. IEEE Trans. Pattern Anal. Mach. Intell. PAMI **3**(3), 324–331 (1981)

Adaptation of Tensor Voting to Image Structure Estimation

Rodrigo Moreno, Luis Pizarro, Bernhard Burgeth, Joachim Weickert, Miguel Angel Garcia, and Domenec Puig

Abstract Tensor voting is a well-known robust technique for extracting perceptual information from clouds of points. This chapter proposes a general methodology to adapt tensor voting to different types of images in the specific context of image structure estimation. This methodology is based on the structural relationships between tensor voting and the so-called structure tensor, which is the most popular technique for image structure estimation. The problematic Gaussian convolution used by the structure tensor is replaced by tensor voting. Afterwards, the results are

---

R. Moreno (✉)
Department of Medical and Health Sciences (IMH), Linköping University, Center for Medical Image Science and Visualization (CMIV), Campus US, 58185, Linköping, Sweden
e-mail: rodrigo.moreno@liu.se

L. Pizarro
Department of Computing, Imperial College London, 180 Queen's Gate, SW7 2AZ, London, UK
e-mail: luis.pizarro@imperial.ac.uk

B. Burgeth
Faculty of Mathematics and Computer Science, Saarland University, Building E2.4, 66041, Saarbrücken, Germany
e-mail: burgeth@math.uni-sb.de

J. Weickert
Faculty of Mathematics and Computer Science, Saarland University, Mathematical Image Analysis Group (MIA), Building E1.1, 66041, Saarbrücken, Germany
e-mail: weickert@mia.uni-saarland.de

M.A. Garcia
Department of Electronic and Communications Technology, Autonomous University of Madrid, Francisco Tomas y Valiente 11, 28049, Madrid, Spain
e-mail: miguelangel.garcia@uam.es

D. Puig
Department of Computer Science and Mathematics, Intelligent Robotics and Computer Vision Group (IRCV), Rovira i Virgili University, Av. Països Catalans 26, 43007, Tarragona, Spain
e-mail: domenec.puig@urv.cat

D.H. Laidlaw and A. Vilanova (eds.), *New Developments in the Visualization and Processing of Tensor Fields*, Mathematics and Visualization,
DOI 10.1007/978-3-642-27343-8_2,

appropriately rescaled. This methodology is adapted to gray-valued, color, vector- and tensor-valued images. Results show that tensor voting can estimate image structure more appropriately than the structure tensor and also more robustly.

1 Introduction

Medioni and colleagues [26] proposed tensor voting as a robust technique for extracting perceptual information from a cloud of points. In 3D, tensor voting estimates saliency measurements of how likely a point lies on a surface, a curve, a junction, or it is an outlier. It is based on the propagation and aggregation of the most likely normal(s) encoded by means of second-order tensors through a convolution-like voting process, assuming that neighboring points belong to the same smooth surface. This technique has been proven versatile, since it has successfully been adapted to problems well beyond the ones to which it was originally applied with excellent results. For example, this method has already been applied to a variety of problems in image and video processing, such as perceptual organization [26, 43], image restoration [15], image segmentation [22, 29], video segmentation [27, 36], mesh analysis [17], 3D reconstruction [49] and dimensionality estimation [28]. Since the input data for most of these applications are not clouds of points, a common approach is to apply tensor voting as described in [26] to clouds of points derived from the original data. Although, in principle, it is more natural to apply tensor voting to the original data, that application requires extensions of tensor voting to different types of data, which, in most cases, have not been proposed so far.

Furthermore, extensions of tensor voting specifically tailored to applications have also been proved effective. They are based on the incorporation of additional application-dependent perceptual rules to the voting process. For example, the use of specifically designed inhibitory voting fields has been reported beneficial for perceptual organization in gray-scale images [25]. In addition, an extension of tensor voting specifically tailored to color image denoising [30], robust color edge detection [31] and color image segmentation [32] has yielded significantly good results. Related to tensor voting, a voting process specifically designed for detecting X- and T-shaped junctions has been proposed in [1].

In a different context, local image structure estimation methods aim at typifying the region around every pixel. These methods estimate similarity measurements of every local region with respect to certain patterns of interest, such as flat and textured regions, and regions that contain edges, lines or corners. These measurements can be used, for example, to steer image processing methods or to extract local features, such as edges, lines and corners, in a further step.

During the last decades, the use of tensors has allowed local image structure estimation methods to represent several types of local patterns with a same mathematical entity. The most popular of these methods is the structure tensor [10], which is able to typify flat regions, regions with edges and regions with corners, through second-order tensors. It has been used in a multitude of applications, such as edge detection [11], corner detection [16, 39], texture analysis [38, 40], image filtering [45], image compression [14], optic flow estimation [2, 24], and detection of X- and T-shaped junctions [1]. It has gained popularity thanks to its robustness, efficiency and easiness of implementation. In addition, it depends on a single parameter, which is usually easy to tune.

The main hypothesis made by the structure tensor is that the orientation of the gradient changes slowly in regions with edges, and quickly in regions with corners. In addition, it assumes the size of the gradient to be small in flat regions, and large in both regions with edges and regions with corners. Thus, the structure tensor estimates local image structure by means of a weighted sum of gradients within a neighborhood. For a gray-scale image, the structure tensor, J, is defined as the convolution of a Gaussian with the tensorized gradient of the image [10]:

$$\mathsf{J} = G_\rho * \nabla u \nabla u^T , \tag{1}$$

where G_ρ is the Gaussian with zero mean and standard deviation ρ, ∇u is the gradient of the image u, and $\nabla u \nabla u^T$ represents the tensorized gradient at every pixel. A related approach based on quadrature filters has been proposed in [12, 19]. Furthermore, extensions using higher-order derivatives [9, 21], and extensions for curved structures [3] have also been proposed.

Despite its popularity, the structure tensor also has important shortcomings, such as detection of features in flat regions, loss of small features, detection of false corners, and misplacement of corners. These shortcomings are mainly related to the use of a Gaussian kernel, since it can propagate the gradient to pixels in flat regions. Thus, the structure tensor can yield similar tensors for flat regions, and regions with edges or corners, leading to errors in the extraction of features. This fact has encouraged researchers to propose alternatives to the structure tensor.

Most of the strategies intend to avoid the integration of different orientations of the gradient by adapting the neighborhood to the data in such a way that only neighbors with similar orientations of the gradient are taken into account in the summation. For example, Nagel and Gehrke [34] and Nath and Palaniappan [35] use adaptive Gaussians instead of a Gaussian convolution; Köthe [20] uses a hourglass-shaped kernel instead of the Gaussian; van de Weijer and van den Boomgaard [44] use robust statistics to choose one of the ambiguous orientations at every pixel; Brox et al. [4] and Hahn and Lee [13] propose non-linear diffusion processes in order to aggregate contributions of the neighbors.

Although tensor voting and the structure tensor have been proposed in different contexts, they have important similarities, as will be shown in Sect. 3. Thus, the aim of this chapter is twofold: First, to propose a general methodology to extend tensor voting to different types of images. This methodology is based on the similarities between the formulations of both tensor voting and the structure tensor. Second, to compare the performance of both methods in the specific context of image structure estimation for different types of images. It is important to remark that application-dependent extensions of tensor voting are not considered in this chapter, since their formulation could not be related to the structure tensor.

Related to this work, two extensions of classical tensor voting in order to directly apply it to gray-scale images have been proposed. First, Tai et al. [42]

encode curveness and regionness in the tensors before applying tensor voting. Unfortunately, discriminating edges from corners is not possible by applying this strategy, since both types of structure will yield high curveness using this approach. Second, Loss et al. [23] initialize the tensors with *ball* tensors (cf. Sect. 2) whose size depends on the gray-scale value of the pixel. However, this strategy cannot be used to extract corners.

The chapter is organized as follows. Section 2 summarizes the tensor voting formalism. Section 3 shows the relationships between tensor voting and the structure tensor. Section 4 describes a general methodology to extend tensor voting to different types of images, in particular to gray-scale, vector- and tensor-valued images in the specific application of image structure estimation. Section 5 shows some results of tensor voting applied to image structure estimation. Finally, Sect. 6 discusses the obtained results and makes some final remarks.

2 Tensor Voting

Medioni et al. [26] proposed tensor voting as a technique for extracting perceptual information from clouds of points, in particular in 3D. The method robustly estimates saliency measurements of how likely a point lies on a surface, a curve, a junction, or it is an outlier. It is based on the propagation and aggregation of the most likely normal(s) encoded by means of second-order tensors modeled by means of symmetric positive semidefinite matrices. In a first stage, a tensor is initialized at every point in the cloud either with a first estimation of the normal, or with a ball-shaped tensor if a priori information is not available. Afterwards, every tensor is decomposed into its three components: a *stick*, a *plate* and a *ball*. Every component casts votes, which are tensors that encode the most likely direction(s) of the normal at a neighboring point taking into account the information encoded by the voter in that component. Finally, the votes are summed up and analyzed in order to estimate surfaceness, curveness and junctionness measurements at every point. Points with low saliency are assumed to be outliers. More formally, the tensor voting at $\mathbf{p}$, $\mathsf{TV}(\mathbf{p})$ is given by:

$$\mathsf{TV}(\mathbf{p}) = \sum_{\mathbf{q} \in neigh(\mathbf{p})} \mathsf{SV}(\mathbf{v}, \mathsf{S}_\mathbf{q}) + \mathsf{PV}(\mathbf{v}, \mathsf{P}_\mathbf{q}) + \mathsf{BV}(\mathbf{v}, \mathsf{B}_\mathbf{q}), \tag{2}$$

where $\mathbf{q}$ represents each of the points in the neighborhood of $\mathbf{p}$, SV, PV and BV are the *stick*, *plate* and *ball* tensor votes cast to $\mathbf{p}$ by every component of $\mathbf{q}$, $\mathbf{v} = \mathbf{p} - \mathbf{q}$, and $\mathsf{S}_\mathbf{q}$, $\mathsf{P}_\mathbf{q}$ and $\mathsf{B}_\mathbf{q}$ are the *stick*, *plate* and *ball* components of the tensor at $\mathbf{q}$ respectively. These components are given by:

$$\mathsf{S}_\mathbf{q} = (\lambda_1 - \lambda_2)\left(\mathbf{e}_1\mathbf{e}_1{}^T\right), \tag{3}$$

$$\mathsf{P}_\mathbf{q} = (\lambda_2 - \lambda_3)\left(\mathbf{e}_1\mathbf{e}_1{}^T + \mathbf{e}_2\mathbf{e}_2{}^T\right), \tag{4}$$

$$\mathsf{B}_\mathbf{q} = \lambda_3\left(\mathbf{e}_1\mathbf{e}_1{}^T + \mathbf{e}_2\mathbf{e}_2{}^T + \mathbf{e}_3\mathbf{e}_3{}^T\right), \tag{5}$$

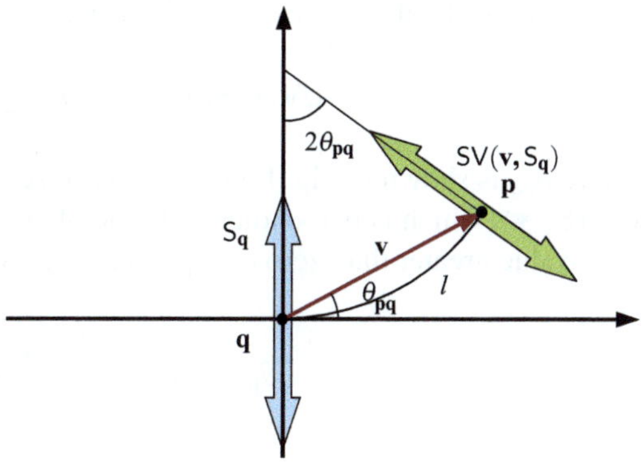

Fig. 1 The *stick* tensor voting. A stick $\mathsf{S}_\mathbf{q}$ casts a stick vote $\mathsf{SV}(\mathbf{v}, \mathsf{S}_\mathbf{q})$ to **p** that corresponds to the most likely tensorized normal at **p**

where λ_i and $\mathbf{e_i}$ are the i-th largest eigenvalue and its corresponding eigenvector of the tensor at **q**.

Saliency measurements can be estimated from an analysis of the eigenvalues of the resulting tensors in (2). Thus, $s_1 = (\lambda_1 - \lambda_2)$, $s_2 = (\lambda_2 - \lambda_3)$, and $s_3 = \lambda_3$ can be used as measurements of surfaceness, curveness and junctionness respectively. Points whose three eigenvalues are small are regarded as outliers. In addition, eigenvector $\pm\mathbf{e_1}$ represents the most likely normal for points lying on a surface, whereas $\pm\mathbf{e_3}$ represents the most likely tangent direction of a curve for points belonging to that curve.

Extensions of tensor voting to N-dimensions are straightforward. In this case, tensors are decomposed into a *stick*, a *ball* and N-2 *plate* components, which are processed through the N-D *stick*, N-D *ball* and N-D *plate* tensor voting respectively [43]. These processes are natural extensions of the 3D case. The next subsections describe how the *stick*, *plate* and *ball* votes are calculated in 3D.

2.1 Stick Tensor Voting

Stick tensors are used by tensor voting in 3D to encode the orientation of the surface normal at a specific point. Tensor voting handles *stick* tensors through the so-called *stick* tensor voting, which aims at propagating surfaceness in a neighborhood by using the perceptual principles of proximity, similarity and good continuation borrowed from the Gestalt psychology [5]. The *stick* tensor voting is based on the hypothesis that surfaces are usually smooth. Thus, tensor voting assumes that normals of neighboring points lying on the same surface change smoothly. This process is illustrated in Fig. 1. Given a known orientation of the normal at a point **q**, which is encoded by $\mathsf{S}_\mathbf{q}$, the orientation of the normal at a neighboring point **p** can be inferred by tracking the change of the normal on a joining smooth curve. Although any smooth curve can be used to calculate *stick* votes, a circumference is usually chosen. A decaying function, w_s, is also used to weight the vote as defined below.

It is not difficult to show from Fig. 1 that for a circumference:

$$\mathsf{SV}(\mathbf{v}, \mathsf{S}_\mathbf{q}) = w_s \left[R_{2\theta_{\mathbf{pq}}} \, \mathsf{S}_\mathbf{q} \, R^T_{2\theta_{\mathbf{pq}}} \right], \tag{6}$$

where $\theta_{\mathbf{pq}}$ is shown in Fig. 1 and $R_{2\theta_{\mathbf{pq}}}$ represents a rotation with respect to the axis $\mathbf{v} \times (\mathsf{S}_\mathbf{q}\ \mathbf{v})$, which is perpendicular to the plane that contains $\mathbf{v}$ and $\mathsf{S}_\mathbf{q}$. Let λ be the eigenvalue greater than zero of $\mathsf{S}_\mathbf{q}$. Angle $\theta_{\mathbf{pq}}$ is given by:

$$\theta_{\mathbf{pq}} = arcsin\left(\sqrt{\frac{\mathbf{v}^T \ \mathsf{S}_\mathbf{q} \ \mathbf{v}}{\lambda \mathbf{v}^T \mathbf{v}}} \right). \tag{7}$$

A point $\mathbf{q}$ can only cast *stick* votes for $\theta_{\mathbf{pq}} \le \pi/4$, since the hypothesis that both points $\mathbf{p}$ and $\mathbf{q}$ belong to the same surface becomes more unlikely for larger values of $\theta_{\mathbf{pq}}$. On the other hand, the weighting function w_s is used to reduce the strength of the vote with the arc length, l, given by:

$$l = \frac{||\mathbf{v}|| \ \theta_{\mathbf{pq}}}{\sin(\theta_{\mathbf{pq}})}, \tag{8}$$

and with its curvature, κ, given by:

$$\kappa = \frac{2 \ \sin(\theta_{\mathbf{pq}})}{||\mathbf{v}||}. \tag{9}$$

Thus, w_s is defined as [33]:

$$w_s(\mathbf{v}, \mathsf{S}_\mathbf{q}) = \begin{cases} e^{-\frac{l^2}{2\rho^2} - b\kappa^2} & \text{if } \theta_{\mathbf{pq}} \le \pi/4 \\ 0 & \text{otherwise,} \end{cases} \tag{10}$$

where ρ is a scale parameter and b can be adjusted to give more importance to the curvature. Following the methodology proposed in [33], b has been set to $||v||^2/4$ in the experiments.

2.2 *The Plate Tensor Voting*

A *plate* tensor is a tensor with $\lambda_1 = \lambda_2 \ge 0$ and $\lambda_3 = 0$. *Plate* tensors are processed through the so-called *plate* tensor voting. The *plate* tensor voting uses the fact that any *plate* tensor, P, can be decomposed into all possible *stick* tensors inside the *plate*. Let $\mathsf{S}_\mathsf{P}(\beta) = R_\beta \mathbf{e_1} \mathbf{e_1}^T R^T_\beta$ be a *stick* inside the *plate* P, with $\mathbf{e_1}$ being its principal eigenvector, and R_β being a rotation with respect to an axis perpendicular to $\mathbf{e_1}$ and $\mathbf{e_2}$. Thus, P can be written as:

$$\mathsf{P} = \frac{\lambda_1 + \lambda_2}{2\pi} \int_0^{2\pi} \mathsf{S}_\mathsf{P}(\beta) \, d\beta. \tag{11}$$

Taking into account that $\mathsf{S_P}(\beta)$ is a *stick* tensor, the *plate* vote is defined as the aggregation of *stick* votes cast by all the *stick* tensors $\mathsf{S_{P_q}}(\beta)$ that constitute $\mathsf{P_q}$. Thus, the *plate* vote is defined as:

$$\mathsf{PV}(\mathbf{v}, \mathsf{P_q}) = \frac{\lambda_1}{\pi} \int_0^{2\pi} \mathsf{SV}(\mathbf{v}, \mathsf{S_{P_q}}(\beta))\, d\beta. \tag{12}$$

Although this integral cannot be simplified, *plate* votes can be computed efficiently by using the method proposed in [33].

2.3 *The Ball Tensor Voting*

A *ball* tensor is a tensor with $\lambda_1 = \lambda_2 = \lambda_3 \geq 0$. The *ball* tensor voting is defined in a similar way as the *plate* tensor voting. Let $\mathsf{S_B}(\phi, \psi)$ be a unitary *stick* tensor oriented in the direction $(1, \phi, \psi)$ in spherical coordinates. Then, any *ball* tensor B can be written as:

$$\mathsf{B} = \frac{\lambda_1 + \lambda_2 + \lambda_3}{4\pi} \int_\Gamma \mathsf{S_B}(\phi, \psi)\, d\Gamma, \tag{13}$$

where Γ represents the surface of the unitary sphere. Using the same argument as in the case of the *plate* tensor voting, the *ball* vote is defined as:

$$\mathsf{BV}(\mathbf{v}, \mathsf{B_q}) = \frac{3\lambda_1}{4\pi} \int_\Gamma \mathsf{SV}(\mathbf{v}, \mathsf{S_{B_q}}(\phi, \psi))\, d\Gamma. \tag{14}$$

Similarly to the *plate* tensor voting, this integral cannot be simplified. However, *ball* votes can be computed efficiently by using the method proposed in [33].

3 Relationships Between the Structure Tensor and Tensor Voting

Although the structure tensor and tensor voting are usually applied to two different scopes, images and clouds of points, both aim at estimating structure, as it will be shown in this section. This section describes the relationships between the structure tensor and tensor voting.

3.1 *Similarities*

With the exception of the rotation term and the restriction of $\theta_{\mathbf{pq}} \leq \pi/4$ in (10), the formulation of the *stick* tensor voting in (6) has a structure similar to that of the structure tensor in (1). In particular, the term $\nabla u \nabla u^T$ in (1) plays a similar role as the term $\mathsf{S}_\mathbf{q}$ in (6), while function w_s of the *stick* tensor voting is closely related to the Gaussian kernel used by the structure tensor. In addition to these structural similarities, both methods have functional connections, since they can be adapted to be applied to the same contexts. Especially, the structure tensor can be adapted to estimation of structures in 3D, and tensor voting can be adapted to estimation of structures in gray-scale images.

On the one hand, the structure tensor can be adapted to estimation of structures in 3D with the help of a norm estimator. For example, the local norm can be estimated by computing the equation of the most likely tangent plane at every point. The norms obtained with such an estimator can be tensorized and convolved with a Gaussian in order to estimate structure in 3D. The resulting tensors yielded by both methods can be analyzed in the same manner. For example, $\lambda_1 - \lambda_2$ can be used as a measure of surfaceness, $\lambda_2 - \lambda_3$ as a measure of curveness, and λ_3 as a measure of junctionness, as in the case of tensor voting [26].

In turn, tensor voting can be adapted to image structure estimation by designing an appropriate encoding step. Taking into account that the normal, $\mathbf{n}_\mathbf{q}$ in a gray-scale image corresponds to the normalized gradient, $\nabla u_\mathbf{q}/||\nabla u_\mathbf{q}||$, the *stick* component $\mathsf{S}_\mathbf{q}$ in (3) can be written as:

$$\mathsf{S}_\mathbf{q} = (\lambda_1 - \lambda_2)\left(\frac{\nabla u_\mathbf{q} \nabla u_\mathbf{q}^T}{||\nabla u_\mathbf{q}||^2}\right), \tag{15}$$

which can be further simplified by choosing $(\lambda_1 - \lambda_2) = ||\nabla u_\mathbf{q}||^2$. Thus:

$$\mathsf{S}_\mathbf{q} = \nabla u_\mathbf{q} \nabla u_\mathbf{q}^T. \tag{16}$$

In addition, if the components $\mathsf{P}_\mathbf{q}$ and $\mathsf{B}_\mathbf{q}$ are set to zero, the input of both, the structure tensor and tensor voting, becomes equivalent for gray-scale images. As in the 3D case, the output of both methods can be analyzed in a similar way, since, in 2D, the shape of the tensors at edges is closer to a *stick*, while the shape tends to a *ball* at corners in both cases (in 2D, the *plate* component is undefined). However, the tensors obtained by means of tensor voting are in a different scale. Hence, it is necessary to apply a rescaling function in order to have comparable results.

3.2 Differences

As already mentioned, both methods have two essential differences: the rotation term in (6) and the restriction of $\theta_{\mathbf{pq}} \leq \pi/4$ in (10). These differences are given by the different assumptions made by both methods. On the one hand, the hypothesis of tensor voting is that $\mathbf{p}$ and $\mathbf{q}$ belong to the same smooth curve and the voting processes are adjusted according to this hypothesis. On the other hand, the hypothesis made by the structure tensor is that the orientation of the normal at

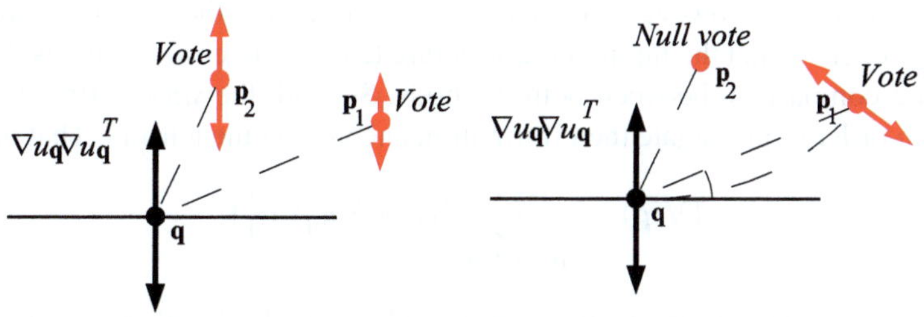

Fig. 2 *Left*: the structure tensor seen as a voting process. *Right*: the *stick* tensor voting. The main differences between both are the rotation term (see the difference of votes at $\mathbf{p}_1$) and the anisotropic behavior of tensor voting (tensor voting does not cast votes to $\mathbf{p}_2$)

neighboring points should be similar, by taking into account that the orientation of the normal in a smooth curve usually changes slowly.

These differences can be seen in Fig. 2. The structure tensor can be modeled as a voting process in which every point votes for its own orientation with a strength given by a Gaussian function. Thus, the structure tensor propagates its own orientation isotropically. This approach can be seen as a displacement to $\mathbf{p}$ of the surface at $\mathbf{q}$. In turn, tensor voting propagates a rotated version of the original orientation when $\theta_{\mathbf{pq}} \leq \pi/4$. It is expected that tensor voting performs better than the structure tensor as it makes stronger assumptions.

4 Tensor Voting for Structure Estimation

The structural relationships shown in Sect. 3 lead to a general methodology to extend tensor voting to different types of images. These extensions can be used to improve the image structure estimation obtained by means of the structure tensor. The methodology comprises three steps. First, tensors are initialized in the same way as for the structure tensor in every different type of images. Second, the Gaussian convolution used by the structure tensor is replaced by tensor voting. Finally, the resulting tensors are rescaled in order to renormalize the total energy stored in the tensors. The following subsections show how this general methodology can be applied to different types of images.

4.1 Gray-Scale Images

From Sect. 3, tensor voting can be directly applied to image structure estimation in gray-scale images by following the next three steps. First, the tensorized gradient, $\nabla u \nabla u^T$, is used to initialize a tensor at every pixel. It is important to remark that other types of tensor can be used in the initialization step, for example, *ball* tensors as proposed in [23]. However, the advantage of initializing the tensors with the tensorized gradient is that the input of the structure tensor and tensor voting is the same, easing the comparison between both methods. Second, the *stick* voting process is applied in order to propagate the information encoded in the tensors. That is:

$$\mathsf{TV}(\mathbf{p}) = \sum_{\mathbf{q}\in neigh(\mathbf{p})} \mathsf{SV}(\mathbf{v}, \nabla u_{\mathbf{q}} \nabla u_{\mathbf{q}}^T). \tag{17}$$

Notice that it is not necessary to apply the *plate* and *ball* voting processes since the *plate* and *ball* components are zero at every pixel due to the initialization step. Finally, the resulting tensors are rescaled by the factor:

$$\xi = \frac{\sum_{\mathbf{p}\in\Omega} trace(\nabla u_{\mathbf{p}} \nabla u_{\mathbf{p}}{}^T)}{\sum_{\mathbf{p}\in\Omega} trace(\mathsf{TV}(\mathbf{p}))}, \tag{18}$$

in order to renormalize the total energy of the tensorized gradient, where Ω refers to the given image. This scaling is applied in order to get comparable results to those obtained with the structure tensor.

4.2 Color and Vector-Valued Images

The structure tensor has already been extended to multivalued images in [7] and in a more general way in [46]:

$$\mathsf{J} = \sum_{k=1}^{d} G_\rho * w_k \nabla u(k) \nabla u(k)^T = G_\rho * \sum_{k=1}^{d} w_k \nabla u(k) \nabla u(k)^T, \tag{19}$$

where d is the number of channels, $\nabla u(k)$ is the gradient at channel k, and w_k are weights used to give different relevance to every channel. From (19), the structure tensor can be equivalently estimated either by adding d structure tensors, one for every channel, or by applying a Gaussian kernel on the (weighted) summation of the tensorized gradients $\nabla u(k) \nabla u(k)^T$. The reason why both alternatives are equivalent for computing the structure tensor is that Gaussian convolution is linear. However, this equivalence does not hold for non-linear averaging methods, including tensor voting. Thus, there are two options to extend tensor voting for this kind of images, considering that tensor voting must replace the Gaussian convolution used in the structure tensor. The first option is to apply the *stick* tensor voting independently to every channel and then adding up the individual results:

$$\mathsf{TV}(\mathbf{p}) = \sum_{k=1}^{d} \sum_{\mathbf{q}\in neigh(\mathbf{p})} w_k\ \mathsf{SV}(\mathbf{v}, \nabla u_{\mathbf{q}}(k) \nabla u_{\mathbf{q}}(k)^T). \tag{20}$$

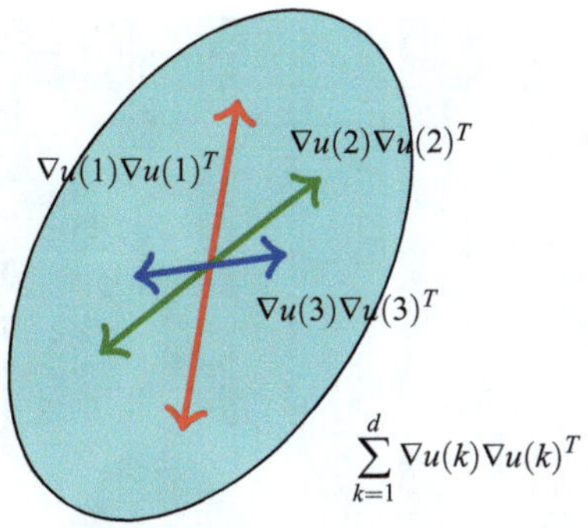

Fig. 3 Tensor voting can be applied to the channels independently (the *red*, *green* and *blue sticks*) or to the sum of the tensorized gradients (the ellipse)

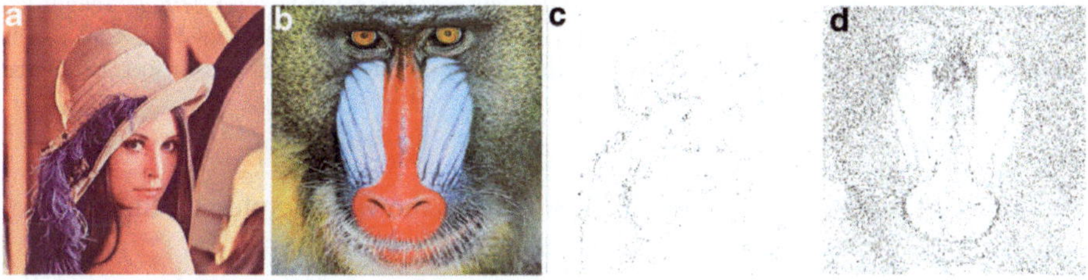

Fig. 4 (**a**) Lenna. (**b**) Mandrill. (**c, d**) Pixels (in *black*) with $\lambda_2 \geq 0.1\,\lambda_1$ of T_q for both images. Processing channels independently is appropriate in most pixels of natural images

The second option is to apply (2) to the sum of tensorized gradients with S_q, P_q and B_q being the *stick*, *plate* and *ball* components of $T_q = \sum_{k=1}^{d} w_k \nabla u_q(k) \nabla u_q(k)^T$. For two-dimensional images, $P_q = 0$. In both options, rescaling the calculated tensors is performed in a similar way as described for the gray-scale images. Figure 3 shows the options described above.

The first option has the advantage that only the application of the *stick* tensor voting is necessary, whereas for the second option, the *stick*, *plate* (for 3D color images) and *ball* tensor voting are required. On the other hand, the second option tends to be more robust since it is less sensitive to bad initial estimations of the gradient. However, in practice, $T_q \approx S_q$ in most pixels of natural images. As an example, in Fig. 4 the number of pixels with λ_2 greater than the 10 % of λ_1 of T_q corresponds to only 0.8 % of the total for Lenna and 12.2 % for the more textured Mandrill. Thus, the first option can be used in most of the pixels and the second one only in those pixels in which the approximation is not valid.

4.3 Tensor-Valued Images

A tensor-valued image is an image in which a tensor is associated with every pixel or voxel. As an example, images acquired through diffusion tensor magnetic resonance

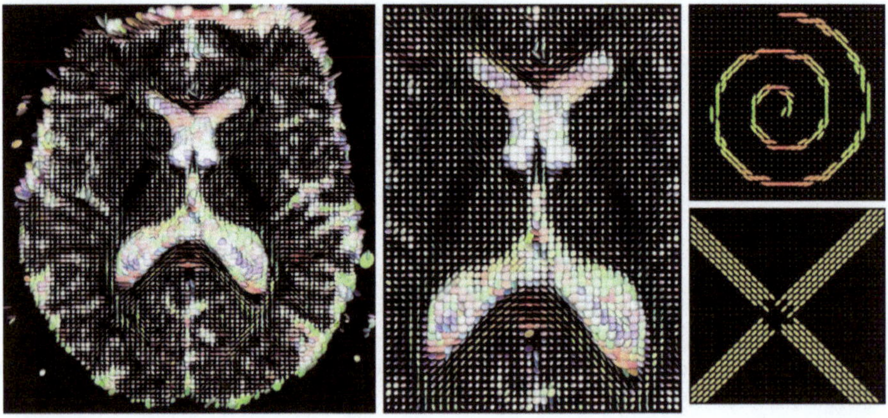

Fig. 5 *Left*: 2D slice extracted from a 3D DT-MRI data set (128 × 128 voxels). *Middle*: magnified slice around the lateral ventricles (40 × 55 voxels). *Right*: two synthetic data sets (32 × 32 voxels each) of a spiral (*top*) and a cross (*bottom*). Ellipsoids are used to represent the tensors associated with every voxel

imaging (DT-MRI) are tensor-valued. Figure 5 shows examples of this kind of images.

Unlike gray-valued and color images, there are several ways to extend the structure tensor concept to tensor-valued images. One of them was proposed by Weickert and Brox [47] in which the structure tensor is calculated through (19), with the channels corresponding to the entries in the tensors. Thus, the same methodology presented in the previous subsection can be used for adapting tensor voting to tensor-valued images by using the entries in the tensors as the channels of a vector-valued image. Moreover, the factors w_k can be set for tensor-valued images by using the fact that any symmetric matrix, M, can be modeled by means of a vector, m, which is given in an orthonormal tensorial basis with respect to the internal product $\langle A, B\rangle = \mathit{trace}(AB^T)$ [18,37]:

$$\mathsf{M} = \begin{bmatrix} m_{11} & m_{12} & m_{13} \\ m_{21} & m_{22} & m_{23} \\ m_{31} & m_{32} & m_{33} \end{bmatrix} \iff m = \begin{bmatrix} m_{11} \\ \sqrt{2}m_{12} \\ \sqrt{2}m_{13} \\ m_{22} \\ \sqrt{2}m_{23} \\ m_{33} \end{bmatrix}. \tag{21}$$

This modeling makes equivalent the Frobenius norm $|\mathsf{M}|_F = \sqrt{\mathit{trace}(\mathsf{M}\mathsf{M}^T)}$ and the norm of m. Thus, tensor voting can be applied to vectors m instead of to tensors M by using the methodology presented in the previous subsection, with $w_k = 1$ for the diagonal entries and $w_k = \sqrt{2}$ for the other entries.

Similarly to the case of color images, there are two options for applying tensor voting: to compute six *stick* votes, one for every channel, or to compute the complete tensor voting framework on the summation of six *stick* tensors, one for every channel (cf. Fig. 3). As in the case of color images, the more expensive second option is only necessary at pixels (voxels) where the gradient computed for one channel is very different from the one computed for one another channel.

An alternative extension of tensor voting for diffusion images can be proposed by taking into account the differential nature of this type of images. In DT-MRI, the eigenvectors of the acquired tensors are tangent to the main diffusivity orientations of the movement of water molecules at every voxel. Instead, normal orientations are required to compute structure. Following the approach in [3], such orientations can be extracted from tensors R computed as:

$$\mathsf{R} = trace(\mathsf{T})\ \mathsf{I} - \mathsf{T}, \tag{22}$$

where T is the acquired tensor and I the identity matrix. This transformation only modifies the eigenvalues, since tensors R and T share the same eigenvectors. The eigendecomposition of these tensors is given by:

$$\mathsf{R} = \sum_{i=1}^{3} \lambda_i \mathbf{e}_i \mathbf{e}_i^T. \tag{23}$$

Factor $\mathbf{e}_i \mathbf{e}_i^T$ can be interpreted as tensorized gradients in the image. Thus, by using this analogy, a structure tensor can be defined as:

$$\mathsf{J} = G_\rho * \sum_{i=1}^{3} \lambda_i \mathbf{e}_i \mathbf{e}_i^T = \sum_{i=1}^{3} G_\rho * \lambda_i \mathbf{e}_i \mathbf{e}_i^T, \tag{24}$$

where the equivalence is given by the linearity of the Gaussian convolution.

Similarly to the case of Fig. 3, there are two approaches to extend tensor voting to this type of images: to apply the *stick* tensor voting to every $\lambda_i \mathbf{e}_i \mathbf{e}_i^T$ or to directly apply the *stick*, *plate* and *ball* tensor voting to the tensors R.

In Chapter 13, the *stick* votes are represented and accumulated as higher-order tensors, whose weight and orientation are derived from second-order tensors. The analysis of these higher-order tensors is then performed through a low-rank approximation, as proposed in [41].

More sophisticated methods have already been proposed for extending the concept of the structure tensor to tensor-valued images. For example, Burgeth et al. [6] use an algebraic approach to deal with the intrinsic third order nature of the gradient of tensor-valued images. Nevertheless, an adaptation of tensor voting based on these methods requires the extension of the voting processes for higher-dimensional tensor-valued images, which is out of the scope of this chapter.

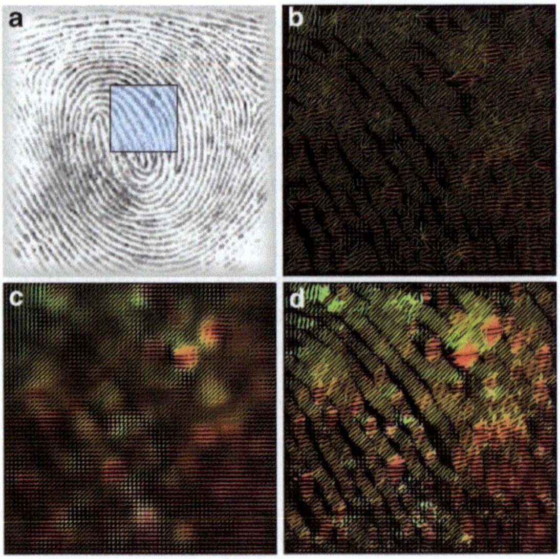

Fig. 6 (**a**) A fingerprint with a region of interest (ROI). (**b**) $\nabla u \nabla u^T$ in the ROI. (**c, d**) The structure tensor and tensor voting in the ROI respectively ($\rho = 2/\sqrt{2}$). Tensor voting preserves gaps

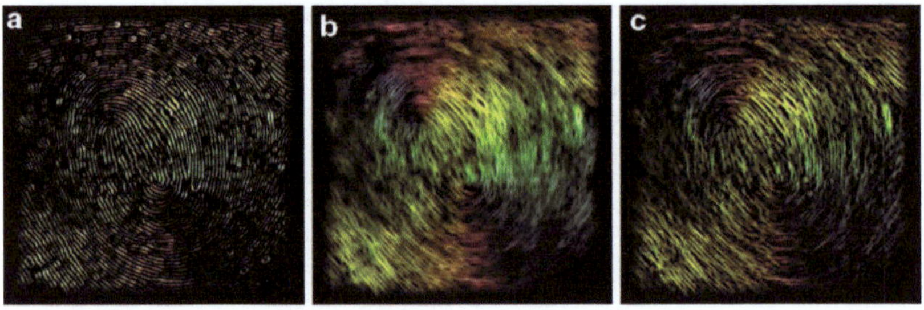

Fig. 7 (**a–c**) Color coded orientation (*green* = 0, *yellow* = $pi/4$, *red* = $\pi/2$, *blue* = $3\pi/4$) of $\nabla u_q \nabla u_q{}^T$, the structure tensor and tensor voting respectively ($\rho = 3/\sqrt{2}$) for the fingerprint of Fig. 6a. Both methods smooth the orientation of the gradient

5 Experimental Results

Figures 6–8 present the structure estimation in a fingerprint by means of both the structure tensor and tensor voting. Figure 6 shows that tensor voting is able to preserve the gaps in the image, while the structure tensor is not. This means that tensor voting avoids estimating structure in unstructured regions, which is one of the known problems of the structure tensor.

Figure 7 shows that the orientation of the gradient is smoothed by both the structure tensor and tensor voting. This is a good property of a structure estimator, since orientation usually changes slowly in an image and is noisy in $\nabla u \nabla u^T$.

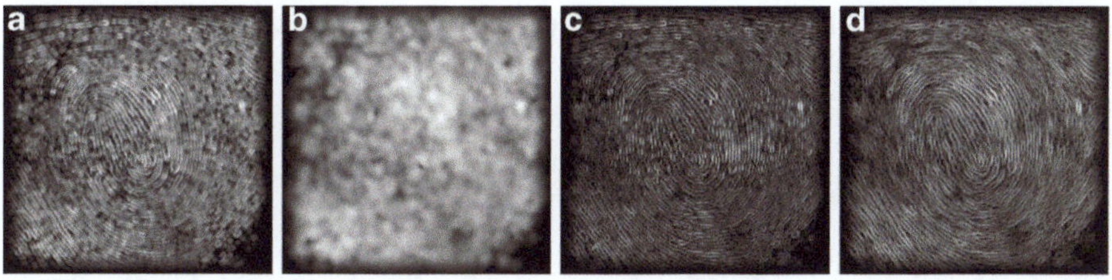

Fig. 8 (**a**, **b**) Map of $\lambda_1 - \lambda_2$ obtained with the structure tensor for $\rho = 1/\sqrt{2}$ and $\rho = 2/\sqrt{2}$ respectively. (**c**, **d**) Map of $\lambda_1 - \lambda_2$ obtained with tensor voting for the same values of ρ. The structure tensor is more sensitive to ρ

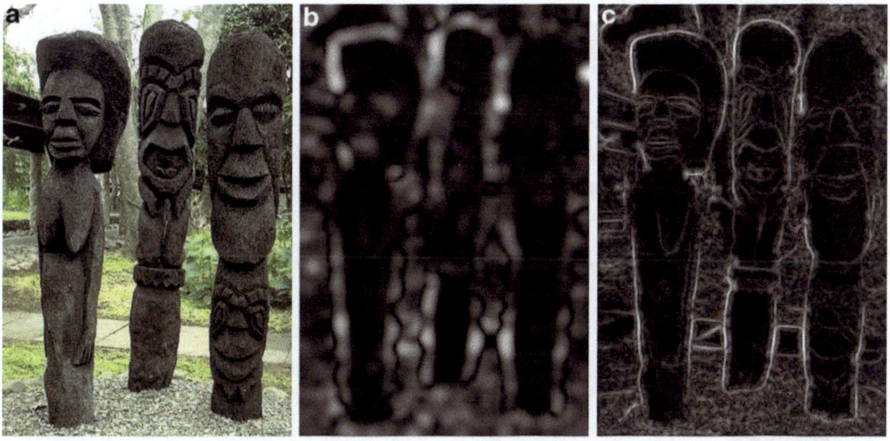

Fig. 9 (**a**) Original image. (**b**, **c**) Map of $\lambda_1 - \lambda_2$ for the structure tensor and tensor voting respectively ($\rho = 3/\sqrt{2}$). The structure tensor blurs the edges

Figure 8 shows the map of $\lambda_1 - \lambda_2$, which can be used to extract edges. It can be seen that the structure tensor is more sensitive to the selection of the parameter ρ, while tensor voting yields similar results for a greater range of values. Thus, it is more difficult to tune the parameter of the structure tensor than the scale parameter of tensor voting.

Figure 9 shows an example for edge detection. Since ideal edges are characterized by *stick* tensors, edges can be obtained by applying non-maximum suppression and hysteresis to the map of $\lambda_1 - \lambda_2$, which measures how far every pixel is from that condition. It can be seen that the structure tensor blurs that map. This can lead to misplacements of the binary edges extracted from these maps and to loss of small edges. For example, edges inside faces are completely lost, and the eyebrow of the totem at the left-hand side is misplaced. Tensor voting is able to keep edges thinner, reducing in that way the problems of the structure tensor.

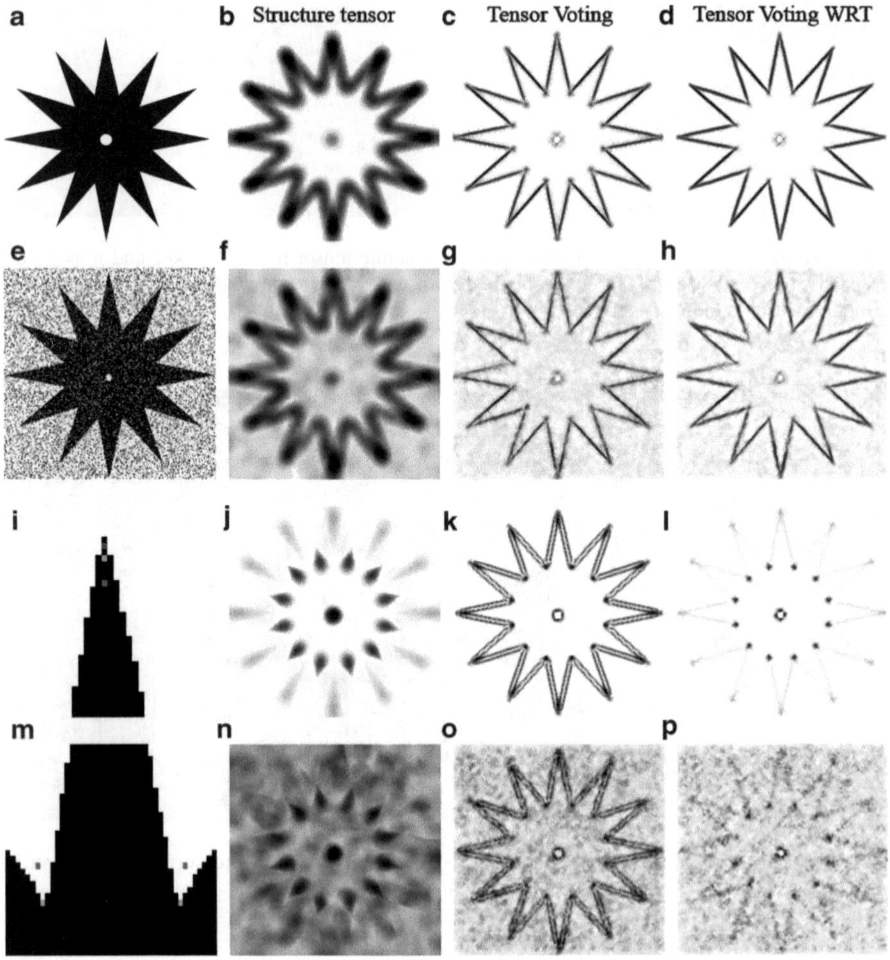

Fig. 10 (**a**) Original image. (**e**) Noisy image (truncated Gaussian noise with $\sigma = 100$). (**b–d**) Maps of λ_1 obtained with the structure tensor, and tensor voting with and without rotation term (WRT) in (6) respectively for the original image ($\rho = 3/\sqrt{2}$). (**f–h**) Maps of λ_1 obtained with the three methods for the noisy image. (**j–l**) Maps of λ_2 obtained with the three methods for the original image. (**n–p**) Maps of λ_2 obtained with the three methods for the noisy image. (**i**) Detected corner at a peak of the star by the structure tensor (*red*), tensor voting (*green*) and tensor voting without the rotation term (*blue*). (**m**) Detected corners at two valleys of the star by the structure tensor (*red*) and both versions of tensor voting (*green*)

Most corner detectors apply a function on the eigenvalues of the structure tensor [16]. Hence, accuracy and robustness in the estimation of eigenvalues are requirements for this application. Figure 10 shows plots of λ_1 and λ_2 from tensors estimated by means of both the structure tensor and tensor voting for a noiseless and a noisy synthetic image. Figure 10 shows that tensor voting is more robust and more

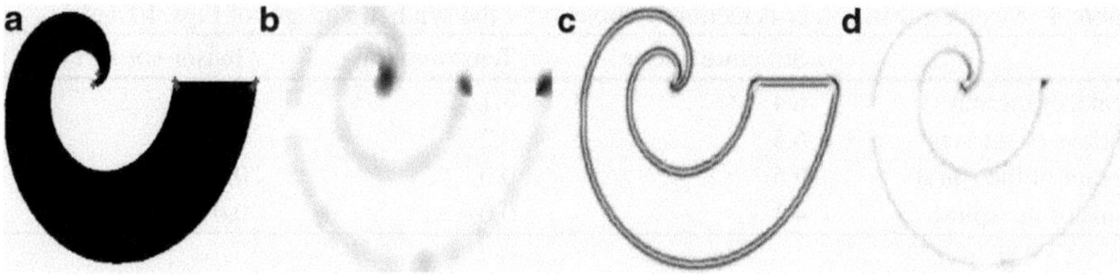

Fig. 11 (**a**) Original image with the detected corners with the structure tensor (in *red*) and tensor voting with and without rotation term in (6) (in *green*). (**b–d**) Maps of λ_2 obtained with the three methods respectively for the original image ($\rho = 3/\sqrt{2}$)

accurate than the structure tensor in the estimation of λ_1. In addition, the structure tensor mistakenly introduces a maximum in λ_1 in the middle of the small hole inside the star, while tensor voting does not.

In addition, Fig. 10 shows that the structure tensor has a bad performance for both noiseless and noisy images. Actually, it blurs λ_2 in such a way that the corners are displaced. In addition, it is very sensitive to noise and generates a false maximum in the hole at the middle of the star. On the other hand, tensor voting has a more consistent performance in estimating λ_2 in both noiseless and noisy images. Although tensor voting generates a halo near edges, it can be filtered out by taking into account that it only appears near edges and has smaller values of λ_2 than in the corners.

Figure 10 also shows the effect of the rotation term in (6). This figure shows plots of λ_1 and λ_2 from tensors estimated by means of tensor voting without the rotation term for the image of the star. It can be seen that the effect of the rotation term in (6) on λ_1 is almost negligible, since the results are similar for both the noiseless and noisy images (see Fig. 10c vs. 10d, and Fig. 10g vs. 10h). Regarding λ_2, tensor voting without the rotation term has a better performance in the noiseless image, since it does not insert halos (see Fig. 10l). However, its performance is not robust, since it is difficult to extract maxima from its estimation for the noisy image (see Fig. 10p). Thus, tensor voting with the rotation term is more robust in the estimation of λ_2. This effect also appears in curved edges, as shown in Fig. 11. In conclusion, the rotation term of (6) robustifies the estimation of λ_2 at a cost of introducing halos that should be filtered out a posteriori. It is noteworthy to remark that the method proposed by Köthe [20] is closely related to tensor voting without the rotation term. The only difference between both methods is the use of a different, but still closely related, weighting function.

Regarding precision, tensor voting both with and without the rotation term is able to detect corners with a smaller error. Corners have been detected by looking at local maxima in the map of λ_2 (see Figs. 10i, m and 11a). Table 1 shows the mean errors yielded by both the structure tensor and tensor voting. The strategies based on tensor voting yield better results than the structure tensor in all cases. Notice that corners at the peaks of the star are more difficult to detect, since the angles between

Table 1 Mean error in corner detection (in pixels) for the synthetic images of Figs. 10 and 11

	Structure tensor	Tensor voting	Tensor voting (WRT)
Peaks of the star	6.4	3.1	0.9
Valleys of the star	6.5	0.2	0.2
Center of the spiral	9.5	0.0	0.0
Ends of the spiral	4.1	0.0	0.0

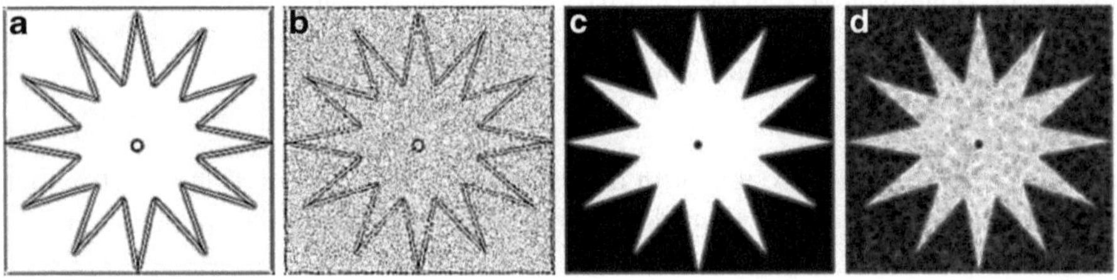

Fig. 12 (**a, b**) Maps of $\lambda_1 - \lambda_2$ calculated with the method by Loss et al. [23] for the images of Fig. 10a, e respectively ($\rho = 3/\sqrt{2}$). (**c, d**) Maps of λ_2 calculated with he method by Loss et al. [23] for the images of Fig. 10a, d respectively ($\rho = 3/\sqrt{2}$). Values have been inverted for a better visualization

the edges that abut at the corner are smaller. In turn, binary edges extracted from the star and the spiral through non-maximum suppression coincide with the ground-truth for both versions of tensor voting. The accuracy of the edges extracted from the structure tensor is also good in regions far away from corners, but it is largely degraded in regions close to corners.

Moreover, the method proposed by Loss et al. [23] has been implemented in order to compare two different approaches for extending tensor voting to gray-scale images. Figure 12 shows the results of applying this method to the images of Fig. 10a, e. As can be seen, $\lambda_1 - \lambda_2$ generates similar responses at both edges and flat regions, making it difficult to detect edges in noisy images. In turn, λ_2 gives no additional information, since it yields a smoothed version of the original image.

Finally, Fig. 13 shows that tensor voting is also a better option to be used instead of the structure estimation for tensor-valued images. This figure shows the results yielded for the images of Fig. 5 by both the structure tensor and tensor voting computed through the two alternatives described in Sect. 4.3, that is, by modeling tensors as vectors, and by applying (24) and its extension to tensor voting. Notice that both alternatives are not comparable since the former estimates structure in the input tensorial image, while the latter estimates structure in an image related to the inverse gradient [8] of the input image. This fact explains, for example, why tensor voting detects two edges in Fig. 13j for every leg of the cross, while it detects only one edge in Fig. 13l. As appreciated in these images, the structure tensor blurs the resulting tensors in such a way that it is difficult to extract edges and corners from them. On the contrary, tensor voting is able to estimate structure in a better way.

Str. tensor (Alt. 1)	Tensor Voting (Alt. 1)	Str. tensor (Alt. 2)	Tensor Voting (Alt. 2)

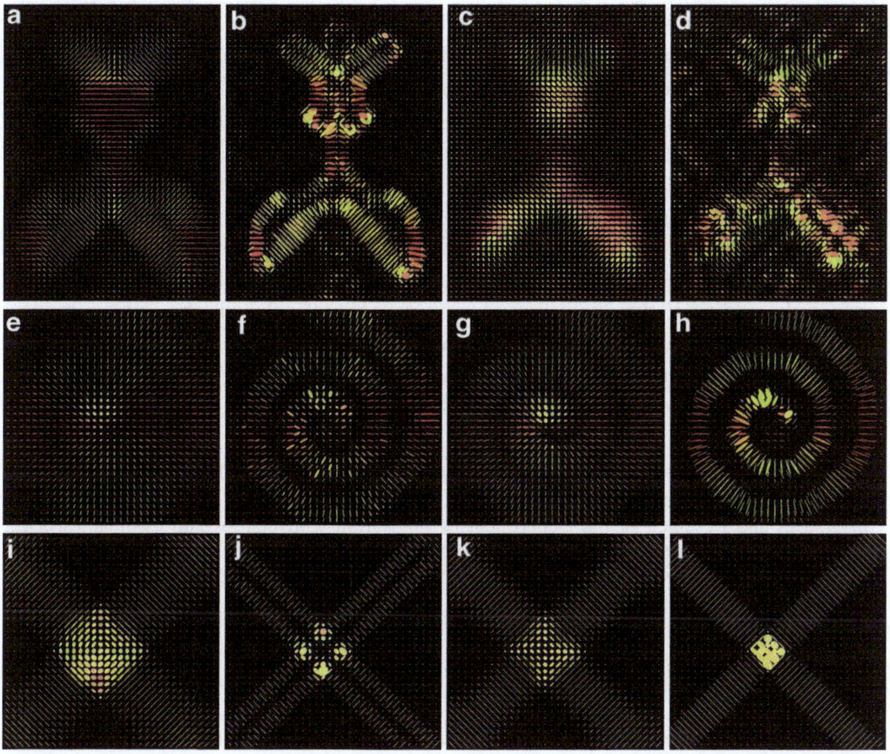

Fig. 13 Resulting tensor fields after applying the structure tensor and the two alternative extensions of tensor voting described in Sect. 4.3 respectively ($\rho = 5/\sqrt{2}$) for the images of Fig. 5. Alt. 1 models tensors as vectors, and Alt. 2 is based on (24)

6 Concluding Remarks

This chapter proposes a general methodology to adapt tensor voting for estimating image structure based on the fact that the *stick* tensor voting and the structure tensor are structurally similar, as shown in Sect. 3. Section 4 has shown how this methodology can be applied to different types of images. Experimental results show that tensor voting can estimate structure more appropriately than the structure tensor. In addition, tensor voting yields more robust estimations of structure than the structure tensor. The rotation term in the stick tensor voting leads to more robust estimations of λ_2 but also generates halos that should be filtered out a posteriori.

It is interesting to remark that the close relationship between the structure tensor and tensor voting has advantages and shortcomings. On the one hand, this relationship can be used to extend tensor voting to different types of images, as proposed in this paper. On the other hand, this relationship also limits the scope of use of tensor voting to structure estimation. Thus, there are three options to extend tensor voting to other applications. The first one is to use tensor voting in the processing step where structure estimation is required, as many previous works have done. The second one is to model the problem in terms of structure estimation, for example, by using different encoding steps. The third one is to adapt the voting process by encoding new application-dependent perceptual rules. Given its recent success [30–32], the third option appears to be the most promising approach for the majority of applications.

Future work includes comparing different ways to perform tensor voting on tensor-valued images and extending the proposed methodology to higher-order tensors. In addition, the inclusion of new perceptual rules in the voting process will be explored in order to eliminate the halos generated by tensor voting in the estimation of λ_2 without a post-processing step. Furthermore, comparisons with other approaches in order to combine tensors locally, e.g. [48], are planned.

Acknowledgements This research has been partially supported by the Spanish Ministry of Science and Technology under project DPI2007-66556-C03-03, by the Commissioner for Universities and Research of the Department of Innovation, Universities and Companies of the Catalonian Government and by the European Social Fund.

References

1. Arseneau, S., Cooperstock, J.R.: An improved representation of junctions through asymmetric tensor diffusion. In: Bebis, G. et al. (eds.) Proc. Int. Symp. Vis. Comp. (ISVC). Lake Tahoe, NV, USA, Lecture Notes in Computer Science, vol. 4291, pp. I:363–372. Springer, Berlin/Heidelberg (2006)
2. Bigün, J., Granlund, G., Wiklund, J.: Multidimensional orientation estimation with applications to texture analysis and optical flow. IEEE Trans. Pattern Anal. Mach. Intell. **13**(8), 775–790 (1991)
3. Bigün, J., Bigun, T., Nilsson, K.: Recognition by symmetry derivatives and the generalized structure tensor. IEEE Trans. Pattern Anal. Mach. Intell. **26**(12), 1590–1605 (2004)
4. Brox, T., Weickert, J., Burgeth, B., Mrázek, P.: Nonlinear structure tensors. Image Vis. Comput. **24**(1), 41–55 (2006)
5. Bruce, V., Green, P.R., Georgeson, M.A.: Visual Perception: Physiology, Psychology and Ecology, 4th edn. Psychology Press, London, UK (2003)
6. Burgeth, B., Didas, S., Weickert, J.: A general structure tensor concept and coherence-enhancing diffusion filtering for matrix fields. In: Laidlaw, D., Weikert, J. (eds.) Visualization and Processing of Tensor Fields: Advances and Perspectives, pp. 305–323. Springer, Berlin/Heidelberg (2009)
7. Di Zenzo, S.: A note on the gradient of a multi-image. Comput. Vis. Graph. Image Process. **33**(1), 116–125 (1986)
8. Farnebäck, G., Rydell, J., Ebbers, T., Andersson, M., Knutsson, H.: Efficient computation of the inverse gradient on irregular domains. In: Int. Conf. Comput. Vis. (ICCV), Rio de Janeiro, pp. 1–8. IEEE Computer Society Press, Los Alamitos (2007)
9. Felsberg, M., Jonsson, E.: Energy tensors: quadratic, phase invariant image operators. In: Kropatsch, W., Sablatnig, R., Hanbury, A. (eds.) Proc. Symp. Ger. Assoc. Pattern Recognit. (DAGM), Vienna. Lecture Notes in Computer Science, vol. 3663, pp. 493–500. Springer, Berlin/New York (2005)

10. Förstner, W.: A feature based correspondence algorithm for image matching. In: Int. Arch. Photogramm. Remote Sens., Rovaniemi, vol. 26, pp. 150–166. ISPRS, Enschede (1986)
11. Förstner, W.: A framework for low-level feature extraction. In: Eklundh, J-O. (ed.) Proc. Eur. Conf. Comput. Vis. (ECCV), Stockholm. Lecture Notes in Computer Science, vol. 801, pp. 383–394. Springer, Berlin/New York (1994)
12. Granlund, G., Knutsson, H.: Signal Processing for Computer Vision. Kluwer, Dordrecht (1995)
13. Hahn, J., Lee, C.O.: A nonlinear structure tensor with the diffusivity matrix composed of the image gradient. J. Math. Imaging Vis. **34**, 137–151 (2009)
14. Hwang, C., Zhuang, S., Lai, S.H.: Efficient intra mode selection using image structure tensor for H.264/AVC. In: Proc. Int. Conf. Image Process. (ICIP), San Antonio, pp. V:289–292. IEEE, Piscataway (2007)
15. Jia, J., Tang, C.K.: Inference of segmented color and texture description by tensor voting. IEEE Trans. Pattern Anal. Mach. Intell. **26**(6), 771–786 (2004)
16. Kenney, C., Zuliani, M., Manjunath, B.: An axiomatic approach to corner detection. In: Proc. Comput. Vis. Pattern Recognit. (CVPR), San Diego, pp. I:191–197. IEEE Computer Society Press, Los Alamitos (2005)
17. Kim, H.S., Choi, H.K., Lee, K.H.: Feature detection of triangular meshes based on tensor voting theory. Comput.-Aided Des. **41**(1), 47–58 (2009)
18. Kindlmann, G., Ennis, D.B., Whitaker, R., Westin, C.F.: Diffusion tensor analysis with invariant gradients and rotation tangents. IEEE Trans. Med. Imaging **26**(11), 1483–1499 (2007)
19. Knutsson, H.: A tensor representation of 3-D structures. In: Proc. Workshop Multidimens. Signal Process. (1987)
20. Köthe, U.: Edge and junction detection with an improved structure tesnsor. In: Michaelis, B., Krell, G. (eds.) Proc. Symp. Ger. Assoc. Pattern Recognit. (DAGM), Magdeburg. Lecture Notes in Computer Science, vol. 2781, pp. 25–32. Springer, Berlin/New York (2003)
21. Köthe, U., Felsberg, M.: Riesz-transforms versus derivatives: on the relationship between the boundary tensor and the energy tensor. In: Kimmel, R., Sochen, N., Weickert, J. (eds.) Scale Space and PDE Methods in Computer Vision. Lecture Notes in Computer Science, vol. 3459, pp. 179–191. Hofgeismar, Germany (2005)
22. Lim, J., Park, J., Medioni, G.: Text segmentation in color images using tensor voting. Image Vis. Comput. **25**(5), 671–685 (2007)
23. Loss, L.A., Bebis, G., Parvin, B.: Iterative tensor voting for perceptual grouping of ill-defined curvilinear structures: application to adherens junctions. IEEE Trans. Med. Imaging 30(8), 1503–1513 (2011)
24. Lucas, B.D., Kanade, T.: An iterative image registration technique with an application to stereo vision. In: Proc. Imagning Underst. Workshop, pp. 121–130 (1981)
25. Massad, A., Babós, M., Mertsching, B.: Application of the tensor voting technique for perceptual grouping to grey-level images. In: Van Gool, L. (ed.) Proc. Symp. Ger. Assoc. Pattern Recognit. (DAGM), Zürich. Lecture Notes in Computer Science, vol. 2449, pp. 306–313. Springer, Berlin/New York (2002)
26. Medioni, G., Lee, M.S., Tang, C.K.: A Computational Framework for Feature Extraction and Segmentation. Elsevier Science, Amsterdam/New York (2000)
27. Min, C., Medioni, G.: Inferring segmented dense motion layers using 5D tensor voting. IEEE Trans. Pattern Anal. Mach. Intell. **30**(9), 1589–1602 (2008)
28. Mordohai, P., Medioni, G.: Dimensionality estimation, manifold learning and function approximation using tensor voting. J. Mach. Learn. **11**, 411–450 (2010)
29. Moreno, R., Garcia, M.A., Puig, D.: Graph-based perceptual segmentation of stereo vision 3D images at multiple abstraction levels. In: Escolano, F., Vento, M. (eds.) Proc. Workshop Graph-Based Represent. Pattern Recognit. (GbRPR), Alicante. Lecture Notes in Computer Science, vol. 4538, pp. 148–157. Springer, Berlin/New York (2007)
30. Moreno, R., Garcia, M.A., Puig, D., Julià, C.: On adapting the tensor voting framework to robust color image denoising. In: Jiang, X., Petkov, N. (eds.) Proc. Comput. Anal. Images Patterns (CAIP), Münster. Lecture Notes in Computer Science, vol. 5702, pp. 492–500. Springer, Berlin/New York (2009)

31. Moreno, R., Garcia, M.A., Puig, D., Julià, C.: Robust color edge detection through tensor voting. In: Proceedings of International Conference on Image Processing (ICIP), Cairo, pp. 2153–2156. IEEE, Piscataway (2009)
32. Moreno, R., Garcia, M.A., Puig, D.: Robust color image segmentation through tensor voting. In: Proc. Int. Conf. Pattern Recognit (ICPR), Istanbul, Turkey, pp. 3372–3375 (2010)
33. Moreno, R., Garcia, M.A., Puig, D., Pizarro, L., Burgeth, B., Weickert, J.: On improving the efficiency of tensor voting. IEEE Trans. Pattern Anal. Mach. Intell. 33(11), 2215–2228 (2011)
34. Nagel, H.H., Gehrke, A.: Spatiotemporally adaptive estimation and segmentation of OF-fields. In: Burkhardt, H., Neumann, B. (eds.) Proc. Eur. Conf. Comput. Vis. (ECCV), Freiburg. Lecture Notes in Computer Science, vol. 1407, pp. 86–102. Springer, Berlin/New York (1998)
35. Nath, S., Palaniappan, K.: Adaptive robust structure tensors for orientation estimation and image segmentation. In: Boyle, R., Koracin, D., Parvin, B. (eds.) Proc. Int. Symp. Vis. Comp. (ISVC), Lake Tahoe. Lecture Notes in Computer Science, vol. 3804, pp. 445–453. Springer, Berlin/New York (2005)
36. Nicolescu, M., Medioni, G.: A voting-based computational framework for visual motion analysis and interpretation. IEEE Trans. Pattern Anal. Mach. Intell. **27**(5), 739–752 (2005)
37. Pajevic, S., Aldroubi, A., Basser, P.J.: A continuous tensor field approximation of discrete DT-MRI data for extracting microstructural and architectural features of tissue. J. Magn. Reson. **154**, 85–100 (2002)
38. Rao, A.R., Schunck, B.G.: Computing oriented texture fields. CVGIP: Graph. Models Image Process. **53**, 157–185 (1991)
39. Rohr, K.: Localization properties of direct corner detectors. J. Math. Imaging Vis. **4**, 139–150 (1994)
40. Rousson, M., Brox, T., Deriche, R.: Active unsupervised texture segmentation on a diffusion based feature space. In: Proc. Comput. Vis. Pattern Recognit. (CVPR), Madison, pp. II-699–704. IEEE, Los Alamitos (2003)
41. Schultz, T., Seidel, H.P.: Estimating crossing fibers: a tensor decomposition approach. IEEE Trans. Vis. Comput. Graph. **14**(6), 1635–1642 (2008)
42. Tai, Y.W., Tong, W.S., Tang, C.K.: Perceptually-inspired and edge-directed color image super-resolution. In: Proc. Comput. Vis. Pattern Recognit. (CVPR), New York, NY, USA, pp. II:1948–1955. IEEE, Los Alamitos (2006)
43. Tang, C.K., Medioni, G., Lee, M.S.: N-Dimensional tensor voting and application to epipolar geometry estimation. IEEE Trans. Pattern Anal. Mach. Intell. **23**(8), 829–844 (2001)
44. van de Weijer, J., van den Boomgaard, R.: Least squares and robust estimation of local image structure. Int. J. Comput. Vis. **64**(2/3), 143–155 (2005)
45. Weickert, J.: Coherence-enhancing diffusion filtering. Int. J. Comput. Vis. **31**(2–3), 111–127 (1999)
46. Weickert, J.: Coherence-enhancing diffusion of colour images. Image Vis. Comput. **17**, 199–212 (1999)
47. Weickert, J., Brox, T.: Diffusion and regularization of vector- and matrix-valued images. In: Nashed, M.Z., Scherzer, O. (eds.) Inverse Problems, Image Analysis, and Medical Imaging, pp. 251–268. AMS, Providence (2002)
48. Westin, C.F., Knutsson, H.: Tensor field regularization using normalized convolution. In: Moreno Diaz, R., Pichler, F. (eds.) Proc. Int. Conf. Comput. Aided Syst. Theory (EUROCAST), Las Palmas de Gran Canaria. Lecture Notes in Computer Science, vol. 2809, pp. 564–572. Springer, Berlin/Heidelberg (2003)
49. Wu, T.P., Yeung, S.K., Jia, J., Tang, C.K.: Quasi-dense 3D reconstruction using tensor-based multiview stereo. In: Proc. Comput. Vis. Pattern Recognit. (CVPR), San Francisco, pp. 1482–1489. IEEE Computer Society Press, Los Alamitos (2010)

Edge-Enhancing Diffusion Filtering for Matrix Fields

Bernhard Burgeth, Luis Pizarro, and Stephan Didas

Abstract The elimination of noise and small details from an image while simultaneously preserving or enhancing the edge structures in an image is a ever-lasting task in image processing. Edge-enhancing anisotropic diffusion is known to tackle this problem successfully. The problem of noise removal and edge enhancement is also a major concern in diffusion tensor magnetic resonance imaging (DT-MRI). This medical image acquisition technique outputs a 3D matrix field of symmetric 3×3-matrices, and it helps to visualise, for example, the nerve fibres in brain tissue. As any physical measurement DT-MRI is subjected to errors causing faulty representations of the tissue structure corrupted by noise. In this paper we address that problem by proposing a edge-enhancing diffusion filtering methodology for matrix fields. The approach is based on a generic structure tensor concept for matrix fields that relies on the operator-algebraic properties of symmetric matrices, rather than their channel-wise treatment of earlier proposals. Numerical experiments with artificial and real DT-MRI data confirm the noise-removing and edge-enhancing qualities of the technique presented.

B. Burgeth (✉)
Faculty of Mathematics and Computer Science, Saarland University, Building E2.4, 66041 Saarbrücken, Germany
e-mail: burgeth@math.uni-sb.de

L. Pizarro
Department of Computing, Imperial College London, 180 Queen's Gate, SW7 2AZ, London, UK
e-mail: luis.pizarro@imperial.ac.uk

S. Didas
Abteilung Bildverarbeitung, Fraunhofer-Institut für Techno- und Wirtschaftsmathematik ITWM, Fraunhofer-Platz 1, 67663 Kaiserslautern, Germany
e-mail: stephan.didas@itwm.fraunhofer.de

D.H. Laidlaw and A. Vilanova (eds.), *New Developments in the Visualization and Processing of Tensor Fields*, Mathematics and Visualization,
DOI 10.1007/978-3-642-27343-8_3,

1 Introduction

Diffusion tensor magnetic resonance imaging (DT-MRI) is a well-established technique to examine in vivo the structure of soft body tissue, such as brain or muscle tissue. However, ubiquitous measurement errors during acquisition produce noise and smeared out edges in the final visualisation. It would be desirable to have a method available that diminishes noise and enhances edge-like structures at the same time. A method achieving this goal for scalar and vector-valued images is already at our disposal: Edge-enhancing diffusion filtering [18]. It is governed by the equation

$$\begin{aligned} \partial_t u - \operatorname{div}\left(D(u) \cdot \nabla u\right) &= 0 \qquad && \text{in } \Omega \times I, \\ \partial_n u &= 0 && \text{on } \partial\Omega \times I, \\ u(x,0) &= f(x) && \text{in } \Omega, \end{aligned} \tag{1}$$

where $\Omega \subset \mathbb{R}^d$ is the image domain and $I = [0, T[$ a potentially unbounded time interval. The Neumann boundary conditions ensure preservation of the total amount of grey-values. An essential part of this equation is the *diffusion tensor* $D : (x,t) \mapsto D(u(x,t))$ of the scalar image $u(\cdot,t)$ with $t \in I$ which steers the diffusion process: It amplifies smoothing of the image within a relatively homogeneous region and inhibits diffusion across edges marking region boundaries. To the authors' best knowledge, no extension of this method to matrix-valued images, matrix fields for short, has been reported in the literature.

Postponing the detailed construction of D to the subsequent Sect. 2, for now we only remark that it is a function of the structure tensor [10] which is given by

$$S_\rho(u(\cdot,t)) := G_\rho * \left(\nabla u(\cdot,t) \cdot (\nabla u(\cdot,t))^\top\right) = \left(G_\rho * \left(\partial_{x_i} u(\cdot,t) \cdot \partial_{x_j} u(\cdot,t)\right)\right)_{i,j=1,\ldots,d}$$

Here $G_\rho *$ indicates a (spatial) convolution with a Gaussian of standard deviation ρ, however, more general averaging procedures can be used. If $\nabla u(x,t) \neq 0$ for the spatial gradient ∇ the matrix

$$\left(\nabla u(x,t) \cdot (\nabla u(x,t))^\top\right)$$

has rank one, the eigenvector $\nabla u(x,t)$ belongs to the only non-zero eigenvalue $|\nabla u(x,t)|^2$. The eigenvalues are a measure for the contrast in the directions of the eigenvectors. The averaging process then creates a matrix with full rank which contains valuable directional information. Note that the averaging of the structure tensor avoids cancellation of directional information. If one would average the gradients instead, neutralisation of vectors with opposite sign would occur. In many applications it is advantageous to use a presmoothed image $u_\sigma := G_\sigma * u$ instead of u in order to reduce the influence of noise for better numerical results. The structure tensor is a classical tool in image processing to extract directional information from an image, for more details the reader is referred to [2] and the literature cited there.

It is not straightforward to generalise both the structure and the diffusion tensor concept to the setting of matrix-valued images. To fix notation in this work matrix-valued images or matrix fields M are considered as mappings from $\mathbb{R}^d$ into the set $\mathrm{Sym}_n(\mathbb{R})$ of symmetric $n \times n$-matrices

$$M : x \mapsto M(x) = \left(m_{i,j}(x)\right)_{i,j=1,\ldots,n} \in \mathrm{Sym}_n(\mathbb{R})$$

and denoted by capital letters while indexed lower case letters indicate their components. In [3, 22] di Zenso's approach to a structure tensor for multi channel images is generalised: Each channel considered as independent scalar image gives rise to a structure tensor, then these structure tensors are summed up to give the structure tensor of Weickert

$$(J_\rho(U(\cdot,t)))(x) := \sum_{i,j=1}^{n} (S_\rho(u_{i,j}(\cdot,t)))(x)\,.$$

In [16] a more sophisticated construction results in a structure tensor that is a weighted sum of tensors of scalar quantities that are now not just the channels, but other meaningful scalar quantities derived from the matrix field. It is important to mention that in case of a 3D matrix field of 3×3 symmetric matrices these concepts yield also 3×3 structure tensors, the very same order as a 3D scalar image. Here we opt for a different approach: We assume an operator-algebraic view on symmetric matrices as finite dimensional instances of selfadjoint Hilbert space operators. The exploitation of the algebraic properties of matrices alone ensures already the proper interaction between different matrix channels.

Promising proposals to generalise nonlinear regularisation methods and related diffusion filters for scalar images to matrix fields have been made in [5, 6]. These approaches are based on a basic differential calculus for matrix fields, which will be useful in this context as well. Other approaches to tensor field regularisation have a more differential geometric background (i.e. [9, 17]) where the set of positive definite matrices is endowed with a Riemannian metric stemming from the DT-MRI field. In this chapter we will present a general concept for a large size structure tensor that carries all the directional information of the matrix field. We will show how this information can be deduced from this large tensor by a reduction process. We will follow the presentation in [7].

The structure of the article is as follows: The next Sect. 2 is devoted to a brief review of edge-enhancing diffusion (EED) filtering of scalar images. Notions necessary to construct the diffusion tensor and a basic differential calculus for matrix fields necessary to construct the diffusion tensor is provided in Sect. 3. In Sect. 4 we present the structure tensor concept for matrix fields, study some of its properties by investigating the connection to already known structure tensors for matrix-valued data. Employing this structure tensor concept we introduce edge-enhancing diffusion for matrix fields in Sect. 5. We then describe in Sect. 6 how explicit schemes for two-dimensional data can be extended by a Sobel-type method to three-dimensional data. The results of our experiments with matrix-valued edge-enhancing diffusion applied to real DT-MRI images are discussed in Sect. 7. Section 8 is made up by concluding remarks.

2 Edge-Enhancing Diffusion

The construction of the diffusion tensor $D(u(\cdot,t))(x)$ as a function of the matrix $(S_\rho(u(\cdot,t)))(x)$ is guided by the following reasoning: The matrix $(S_\rho(u(\cdot,t)))(x)$ is the positive average of different symmetric positive semidefinite matrices, and as such is positive definite as well. Hence $(S_\rho(u(\cdot,t)))(x)$ possesses an orthonormal system $\{w_1,\ldots,w_d\}$ of eigenvectors corresponding to the non-negative eigenvalues $\mu_1 \geq \mu_2 \geq \ldots \geq \mu_d \geq 0$ indicating the contrast in each direction. Although not indicated to avoid notational clutter these eigenvectors and eigenvalues depend on x and t. In the line defined by w_1, the contrast is the highest compared to other orientations, since w_1 belongs to the highest eigenvalue μ_1. This indicates an edge or a boundary between image regions perpendicular to w_1. Hindering diffusion across image borders the more the higher the contrast, and leaving it basically unaltered in other directions suggests the following design for D: The matrix D has the same eigenvectors as S_ρ, however, its eigenvalues λ_i are functions of the μ_i according to

$$\lambda_1 := g(\mu_1)$$

and

$$\lambda_i := 1 \qquad \text{for } i = 2,\ldots,d$$

where g stands for a diffusivity function. We use either the Perona-Malik diffusivity [15]

$$g(s^2) = \frac{1}{1+\frac{s^2}{c^2}} \tag{2}$$

or Weickert's diffusivity [21]

$$g(s^2) = \begin{cases} 1 & \text{if } s = 0, \\ 1 - \exp\left(-\dfrac{-3.31488}{(s/c)^8}\right) & \text{else.} \end{cases} \tag{3}$$

with a threshold parameter $c > 0$. The procedure described above transforming S_ρ into D induces a mapping $H : \mathrm{Sym}_n(\mathbb{R}) \longrightarrow \mathrm{Sym}_n(\mathbb{R})$ which we call a tensor map for future reference, and we write

$$H(S_\rho) = D. \tag{4}$$

In this chapter we show how edge-enhancing diffusion filtering can be extended to matrix fields. We follow the exposition in [7] and provide basic notions of a calculus for matrix fields in the next section.

3 Basic Differential Calculus for Matrix Fields

This section contains a brief account of the basic definitions for the formulation of a differential calculus for matrix fields. A previous version of this material can be found in [4] but for a more detailed exposition the reader is referred to [5].

1. **Functions of matrices.** Given a real-valued function h of one variable the standard definition for its extension h to $\mathrm{Sym}_n(\mathbb{R})$ (usually denoted with the same letter) is given by [11]:

$$h(M) = V^\top \mathrm{diag}(h(\lambda_1), \ldots, h(\lambda_n))V \in \mathrm{Sym}_n(\mathbb{R}),$$

   if $M = V^\top \mathrm{diag}(\lambda_1, \ldots, \lambda_n)V$ is the spectral/eigen decomposition of the symmetric matrix M, and if $\lambda_1, \ldots, \lambda_n$ lie in the domain of definition of h.
2. **Partial derivatives.** Let $\omega \in \{x_1, \ldots, x_d, t\}$ stand for a spatial or temporal variable, and set $(x,t) = (x_1, \ldots, x_d, t)$. The partial derivative for matrix fields is naturally defined *componentwise* as the limit of a difference quotient:

$$\begin{aligned} \overline{\partial}_\omega U(x,t) &= \lim_{h\to 0} \frac{U((x,t) + h\cdot e_k) - U(x,t)}{h} \\ &= \left(\lim_{h\to 0} \frac{u_{ij}((x,t) + h\cdot e_k) - u_{ij}(x,t)}{h}\right)_{i,j} \\ &= \left(\partial_\omega u_{ij}(x,t)\right)_{i,j} \end{aligned}$$

   where $e_k := (0, \ldots, 0, 1, 0, \ldots, 0) \in \mathbb{R}^{d+1}$ stands for the kth unit vector of space-time $\mathbb{R}^{d+1}$. The generalisation to directional derivatives is straight forward, then ω would denote an appropriate unit vector, for example, a normal vector n to the boundary of the domain $\partial\Omega$:

$$\overline{\partial}_n U(x,t) = n_1\, \overline{\partial}_{x_1}\, U(x,t) + \ldots + n_d\, \overline{\partial}_{x_d}\, U(x,t).$$

   Higher order partial differential operators, such as the Laplacian, or other more sophisticated operators, find their natural counterparts in the matrix-valued framework in this way as well.
3. **Generalised gradient of a matrix field.** The (spatial) gradient of a matrix field with sufficiently smooth component functions is defined via

$$\overline{\nabla} U(x,t) := (\overline{\partial}_{x_1}\, U(x,t), \ldots, \overline{\partial}_{x_d}\, U(x,t))^\top \in (\mathrm{Sym}_n(\mathbb{R}))^d.$$

Table 1 Extensions of elements of scalar valued calculus (*middle*) to the matrix-valued setting (*right*)

Setting	Scalar valued	Matrix-valued
Function	$h : \begin{cases} \mathbb{R} \longrightarrow \mathbb{R} \\ x \mapsto h(x) \end{cases}$	$h : \begin{cases} \mathrm{Sym}_n(\mathbb{R}) \longrightarrow \mathrm{Sym}_n(\mathbb{R}) \\ U \mapsto V^\top \mathrm{diag}(h(\lambda_1), \dots, h(\lambda_n))V \end{cases}$
Partial derivatives	$\partial_\omega u$, $\omega \in \{t, x_1, \dots, x_d\}$	$\overline{\partial}_\omega U := \left(\partial_\omega u_{ij}\right)_{ij}$, $\omega \in \{t, x_1, \dots, x_d\}$
Gradient	$\nabla u(x,t) := (\partial_{x_1} u(x,t), \dots, \partial_{x_d} u(x,t))^\top$, $\nabla u(x,t) \in \mathbb{R}^d$	$\overline{\nabla} U(x,t) := (\overline{\partial}_{x_1} U(x,t), \dots, \overline{\partial}_{x_d} U(x,t))^\top$, $\overline{\nabla} U(x,t) \in (\mathrm{Sym}_n(\mathbb{R}))^d$
Structure tensor	$G_\rho * \left(\nabla u(x,t) \cdot (\nabla u(x,t))^\top\right)$	$\overline{\mathscr{S}}_G\,(U(x,t)) := G_\rho * \left(\overline{\nabla} U(x,t) \cdot (\overline{\nabla} U(x,t))^\top\right)$
Product	$a \cdot b$	$A \bullet_J B := \frac{1}{2}(AB + BA)$

Hence, the generalised gradient $\overline{\nabla} U(x,t)$ at a voxel x and time t is regarded as an element of the module $(\mathrm{Sym}_n(\mathbb{R}))^d$ over $\mathrm{Sym}_n(\mathbb{R})$ in close analogy to the scalar setting where for the spatial gradient one has $\nabla u(x,t) \in \mathbb{R}^d$. In the sequel we will call a mapping from $\mathbb{R}^d$ into $(\mathrm{Sym}_n(\mathbb{R}))^d$ a *module field* rather than a vector field.

4. For the sake of completeness we include the formal definition of the **generalised structure tensor of a matrix field** here. We will discuss its derivation, properties and application in the next section. The structure tensor for a matrix field is given by

$$\overline{\mathscr{S}}_G\,(U(x,t)) := G_\rho * \left(\overline{\nabla} U(x,t) \cdot (\overline{\nabla} U(x,t))^\top\right)$$
$$= \left(G_\rho * \left(\overline{\partial}_{x_i} U(x,t) \cdot \overline{\partial}_{x_j} U(x,t)\right)\right)_{i,j=1,\dots,d}. \tag{5}$$

5. **Symmetrised product of symmetric matrices.** The product of two symmetric matrices $A, B \in \mathrm{Sym}_n(\mathbb{R})$ is not symmetric unless the matrices commute. However, it is vital to our interests to have a symmetrised matrix product at our disposal. There are numerous options to define a symmetrised matrix product, however, we concentrate on a specific one known from algebra and called Jordan product:

$$A \bullet_J B = \frac{1}{2}(AB + BA) \qquad \text{for } A, B \in \mathrm{Sym}_n(\mathbb{R})\,. \tag{6}$$

For commuting A and B we have $A \bullet_J B = A \cdot B$. This product is commutative and distributive but not associative. Most important, it does not preserve the positive semidefinitness of its arguments [7].

We summarised the definitions from above and juxtapose them with their scalar counterparts in the subsequent small Table 1. The matrix field $U(x,t)$ is assumed to be diagonisable with $U = (u_{ij})_{ij} = V^\top \mathrm{diag}(\lambda_1, \dots, \lambda_n)V$, where $V \in O(n)$, the set of all orthogonal $n \times n$-matrices, and $\lambda_1, \dots, \lambda_n \in \mathbb{R}$.

4 The Generalised Structure Tensor $\overline{\mathcal{S}}_G$ for Matrix Fields

In order to extract d-dimensional information we *reduce* the large structure tensor $\overline{\mathcal{S}}_G(U) \in \mathrm{Sym}_{nd}(\mathbb{R})$ to a structure tensor $S(U) \in \mathrm{Sym}_n(\mathbb{R})$ in a generalised projection step employing the block operator matrix

$$\mathrm{Tr}_A := \begin{pmatrix} \mathrm{tr}_A & \cdots & 0 \\ \vdots & \ddots & \vdots \\ 0 & \cdots & \mathrm{tr}_A \end{pmatrix} \tag{7}$$

containing the trace operation $\mathrm{tr}_A : \mathrm{Sym}_n(\mathbb{R}) \longrightarrow \mathbb{R},\ M \mapsto \mathrm{tr}(A \cdot M)$ with a fixed user-defined $d \times d$-matrix A. The operator matrix Tr_A acts on elements of the space $(\mathrm{Sym}_n(\mathbb{R}))^d$ as well as on block matrices via formal blockwise matrix multiplication:

$$\begin{pmatrix} \mathrm{tr}_A & \cdots & 0 \\ \vdots & \ddots & \vdots \\ 0 & \cdots & \mathrm{tr}_A \end{pmatrix} \begin{pmatrix} M_{11} & \cdots & M_{1n} \\ \vdots & \ddots & \vdots \\ M_{n1} & \cdots & M_{nn} \end{pmatrix} = \begin{pmatrix} \mathrm{tr}_A(M_{11}) & \cdots & \mathrm{tr}_A(M_{1n}) \\ \vdots & \ddots & \vdots \\ \mathrm{tr}_A(M_{n1}) & \cdots & \mathrm{tr}_A(M_{nn}) \end{pmatrix} \in \mathrm{Sym}_n(\mathbb{R}),$$

provided that the square blocks M_{ij} are compatible with tr_A, that means here, have the same size as A. The reason for choosing tr_A as reduction operators is their linearity:

$$\mathrm{tr}_A(M_1 + t\ M_2) = \mathrm{tr}_A(M_1) + t\ \mathrm{tr}_A(M_2)$$

for all $t \in \mathbb{R}$ and $M_1, M_2 \in \mathrm{Sym}_n(\mathbb{R})$, which is usefull to infer directional information from $\overline{\mathcal{S}}_G$, see [7] for details. As remarked in [7] the Weickert tensor J_ρ is a elementarily $(A = I)$ reduced version of $\overline{\mathcal{S}}_G$,

$$\mathrm{Tr}_I\ \overline{\mathcal{S}}_G(U) = J_\rho(U) \in \mathrm{Sym}_n(\mathbb{R}).$$

The reduction operation is accompanied by an extension operation defined via the so-called Kronecker product: The I_n-extension operation is the mapping from $\mathrm{Sym}_d(\mathbb{R})$ to $\mathrm{Sym}_{nd}(\mathbb{R})$ given by the Kronecker product $\otimes$:

$$\begin{pmatrix} v_{11} & \cdots & v_{1d} \\ \vdots & \ddots & \vdots \\ v_{d1} & \cdot & v_{dd} \end{pmatrix} \longmapsto \begin{pmatrix} v_{11} & \cdots & v_{1d} \\ \vdots & \ddots & \vdots \\ v_{d1} & \cdots & v_{dd} \end{pmatrix} \otimes I_n := \begin{pmatrix} v_{11}I_n & \cdots & v_{1d}I_n \\ \vdots & \ddots & \vdots \\ v_{d1}I_n & \cdots & v_{dd}I_n \end{pmatrix}.$$

If the $d \times d$-matrix $(v_{ij})_{ij}$ is Kronecker-multiplied with a matrix C we speak of a C-extension.

4.1 A Diffusion Tensor $\overline{D}$ for Matrix Fields

Now it is possible to give an analog $\overline{D}$ to the diffusion tensor D in the framework of matrix fields. We proceed in four steps:

1. The matrix field $\mathbb{R}^d \ni (x,t) \mapsto U(x,t)$ provides us with an module field of generalised gradients $\overline{\nabla} U(x,t)$ from which we construct the generalised structure tensor $\overline{\mathscr{S}}_G\, U(x,t)$ possibly with a certain integration scale ρ. This step corresponds exactly to the scalar case.
2. We infer reliable d-dimensional directional information by reducing $\overline{\mathscr{S}}_G\, U(x,t)$ with tr_A with the help of the block operator matrix given in (7) leading to a symmetric $d \times d$-matrix S, for example $S = J_\rho$ if $A = I_n$.

$$S := \begin{pmatrix} \mathrm{tr}_A & \cdots & 0 \\ \vdots & \ddots & \vdots \\ 0 & \cdots & \mathrm{tr}_A \end{pmatrix} \overline{\mathscr{S}}_G\, U(x,t)\,.$$

3. The symmetric $d \times d$-matrix S is spectrally decomposed, and the tensor map H is applied to S yielding the diffusion tensor D,

$$D := H(S)\,.$$

4. Finally we enlarge the $d \times d$-matrix D to a $nd \times nd$-matrix $\overline{D}$ by the extension operation:

$$\overline{D} = D \otimes C$$

This last step gives another possibility to steer the filter process by the choice of the matrix C. However, this is the subject of current research. For this work we restricted ourselves to $C = I_n$.

5 Edge-Enhancing Diffusion Filtering for Matrix Fields

Now we have gathered the necessary ingredients to formulate the matrix-valued equivalent to the scalar edge-enhancing diffusion as expressed in Eq. (1):

$$\begin{aligned} \overline{\partial}_t U - \sum_{i=1}^{d} \overline{\partial}_{x_i} \left(\overline{D}(U) \bullet \overline{\nabla} U \right) &= 0 \qquad \text{in } \Omega \times I, \\ \overline{\partial}_n U &= 0 \qquad \text{on } \partial\Omega \times I, \\ U(x,0) &= F(x) \qquad \text{in } \Omega. \end{aligned} \tag{8}$$

Note that the Jordan-multiplication in $\overline{D} \bullet \overline{\nabla} U$ is understood in the blockwise sense of partitioned matrices. Moreover, we translated the divergence differential operator acting on a vector-valued function $u = (u_1, \ldots, u_d)$ div $u = \sum_{i=1}^{d} \partial_{x_i} u$ into its matrix-valued counterpart acting on a module field $W \in \mathrm{Sym}_n(\mathbb{R})^d$ by

$$\overline{\mathrm{div}}\ W = \sum_{i=1}^{d} \overline{\partial}_{x_i} W\,.$$

Remark. It becomes apparent that the PDE describing coherence-enhancing diffusion filtering (CED) has the same form as the equation above, see [7]. However, for CED a different tensor map is employed [8], namely the one that alters the eigenvalues $\mu_1, \ldots, \mu_d$ of S into

$$\lambda_i := \alpha \qquad \text{for } i = 1, \ldots, d-1$$

and

$$\lambda_d := \begin{cases} \alpha & \text{if } \kappa = 0, \\ \alpha + (1-\alpha)\exp\left(-\frac{c}{\kappa}\right) & \text{else,} \end{cases}$$

with a threshold $c > 0$, the eigenvalues of the diffusion marix $G(S)$. The quantity

$$\kappa := \sum_{i=1}^{d-1} \sum_{j=i+1}^{d} (\mu_i - \mu_j)^2$$

has been proposed in [20] to measure coherence.

6 Numerical Issues

In the scalar, two-dimensional case an explicit scheme can be found in [19]. We used a matrix-valued version employing the calculus framework for matrix fields as presented before. This provides us with a matrix-valued solution scheme for matrix fields defined over a two-dimensional image domain Ω.

We derive a three-dimensional scheme by employing a Sobel-type construction using the 2D discrete divergence approximation. The idea is based on the following decomposition:

$$\begin{aligned} &\mathrm{div}\left(\begin{pmatrix} 2a & b & c \\ b & 2d & e \\ c & e & 2f \end{pmatrix} \nabla u\right) \\ &= \mathrm{div}\left(\begin{pmatrix} a & b & 0 \\ b & d & 0 \\ 0 & 0 & 0 \end{pmatrix} \nabla u\right) + \mathrm{div}\left(\begin{pmatrix} a & 0 & c \\ 0 & 0 & 0 \\ c & 0 & f \end{pmatrix} \nabla u\right) + \mathrm{div}\left(\begin{pmatrix} 0 & 0 & 0 \\ 0 & d & e \\ 0 & e & f \end{pmatrix} \nabla u\right). \qquad (9) \end{aligned}$$

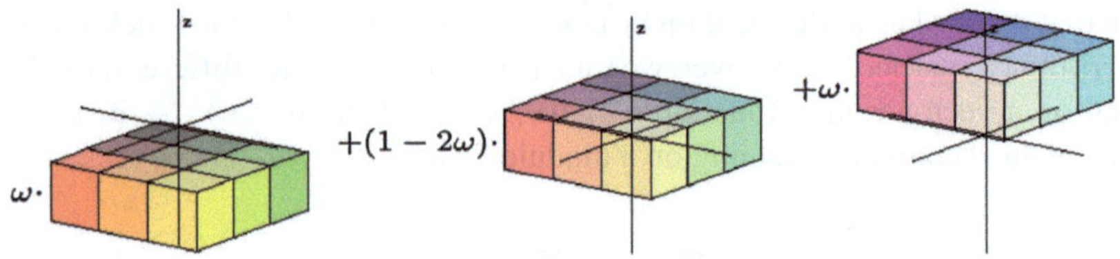

Fig. 1 Construction of a 3D-stencil specific for the $z-$direction. Each layer represents a discrete approximation of an essentially 2D divergence term

Each term on the right hand side of (9) can be approximated by a two dimensional discretisation where we employ additionally a Sobel-type construction with a weight $\omega \in [0, \frac{1}{2}]$ as indicated in Fig. 1 (however, only $\omega < \frac{1}{3}$ leads to good results). For each of the three summands associated with the $z-, y-$ and $x-$ directions a $3 \times 3 \times 3$-stencil is obtained which, when added give the final stencil. We used the matrix-valued version of this stencil. Note that this construction can be applied to more sophisticated two-dimensional stencils.

7 Experiments

In our experiments we use both synthetic and real-world tensor fields of 3×3-symmetric matrices. The data are represented as ellipsoids via the level sets of the quadratic form $\{v^\top U^{-2}(x,t)v = const. : v \in \mathbb{R}^3\}$ associated with a matrix $U(x,t) \in \mathrm{Sym}^+(3)$. By using U^{-2} the lengths of the semi-axes of the ellipsoid correspond directly with the three eigenvalues of the matrix. In the following, we demonstrate the denoising and enhancing capabilities of our edge-enhancing filtering (EED) compared to coherence-enhancing filtering (CED) [8]. Both approaches are implemented according to the 3D numerical scheme described in Sect. 6 with $\omega = \frac{1}{4}$, unless stated otherwise. Before processing, the matrix fields will be presmoothed with a Gaussian kernel of width σ. In EED, we map the largest eigenvalue λ_1 using the Perona-Malik diffusivity (2) with manually adjusted parameter c. In CED, we map the smallest eigenvalue λ_d with parameters $\alpha = 0.001$ and $c = 1$.

We have added (noise) random positive definite matrices to our synthetic data sets. The eigenvectors of the noisy matrices were obtained by choosing Gaussian-distributed numbers with standard deviation γ and taking the absolute value for positive definiteness. The eigenvectors of the noise result in choosing three uniformly distributed angles and rotating the matrix by these angles around the coordinate axes. The top row of Fig. 2 displays an original matrix field (maximum eigenvalue 2,500) and its noisy version ($\gamma = 1{,}000$) with simple line-like structures imitating a crossing of nerve fibres. Depending on the choice of the reduction matrix A in tr_A different filtering effects can be achieved. When no directional preference is sought, i.e. $A = I$, the middle row of Fig. 2 shows that both EED and CED can

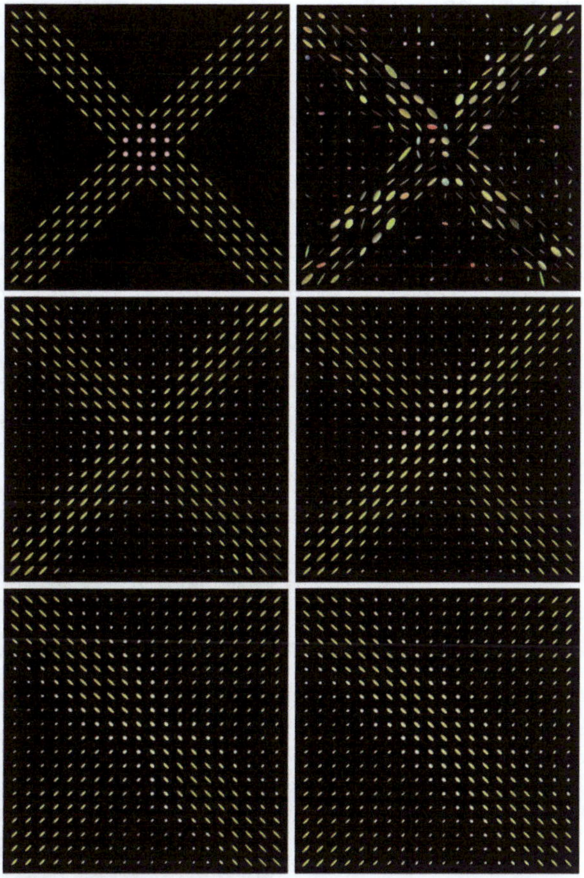

Fig. 2 From *left* to *right*: *Top row:* Synthetic 2D data set and its noisy version. *Middle row:* EED ($c = 100$) and CED ($\rho = 3$) filtering at diffusion time $t = 5$. No specified preferred direction. *Bottom row:* EED ($c = 100$) and CED ($\rho = 3$) filtering at diffusion time $t = 15$. Diagonal-downwards preferred direction. All cases with presmoothing $\sigma = 0.5$

preserve and enhance edges. However, EED has a more local effect than CED. This is expected as the latter integrates directional information from its neighbouring voxels—here we set $\rho = 2$ as the width of a Gaussian neighbourhood [8]. The bottom row of Fig. 2 shows the effect of choosing $A = \begin{pmatrix} 1 & 1 \\ 1 & 1 \end{pmatrix}$ to give preference to diagonal structures directed downwards. Also due to the argument above, CED is able to complete the central isotropic region in a directional fashion faster than EED. As a general remark, note that these diffusion-like approaches decrease the overall size of matrices since the total mass, that is, the volume of the ellipsoids is only redistributed due to the property of mass conservation of diffusion processes.

The top two rows of Fig. 3 display a synthetic 3D spiral (maximum eigenvalue 9,000) and two noisy versions. The first one was generated by adding Gaussian noise with $\gamma = 6{,}000$ to the matrices on the spiral's surface, while in the second one further Gaussian noise with $\gamma = 3{,}000$ was added to 20 % of the background voxels. The bottom two rows of Fig. 3 show the corresponding filtering results using EED and CED without directional preference. It is clear that both approaches are

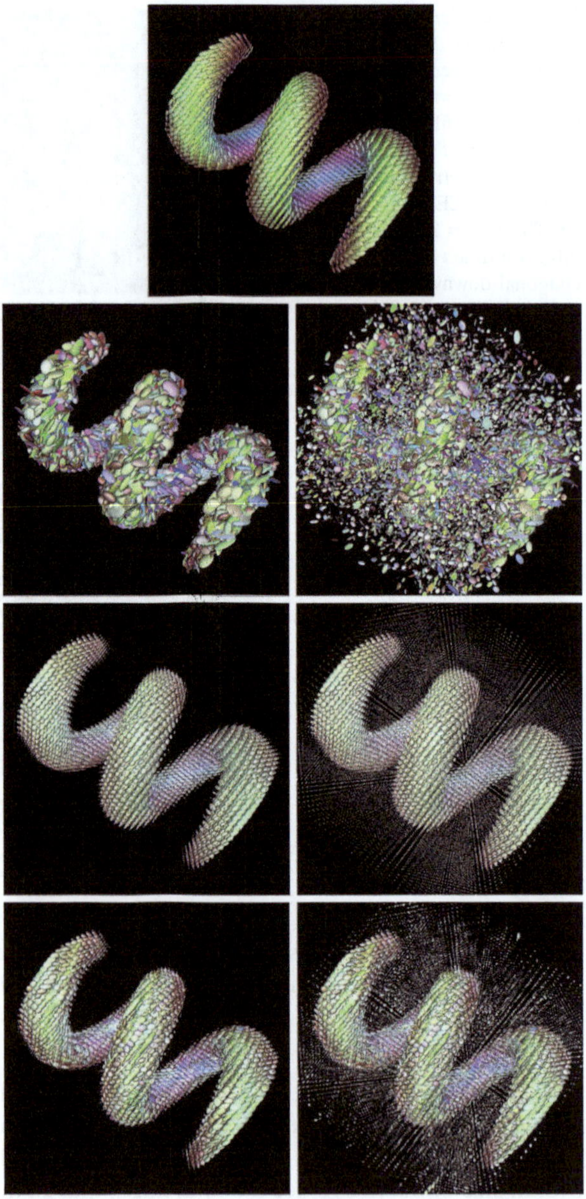

Fig. 3 *Top:* 3D synthetic spiral. *Left column*, from *top* to *bottom*: Noisy spiral (noise on its surface) and its filtered versions with EED ($c = 100$) and CED ($\rho = 2$) at diffusion time $t = 2$. *Right column*, from *top* to *bottom*: Noisy spiral (noise on its surface and background) and its filtered versions with EED ($c = 100$) and CED ($\rho = 2$) at diffusion time $t = 2$. All cases with presmoothing $\sigma = 2.0$ and no specified preferred direction

capable to smoothly restore the spiral's surface while keeping its sharp boundaries. Note that at the same diffusion time ($t = 2$) EED produces slightly more convincing (smoother) results than CED. The latter should thus be run a little longer for a better result, fact that we will take into account later in our experiments with real-world data. Figure 4 shows the effect of using different values of ω in our numerical

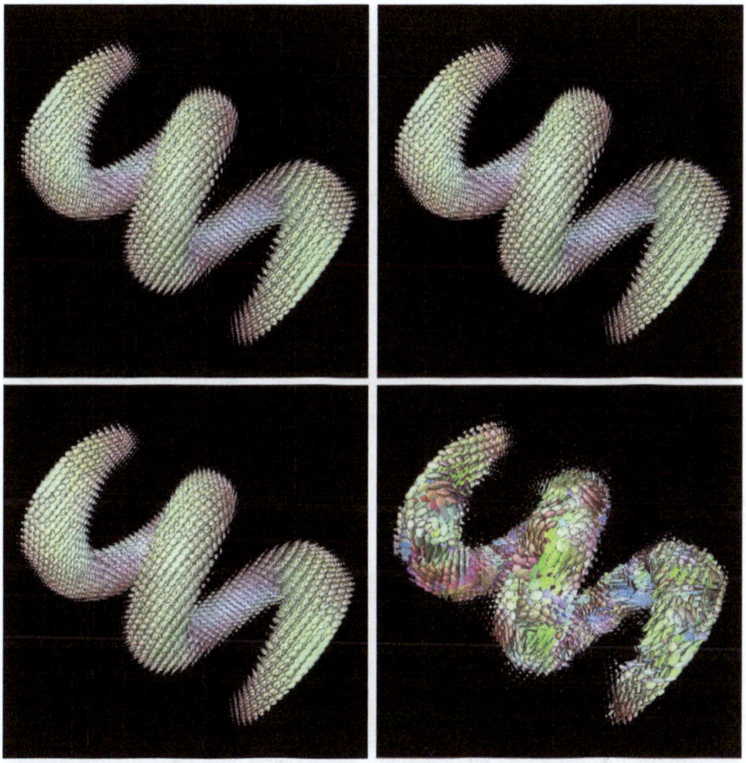

Fig. 4 From *top* to *bottom*, *left* to *right*: EED ($c = 100$) filtering of the noisy spiral shown in Fig. 3 using $\omega = 0.0, 0.1, 0.2, 0.3$ in the numerical scheme at diffusion time $t = 2$. All cases with presmoothing $\sigma = 2.0$ and no specified preferred direction

scheme. A small ω gives more importance to the central layer (cf. Fig. 1). It can be noted that our scheme produces good results with $\omega < \frac{1}{3}$.

Figure 5 displays an original 3D data set obtained by diffusion tensor magnetic resonance imaging (DT-MRI). A close-up around the lateral ventricles shows isotropic (round) tensors as well as anisotropic (elongated) tensors in the area of the genu and the corpus callosum. It has been proven [14] that this type of data are corrupted with multivariate Gaussian noise. Our previous noise model is therefore consistent with these real-world data and we can similarly test our EED and CED filtering approaches on it. Figure 6 shows these results, where EED was stopped at $t = 1$ and CED at $t = 2$ so that both methods achieve similar smoothness. Both approaches produce pleasant results, denoising isotropic and anisotropic structures, preserving and enhancing sharp boundaries, and reconstructing areas with missing tensors. The main difference between both methods can be observed in the upper region, where CED tends to elongate the ventricles as they are considered coherent structures, which changes the original ventricular shape. This effect could have been stronger with $\rho > 1$. Figure 6 also displays the *fractional anisotropy* (FA)

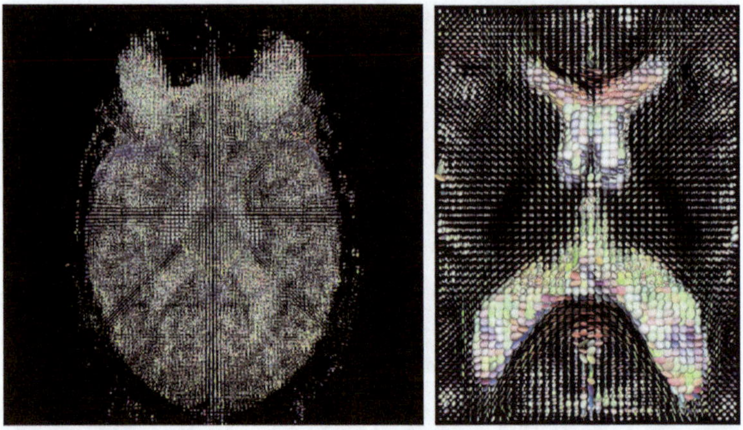

Fig. 5 *Left:* Original 3D DT-MRI data set of a human head, 128×128×30 voxels. *Right:* Close-up region around the lateral ventricles, 40 × 55 × 3 voxels

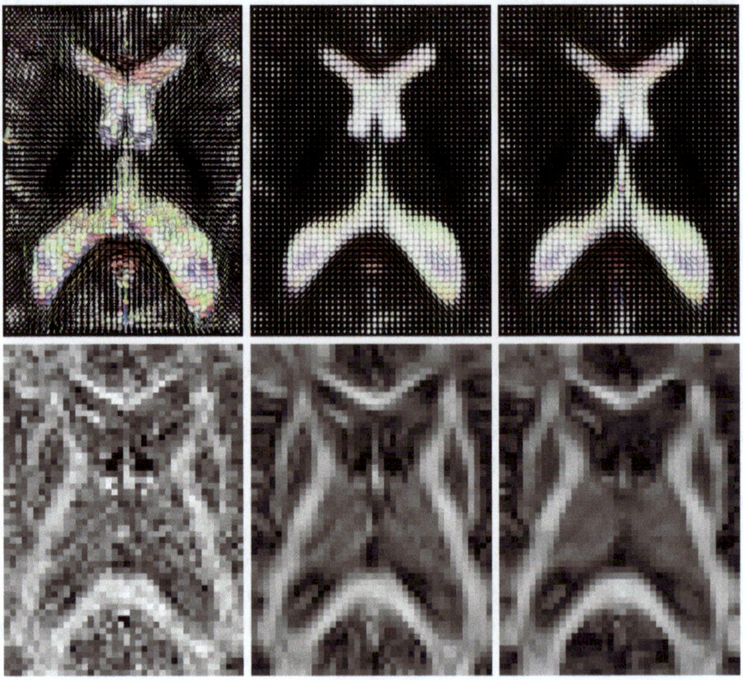

Fig. 6 From *top* to *bottom*: *Left column*: Original 2D slice, 40 × 55 × 1 voxels, and its corresponding FA map. *Middle column*: EED ($c = 50$) filtering at diffusion time $t = 1$ and its corresponding FA map. *Right column*: CED ($\rho = 1$) filtering at diffusion time $t = 2$ and its corresponding FA map. Both approaches were run on the full 3D data set, but to avoid clutter we visualised the results on 2D slices. All cases with presmoothing $\sigma = 0.5$ and no specified preferred direction

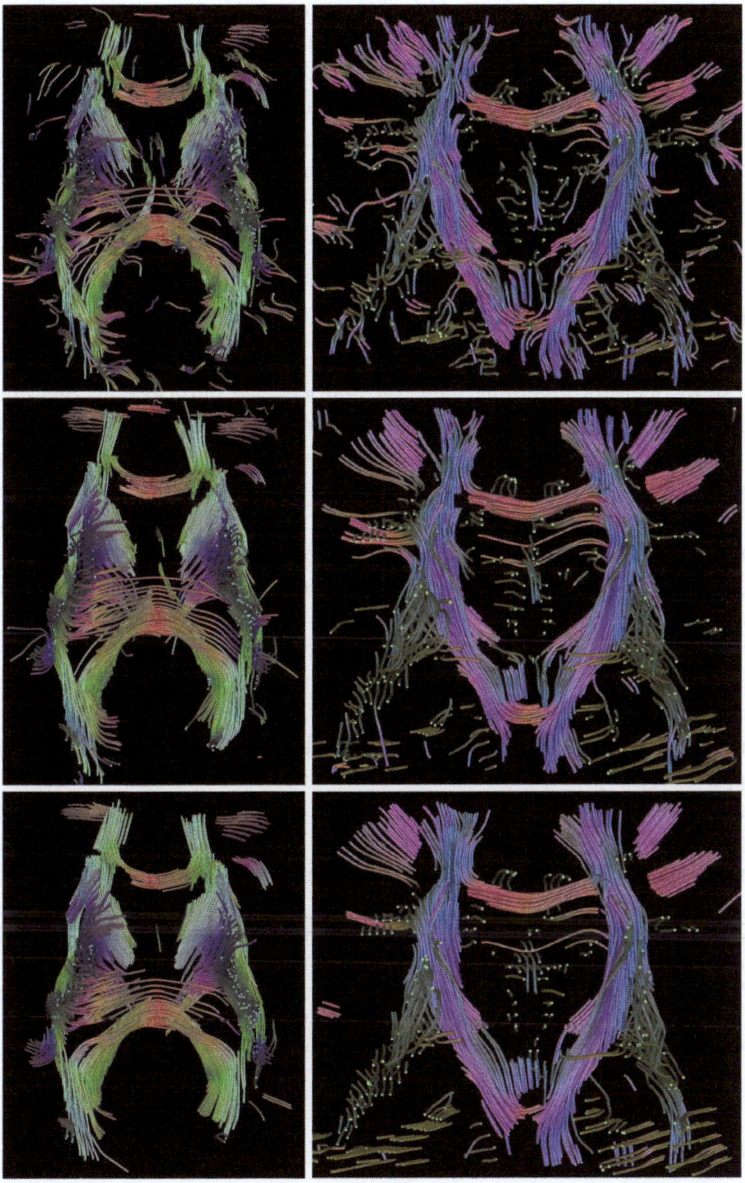

Fig. 7 Two viewpoints of 3D tractography results obtained from the data set shown in Fig. 5. *Top row*: without filtering. *Middle row*: with CED filtering. *Bottom row*: with EED filtering. The paremeters were chosen as in the experiment of Fig. 6.

maps [1] of the shown 2D DT-MRI slices. Note that FA $\in [0, 1]$: FA $= 0$ at voxels with isotropic tensors ($\lambda_1 \approx \lambda_2 \approx \lambda_3$), while FA $= 1$ at voxels with anisotropic tensors ($\lambda_1 \gg \lambda_2 \approx \lambda_3 \approx 0$). These maps exhibit the regularising and edge-enhancing properties of both diffusion approaches: EED seems to produce more homogeneous results at regions with different degrees of anisotropy; and CED highlights anisotropic regions even if they are composed of only a few coherent voxels. These properties are also observed in Fig. 7, which shows that our EED approach is a powerful preprocessing technique that provides algorithms such as tractography [12] with filtered and gap-completed tensors fields, which results in better fibre tracts. These characteristics are important for the study of diseases associated with certain abnormalities in brain anatomy [13].

8 Concluding Remarks

Based on an operator-algebraic view on matrices we described in this chapter a structure tensor concept for matrix fields that allows for edge-enhancing diffusion (EED) filtering of matrix fields by employing a generic differential calculus framework for matrices. To solve the underlying matrix-valued partial differential equation we have described a 3D numerical solution scheme that can be realised as a Sobel-type construction that uses 2D discrete divergence approximations. This allows us to employ all 27 voxels in the numerical discretisation rather than only seven voxels as in standard 3D schemes.

Our matrix-valued EED filtering exhibits behaviour similar to its scalar counterpart. As confirmed with experiments on synthetic and DT-MRI data sets, the expected strength of EED is edge enhancement rather than line completion, the latter being a typical feature of coherence enhancing diffusion (CED) filtering. In general, both approaches represent suitable alternatives for denoising matrix fields with different degree of anisotropy, preserving sharp edges and reconstructing areas with missing tensors. Future research will focus on further applications of the generic differential calculus framework for matrices in image processing for matrix fields.

Acknowledgements We are grateful to Anna Vilanova i Bartrolí (Eindhoven Institute of Technology) and Carola van Pul (Maxima Medical Center, Eindhoven) for providing us with the DT-MRI data set and for discussing questions concerning data conversion. The original helix data set is by courtesy of Gordon Kindlmann (University of Chicago). The thank Thomas Schultz (University of Chicago) for providing the tractography results.

References

1. Basser, P.J., Pierpaoli, C.: Microstructural and physiological features of tissues elucidated by quantitative-diffusion-tensor MRI. J. Magn. Reson. **11**, 209–219 (1996)
2. Bigun, J.: Vision with Direction. Springer, Berlin (2006)
3. Brox, T., Weickert, J., Burgeth, B., Mrázek, P.: Nonlinear structure tensors. Technical report 113, Department of Mathematics, Saarland University, Saarbrücken, October 2004
4. Burgeth, B., Bruhn, A., Didas, S., Weickert, J., Welk, M.: Morphology for matrix-data: ordering versus PDE-based approach. Image Vis. Comput. **25**(4), 496–511 (2007)

5. Burgeth, B., Didas, S., Florack, L., Weickert, J.: A Generic Approach for Singular PDEs for and Variational Methods in Computer Vision. Lecture Notes in Computer Science, vol. 4485, pp. 556–567. Springer, Berlin (2007)
6. Burgeth, B., Didas, S., Florack, L., Weickert, J.: A generic approach to diffusion filtering of matrix-fields. Computing **81**, 179–197 (2007)
7. Burgeth, B., Didas, S., Weickert, J.: A general structure tensor concept and coherenceenhancing diffusion filtering for matrix fields. In: Laidlaw, D., Weickert, J. (eds.) Visualization and Processing of Tensor Fields, pp. 305–323. Springer, Berlin (2009)
8. Burgeth, B., Pizarro, L., Didas, S., Weickert, J.: 3D-coherence-enhancing diffusion filtering for matrix fields. In: Florack, L., Duits, R., Jongbloed, G., van Lieshout, M.-C., Davies, L. (eds.) Locally Adaptive Filters in Signal and Image Processing, pp. 55–70. Springer (2011)
9. Chefd'Hotel, C., Tschumperlé, D., Deriche, R., Faugeras, O.: Constrained flows of matrix-valued functions: Application to diffusion tensor regularization. In: Heyden, A., Sparr, G., Nielsen, M., Johansen, P. (eds.) Computer Vision – ECCV 2002. Lecture Notes in Computer Science, vol. 2350, pp. 251–265. Springer, Berlin (2002)
10. Förstner, W., Gülch, E.: A fast operator for detection and precise location of distinct points, corners and centres of circular features. In: Proceedings of ISPRS Intercommission Conference on Fast Processing of Photogrammetric Data, Interlaken, pp. 281–305 (1987)
11. Horn, R.A., Johnson, C.R.: Matrix Analysis. Cambridge University Press, Cambridge (1990)
12. Mori, S., van Zijl, P.C.: Fiber tracking: principles and strategies – a technical review. NRM Biomed **15**, 468–480 (2002)
13. Nucifora, P.G., Verma, R., Lee, S.-K., Melhem, E.R.: Diffusion-tensor MR imaging and tractography: exploring brain microstructure and connectivity. Radiology **245**, 367–384 (2007)
14. Pajevic, S., Basser, P.J.: Parametric and non-parametric statistical analysis of DT-MRI data. J. Magn. Reson. **161**(1), 1–14 (2003)
15. Perona, P., Malik, J.: Scale space and edge detection using anisotropic diffusion. IEEE Trans. Pattern Anal. Mach. Intell. **12**, 629–639 (1990)
16. Schultz, T., Burgeth, B., Weickert, J.: Flexible segmentation and smoothing of DT-MRI fields through a customizable structure tensor. In: Bebis, G., Boyle, R., Parvin, B., Koracin, D., Remagnino, P., Nefian, A., Meenakshisundaram, G., Pascucci, V., Zara, J., Molineros, J., Theisel, H., Malzbender, T. (eds.) Advances in Visual Computing. Lecture Notes in Computer Science, vol. 4291, pp. 454–464. Springer, Berlin (2006)
17. Tschumperlé, D., Deriche, R.: Regularization of orthonormal vector sets using coupled PDE's. In: Proceedings of First IEEE Workshop on Variational and Level Set Methods in Computer Vision, Vancouver, pp. 3–10. IEEE Computer Society Press (2001)
18. Weickert, J.: Theoretical foundations of anisotropic diffusion in image processing. Comput. Suppl. **11**, 221–236 (1996)
19. Weickert, J.: Anisotropic Diffusion in Image Processing. Teubner, Stuttgart (1998)
20. Weickert, J.: Coherence-enhancing diffusion filtering. Int. J. Comput. Vis. **31**(2/3), 111–127 (1999)
21. Weickert, J.: Design of nonlinear diffusion filters. In: Jähne, B., Haußecker, H. (eds.) Computer Vision and Applications, pp. 439–458. Academic, San Diego (2000)
22. Weickert, J., Brox, T.: Diffusion and regularization of vector- and matrix-valued images. In: Nashed, M.Z., Scherzer, O. (eds.) Inverse Problems, Image Analysis, and Medical Imaging. Contemporary Mathematics, vol. 313, pp. 251–268. AMS, Providence (2002)

[illegible]

Part II
Tensor-Field Visualization

Fabric-Like Visualization of Tensor Field Data on Arbitrary Surfaces in Image Space

Sebastian Eichelbaum, Mario Hlawitschka, Bernd Hamann, and Gerik Scheuermann

Abstract Tensors are of great interest to many applications in engineering and in medical imaging, but a proper analysis and visualization remains challenging. It already has been shown that, by employing the metaphor of a fabric structure, tensor data can be visualized precisely on surfaces where the two eigendirections in the plane are illustrated as thread-like structures. This leads to a continuous visualization of most salient features of the tensor data set.

We introduce a novel approach to compute such a visualization from tensor field data that is motivated by image space line integral convolution (LIC). Although our approach can be applied to arbitrary, non-self-intersecting surfaces, the main focus lies on special surfaces following important features, such as surfaces aligned to the neural pathways in the human brain. By adding a postprocessing step, we are able to enhance the visual quality of the results, which improves perception of the major patterns.

S. Eichelbaum (✉) · G. Scheuermann
Abteilung für Bild- und Signalverarbeitung, Institut für Informatik, Universität Leipzig, Leipzig, Germany
e-mail: eichelbaum@informatik.uni-leipzig.de; scheuermann@informatik.uni-leipzig.de

M. Hlawitschka
Scientific Visualization Group, Institut für Informatik, Universität Leipzig, Leipzig, Germany
e-mail: hlawitschka@informatik.uni-leipzig.de

B. Hamann
Department of Computer Science, Institute for Data Analysis and Visualization (IDAV), University of California, Davis, CA, USA
e-mail: bhamann@ucdavis.edu

D.H. Laidlaw and A. Vilanova (eds.), *New Developments in the Visualization and Processing of Tensor Fields*, Mathematics and Visualization,
DOI 10.1007/978-3-642-27343-8_4, © Springer-Verlag Berlin Heidelberg 2012

1 Motivation and Related Work

Since the introduction of tensor lines and hyperstreamlines [6], there have been many research efforts directed at the continuous representation of tensor fields, including research on tensor field topology [11, 22, 23]. Zheng and Pang introduced HyperLIC [31], which makes it possible to display a single eigendirection of a tensor field in a continuous manner by smoothing a noise texture along integral lines, while neglecting secondary directions. Recent approaches to visualize Lagrangian structures on tensor fields [12] provide information on one chosen tensor direction and are especially useful for diffusion tensor data, where the main tensor direction can be correlated to neural fibers or muscular structures, whereas the secondary direction only plays a minor role. More recently, Dick et al. [7] published an interactive approach to visualize a volumetric tensor field for implant planning.

Hotz et al. [13] introduced Physically Based Methods (PBM) for tensor field visualization in 2004 as a means to visualize stress and strain tensors arising in geomechanics. A positive-definite metric that has the same topological structure as the tensor field is defined and visualized using a texture-based approach resembling LIC [4]. Besides other information, eigenvalues of the metric can be encoded by free parameters of the texture definition, such as the remaining color space. Whereas the method's implementation for parameterizable surfaces that are topologically equivalent to discs or spheres is straightforward, implementations for arbitrary surfaces remains computationally challenging. In 2009, Hotz et al. [14] enhanced their approach to isosurfaces in three-dimensional tensor fields. A three-dimensional noise texture is computed in the data set and a convolution is performed along integral lines tangential to the eigenvector field. LIC has been used in vector field visualization methods to imitate *Schlieren* patterns on surfaces that are generated in experiments where a thin film of oil is applied to surfaces, which show patterns caused by the air flow. In vector field visualization, image space LIC is a method to compute *Schlieren*-like textures in image space [9, 17, 27, 28], intended for large and non-parameterized geometries. Besides the non-trivial application of image space LIC to tensor data, image space LIC has certain other drawbacks. Mainly because the noise pattern is defined in image space, it does not follow the movement of the surface and, therefore, during user interaction, the three-dimensional impression is lost. A simple method proposed to circumvent this problem is animating the texture pattern by applying randomized trigonometric functions to the input noise. Weiskopf and Ertl [26] solved this problem for vector field visualization by generating a three-dimensional texture that is scaled appropriately in physical space.

In contrast to other real-time tensor-field visualizations like [30], we developed and implemented an algorithm similar to the original PBM but for arbitrary non-intersecting surfaces in image space. Our algorithm can perform at interactive frame rates for large data sets on current desktop PCs. We overcome the drawbacks present in image space LIC implementations by defining a fixed parameterization on the surface. Thus, we do not require a three-dimensional noise texture representation defined at sub-voxel resolution for the data set. Our approach is

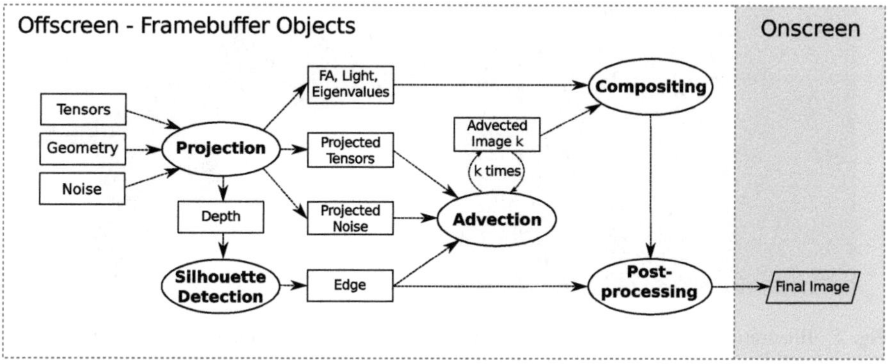

Fig. 1 Flowchart indicating the five major steps (bold font) of the algorithm: **projection**, which transforms the data set in an image space representation and produces the initial noise texture on the geometry; **silhouette detection**, required for the advection step and the final rendering; **advection**, which produces the two eigenvector textures; **compositing**, which combines intermediate textures; and the **postprocessing**, which adds additional shading and improves the perceptual quality of the final visualization. Between consecutive steps, the data is transferred using textures

capable of maintaining local coherence of the texture pattern between frames when (1) transforming, rotating, or scaling the visualization, (2) changing the surface by, e.g., changing isovalues or sweeping the surface through space, and (3) changing the level of detail. In addition, we implemented special application-dependent modes to ensure our method integrates well with existing techniques. Besides this, we also apply several postprocessing steps to further increase the visual quality and clarity of the shown structures.

2 Method

We employ a multi-pass rendering technique that consists of five major rendering passes as outlined in Fig. 1 using bold typesetted boxes. The complete pipeline is processed in one single render-frame and offscreen. In the following sections we describe each single step in our pipeline and imply that all operations are done on a per-pixel basis, if not denoted differently. We additionally rely on Fig. 1 in many sections, as this figure shows all needed inputs and generated outputs of each step.

After generating the basic input textures once, the first pass calculates and projects all required data into image space. This encompasses the eigenvector-decomposition and projection, the noise projection to the surface, lighting and further color-mapping. Using the calculated eigenvalues, the fractional anisotropy is calculated as well. It is used later for clipping and color-mapping. Pass two performs a silhouette detection on the depth-buffer if the rendered geometry. That is used to guarantee integrity of the advection image, computed by multiple iterations of pass three. Pass three uses the projected eigenvectors to smear the projected noise on

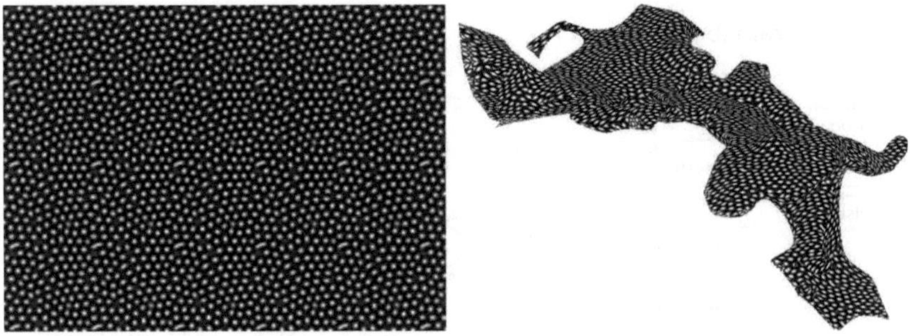

Fig. 2 Illustration of the reaction diffusion texture used (*right*) and the noise texture mapped to geometry (*left*)

the surface in k iterations along the eigenvectors. Eventually, pass four composes the intermediate textures in an image, which is then post-processed by step five and rendered on-screen.

2.1 Initial Noise Texture Generation

In contrast to standard LIC approaches, to achieve a proper visual representation of the data, high-frequency noise textures, such as white noise, are not suitable for the compositing of multiple textures. Therefore, we compute the initial noise texture using the reaction diffusion scheme first introduced by Turing [24]. It simulates the mixture of two reacting chemicals, which leads to larger but smooth "spots" that are randomly and almost uniquely distributed (cf. Fig. 2, right). This can be pre-computed on the CPU once. The created texture can then be used for all consecutive frames. For the discrete case, the governing equations are:

$$\begin{aligned} \Delta a_{i,j} &= F(i,j) + D_a \cdot (a_{i+1,j} + a_{i-1,j} + a_{i,j+1} + a_{i,j-1} - 4 \cdot a_{i,j}), \\ \Delta b_{i,j} &= G(i,j) + D_b \cdot (b_{i+1,j} + b_{i-1,j} + b_{i,j+1} + b_{i,j-1} - 4 \cdot b_{i,j}), \text{ where} \\ F(i,j) &= s(16 - a_{i,j} \cdot b_{i,j}) \text{ and } G(i,j) = s(a_{i,j} \cdot b_{i,j} - b_{i,j} - \beta_{i,j}). \end{aligned} \tag{1}$$

Here, we assume continuous boundary conditions to obtain a seamless texture in both directions. The scalar s allows one to control the size of the spots where a smaller value of s leads to larger spots. The constants D_a and D_b are the diffusion constants of each chemical. We use $D_a = 0.125$ and $D_b = 0.031$ to create the input textures. We gained both constants empirically. They directly influence the shape and density of the created spots.

2.2 Projection Step

The first step of our algorithm is the projection pass. Its purpose is to project the geometry as well as the tensorial data to image space. Besides this, the initial noise texture created earlier is mapped to the surface. After the projection step, the tensors, the noise and the rendered geometry are available in image space and can be used by the consecutive passes. As the next steps are all operating in image space, we state that all consecutive steps are done on a per-pixel basis. The operations of the projection step are done on a per-fragment basis. Only the projection of the geometry is done vertex-wise.

2.2.1 Projection into Image Space

In the first step, we project the data into image space by rendering the surface using the default OpenGL rendering pipeline. Notably, the surface does not need to be represented by a surface mesh. Any other representation that provides proper depth and surface normal information works just as well (e.g., ray-casting methods for implicit surfaces, cf. Knoll et al. [16]). In the same rendering step, the tensor field is transformed from world space to object space, i.e., each tensor T, that is interpolated at the point on the surface from the surrounding two- or three-dimensional tensor field is projected onto the surface by

$$T' = P \cdot T \cdot P^T, \tag{2}$$

with a matrix P defined using the surface normal n as

$$P = \begin{pmatrix} 1 - n_x^2 & -n_y n_x & -n_z n_x \\ -n_x n_y & 1 - n_y^2 & -n_z n_y \\ -n_x n_z & -n_y n_z & 1 - n_z^2 \end{pmatrix}. \tag{3}$$

The camera viewing system configuration and the available screen resolution imply a super- or sub-sampling of the data. We obtain an interpolated surface tensor in every pixel, which is decomposed into the eigenvector/eigenvalue representation using a method derived from the one presented by Hasan et al. [10], only using iteration-free math functions. This causes a tremendous acceleration on the GPU. With this method, we calculate the three real-valued, orthogonal eigenvectors $v_{\lambda_{1-3}}$ and the corresponding eigenvalues $\lambda_1 \geq \lambda_2 \geq \lambda_3$. In our method, we are only using the first two eigenvectors, showing the two main directions. The eigenvectors, still defined in object space, are projected into image space using the same projection matrices M_M and M_P used for projecting the geometry to image space. These usually are the standard *modelview* and *projection* matrices OpenGL offers:

$$v'_{\lambda_i} = M_P \times M_M \times v_{\lambda_i}, \text{ with } i \in \{1, 2\}. \tag{4}$$

After the projection, the two eigenvectors are not necessarily orthogonal anymore.

2.2.2 Noise Texture Transformation

Mapping the initial texture to the geometry is a difficult and application-dependent task. Even though there exist methods to parameterize a surface, they employ restrictions to the surface (such as being isomorphic to discs or spheres), require additional storage for texture atlases (cf. [15, 19]) and, in general, require additional and often time-consuming pre-processing.

Another solution, proposed by Turk et al. [25], calculates the reaction diffusion texture directly on the surface. A major disadvantage of this method is the computational complexity. Even though these approaches provide almost distortion-free texture representations, isosurfaces, for example, may consist of a large amount of unstructured primitives, which increases the pre-processing time tremendously.

Whereas previously published approaches for image space LIC either use parameterized surfaces to apply the initial noise pattern to the surface or use locally or globally defined three-dimensional textures [26], we define an implicit parameterization of the surface that provides an appropriate mapping of the noise texture to the surface.

For our purpose, a simple, yet fast and flexible mapping strategy is used. We implicitly split the world space in voxels of equal size. These voxels fill the bounding volume of the geometry but are never created explicitly. The seamless noise texture is mapped onto each side of each voxel exactly once (no tiling). This creates a seamless mapping of the noise onto the surface of any connected block of voxels. During rendering, each point p of the surface can then be classified to belong to one certain voxel. This can be interpreted as discretization of the surface with the help of the implicit voxels. The normal at p on the surface is then used to find the most similar side of the voxel associated with p. Therefore, the scalar product between the surface normal and the normals of each side are compared. Once the side-plane is found, the following table determines the point's p texture coordinates:

Side-normal	Texture coordinates
$(1,0,0)$ or $(-1,0,0)$	(p_y, p_z)
$(0,1,0)$ or $(0,-1,0)$	(p_x, p_z)
$(0,0,1)$ or $(0,0,-1)$	(p_x, p_y)

Please note, that we assume the texture coordinates to be defined in a wrapped and continuously defined coordinate system, which is common in OpenGL. This allows the seamless tiling of the input noise texture on each voxel surface, which then is mapped to the surface. This can be interpreted as an orthographic projection of the voxel side plane onto the surface along the plane's normal vector.

Regardless of its simplicity, this method supports a fast and flexible parameterization of the surface space that only introduces irrelevant distortions (cf. Fig. 2), which vanish during the advection step.

By changing the size of voxels during the calculation, different frequencies of patterns can easily be produced and projected onto the geometry. This capability

Fig. 3 Comparison of two different voxel sizes during noise mapping. This demonstrates the possibility for dynamic refinement of the input noise to achieve different levels of detail

allows one to change the resolution of the texture as required for automatic texture refinement when zooming. A comparison of two different levels of detail is shown in Fig. 3.

2.3 Silhouette Detection

Following the projection pass, the silhouette detection pass uses the rendered surface's depth as input. To avoid advection over geometric boundaries, a silhouette of the object is required to stop advection in these areas [17]. Otherwise, tensor advection would lead to a constant flow of "particles" across surface boundaries which makes the surface's geometry and topology unrecognizable.

A standard three-by-three Laplacian filter, defined by the convolution mask

$$\begin{bmatrix} 0 & 1 & 0 \\ 1 & -4 & 1 \\ 0 & 1 & 0 \end{bmatrix} \tag{5}$$

applied to the depth values followed by thresholding, has proven to be suitable for our purposes. The silhouette image e for each pixel (x, y) is then provided to the next pass.

2.4 Advection

We have discussed how to project the geometry and the corresponding tensor field to image space. With the prepared image space eigenvectors and the input noise texture on the geometry, the advection can be done. Another important input is the advected image of the previous advection pass, created during the last render-frame. For the first frame, the geometry mapped noise is used as initialization.

Fig. 4 Advection texture after ten iterations. *Left*: red channel containing advected noise along the eigenvectors v'_{λ_1}; *Right*: green channel containing the advected noise along the second eigenvectors v'_{λ_2}

In the advection step, an Euler integration is applied to both vector fields separately. In our case, we do not calculate streamlines at each position of both vector fields, as normally done in LIC. We directly advect the noise input texture with the given vector fields, which provides the same results as locally filtering the data along pre-computed streamlines. During the advection pass, the previous advection results are, again, advected along the both eigenvector-fields separately. Each pass thereby only does one step along the two vector-fields. This decision was based on the fact that massively parallel architectures like modern GPUs are able to perform this task in parallel for each pixel several hundred times per second.

An important abortion-criteria here is the silhouette image e. If an edge is crossed, integration is stopped. The advection iteration can also be stopped if the advection reaches a saturation; i.e. the resulting advected images do not differ from the previous ones. Due to this saturation effect, we do not use the iteration count k as abortion criterion. Since the eigenvectors do not have an orientation, the advection needs to be done in direction of v'_{λ_1} and $-v'_{\lambda_1}$ and for v'_{λ_2} and $-v'_{\lambda_2}$ respectively. At this point, we have two advected images for each eigenvector. These get composited equally to for both eigenvectors at this pixel. Furthermore, the advection results for each eigenvector again get blended with the input noise. The blending ratio between noise and the advected images determines how crisp the results are. Lower ratios produce crispier images. Higher ratios produce more smooth and more smeared images. Throughout this paper, we use a ratio of $\frac{1}{10}$.

The resulting images, one for each eigenvector, are then used as input during the next render-frame. They can be stored in one single texture in different color channels. We use the red color channel for the first eigenvector and the green color channel for the second one. Figure 4 shows the resulting images of the advection step after ten iterations ($k = 10$). For later reference, we denote these advected images after k steps with $A^k_{\lambda_1}$ and $A^k_{\lambda_2}$. The number of iterations k hereby equals the number of rendered frames as we do only one advection step per frame.

2.5 Compositing

In a subsequent rendering pass, an initial fabric-like texture is composed. For the sake of simplicity and the limitations of some graphics boards, we split the final image creation step in an initial compositing followed by a postprocessing step described in the next section. The compositing step combines the advection results $A^k_{\lambda_1}$ and $A^k_{\lambda_2}$ into one image, whereas the postprocessor mainly improves visual quality. The input of the compositing step are the both advected images $A^k_{\lambda_1}$ and $A^k_{\lambda_2}$, the depth-buffer, the silhouette e as well as the light and colormap information from the projection pass. On the GPU, these inputs get composed to the final RGB triple for each pixel:

$$R = \frac{r \cdot A^k_{\lambda_2}}{8 \cdot (A^k_{\lambda_1})^2},$$

$$G = \frac{(1-r) \cdot A^k_{\lambda_1}}{8 \cdot (A^k_{\lambda_2})^2},$$

$$B = 0. \tag{6}$$

Equation 6 is a weighting function between the two advected images for both eigenvectors. The scalar factor r is used to blend between the two tensor directions. If both directions are equally important, a value of 0.5 ensures an equal blending of both directions. To explain the above compositing scheme, we are using the red component as an example. The red color should represent the main tensor direction. We therefore reduce the intensity of the second eigenvector image $A^k_{\lambda_2}$ using the over-emphasized first eigenvector image $A^k_{\lambda_1}$. To furthermore emphasize the influence of a high intensity in the advected image for the first eigenvector, the denominator is squared. This way, pixels with a high intensity in the first eigenvector direction get a high red intensity. This is done vice versa for the green channel. The compositing implicitly utilizes the clamping to [0, 1] which is done for colors on the GPU. This approach creates a mesh resembling the tensor field's structure. The light information is not yet composited. This is done during postprocessing.

2.6 Postprocessing

Additional filters can be applied to the composed image, such as contrast enhancement or sharpening filters, which are commonly used in vector field LIC [9, 26]. Figure 5 shows the result of Eq. 6 combined with Blinn-Phong shading. Even though Blinn-Phong shading [2] provides the required depth cues, additional emphasis of the third dimension using depth-enhancing color coding has proven to provide a

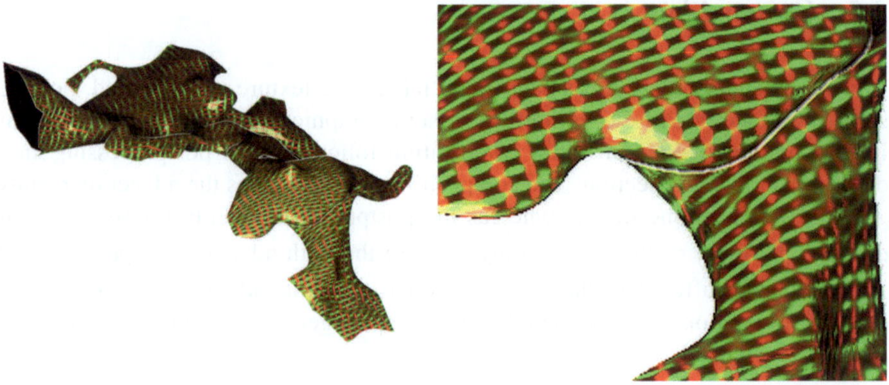

Fig. 5 The composited image produced by the compositing shader. Please note, that we added the light for illustration only. *Left*: the whole geometry. *Right*: a zoomed part of the geometry to show the still blurry fabric structure on the surface

better overall understanding of the data [5]. These techniques can be incorporated in our compositing scheme easily. Anyhow, the results still look blurry and justify the need for additional postprocessing. We have implemented two postprocessings which reduce blur and create crisp and appealing images.

Bump mapping, first introduced by Blinn [3] to simulate three-dimensionality in planar surfaces, can be used to improve spatial perception of the fabric surface. Bump mapping is normally done in world space, where the three-dimensional tangent space is known. In image space, this information is not available anymore. Therefore, we use a modified approach that can be applied in image space. Bump mapping requires the surface normal at each point of the surface. We obtain this normal by estimating the normalized gradient on the two-dimensional texture:

$$g = ||\nabla(R + G)||. \tag{7}$$

The resulting two-dimensional vector g describes the gradient on the image plane using each pixel's intensity. The blue color channel is not used as it does not contain relevant information. It is also worth noting that Eq. 6 did not include edge and light information to the red and green channel. This ensures that the gradient can be estimated on basis of the tensor fabric intensity only. Besides the gradient, the surface-normal in image space is needed too. The image space surface normal can be retrieved in projection step by projection each normal the same way a vertex gets projected. Using this gradient, the new normal is a weighted sum of the surface normal and the gradient and is used for calculating per-pixel Phong lighting as seen in Fig. 6. So, we achieve the bump mapping effect by doing per-pixel Phong lighting on the surface with normals modified by the fabric pattern. For later reference, we call this $\mathscr{B}$.

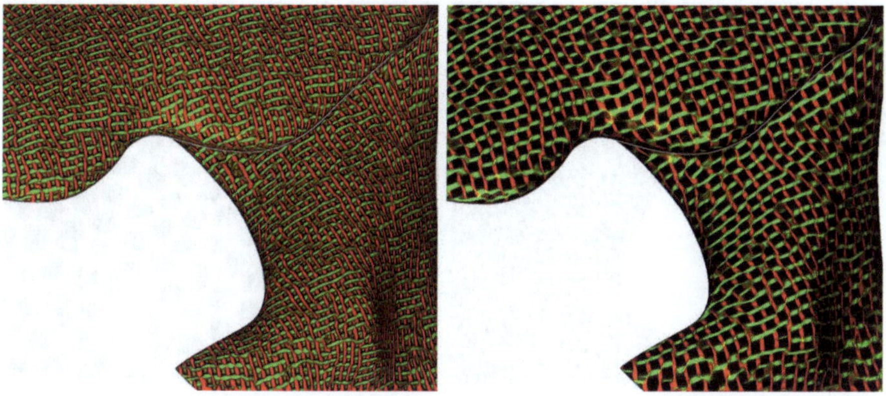

Fig. 6 The final image produced by the postprocessing shader in combination with bump mapping, the geometry's Phong shading and combined edges. *Left*: standard bump mapping. *Right*: the same zoomed part of the original geometry to show the effect of weighting the resulting Phong intensities by the original $R(x, y)$ and $G(x, y)$ intensities. This approach creates a more fabric-like impression that can be misunderstood as rotating ribbons similar to stream ribbons

Figure 6 (right) shows the additional scaling of the red and green color channels by the original color intensities, to lead to a more fabric-like impression of the lines. Equation 8 shows this in more detail and defines the final output color triple as:

$$\begin{aligned} R_{final} &= \mathcal{B} * (R * G + R^2) + e, \\ G_{final} &= \mathcal{B} * (R * G + G^2) + e, \\ B_{final} &= e. \end{aligned} \tag{8}$$

Again, this utilizes the clamping to [0, 1] done on the GPU for colors automatically.

With the help of bump mapping, we achieve a better spatial impression of the fabric-like pattern. A further visual improvement can be achieved by interpreting the structure on the surface as streamtubes [29] along the surface. Therefore, an approach similar to the ones in [18, 20] is appropriate to create the visual effect of streamtubes on the geometry's surface, without actually creating tubes. First, we need to have a tangential coordinate system, similar to the one needed for bump mapping. The eigenvectors v'_{λ_1} and v'_{λ_2} from Eq. 4 are interpreted as the tube tangents for the first and second eigenvector field. These tangents denote the direction of the tube along the surface and, together with the trivial surface normal of $(0, 0, 1)^T$, define the bi-normal vector. The bi-normal b for each eigenvector field i:

$$b_{\lambda_1} = ||(0, 0, 1)^T \times v'_{\lambda_1}|| \text{ and } b_{\lambda_2} = ||(0, 0, 1)^T \times v'_{\lambda_2}|| \tag{9}$$

To simplify the further descriptions, we describe the next steps for rendering the tubes of v'_{λ_1}. These steps need to be done for the second eigenvector too.

The area of the tube on the surface itself is already drawn in the composited image red channel, but their proper shading is missing. The bi-normal b_{λ_1} is now

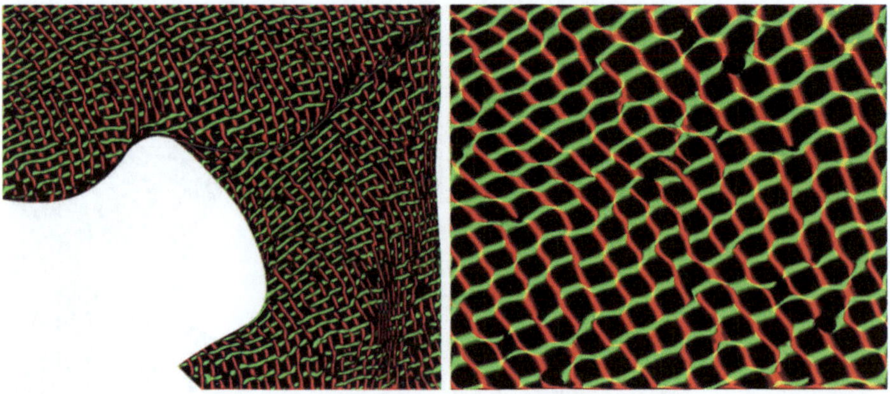

Fig. 7 *Left*: Interpreting the final image from Fig. 5 as streamtubes along the geometry's surface, and lighting them accordingly, results in a less blurry surface. *Right*: zoomed part of the left geometry to show the tube effect. Although there are plenty of artifacts in the zoomed image, they do not influence the overall impression of images not zoomed as much. Especially, such strongly zoomed images are not useful for gathering an overview over the tensor field's structure

used to find the correct normal on the tube-surface at the current pixel for lighting. Therefore, we sample in very small steps in direction of the bi-normal, to find the border of the tube. This check can be done using the red color channel from the composited image, since it is 0 if not on the respective tube anymore. In other words, we search for the smallest, positive scaling factors a_p and a_n which scale the bi-normal b_{λ_1} and $-b_{\lambda_1}$ to point to the nearest 0 in the red channel. If both factors are 0, the current pixel is not on a tube and further calculations can be skipped. The width of the tube passing the current pixel is defined by $a_p + a_n$ and

$$p = 2 * \left(0.5 - \frac{a_p}{a_p + a_n}\right) \in [-1, 1] \tag{10}$$

defines the relative position of the current pixel on this tube regarding the (normalized) bi-normal, where 0 is the middle of the tube. Also note that a_p and a_n are both positive.

This information is enough to define a diffuse shaded surface. However, we want proper per-pixel Phong shading and therefore need to use this to calculate the tube normal at the current pixel:

$$n_{tube} = (1 - p^2)(0, 0, 1)^T + p^2 b_{\lambda_1}. \tag{11}$$

The value of p is additionally squared to achieve the effect of a round surface. The normal n_{tube} is then used to calculate the Phong shading on the surface and produces the tube-like effect with proper spatial impression on the surface, as can be seen in Fig. 7. Again, this needs to be done for both eigenvectors separately. The resulting Phong intensities are then used as red and green color of the pixel in the final image and can be enhanced with the edge, as done in Eq. 8.

The artifacts seen in Fig. 7 result from the local approach we are using to calculate the tubes. As we do not integrate along the eigenvector-field, there may be discontinuities along a tube in the produced image. There are also artifacts caused by a blurry input field, where borders cannot be found clearly. But, since the frequency of the fabric structure is normally much higher, these effects are not visible anymore, as can be seen in Fig. 7, left.

2.7 Implementation

The implementation of the pipeline shown in Fig. 1 is straight forward. The figure clearly shows the input and output textures of each step and their execution order. The whole pipeline is implemented using OpenGL and framebuffer objects (FBO), which allow the efficient offscreen rendering and image space based processing we need. The projection step is the beginning of the pipeline and the only step which is not in image space. For the consecutive steps, we render a quad, filling the whole viewport of the FBO. The inputs and outputs are then bound as textures to the FBO and the quad respectively. Since texture space is limited on the hardware, it is important to store as much information as possible in each texture (four channels per texture available). The steps itself are all implemented as fragment shaders using GLSL. This way, we can work on a per-pixel basis easily. There are only some implementation specifics we want to mention here.

2.7.1 Projection Step

Our implementation is not limited to a special kind of geometry. It is able to handle almost every tensor field defined on a surface. It is, for example, possible to calculate an isosurface on a derived scalar metric, like fractional anisotropy, or on a second data set to generate a surface in a three-dimensional data domain. Other methods include hyper-stream surfaces [6], wrapped streamlines [8], or domain-dependent methods like dissection-like surfaces presented in [1]. The only requirement for the surface is that it is non-self-intersecting and that smooth normals are provided as they are required for the projection step and for proper lighting.

As the tensors are symmetric, it is sufficient to transfer six floating-point values per vertex to the GPU. In our case, two three-dimensional texture coordinates are used per vertex to upload the tensor information along with the geometry. Assuming the tensor T is available on the GPU, it is possible to map the two main directions to the surface described by the normal n at the current vertex using Eq. 2. This projection is implemented in a per-vertex manner in the vertex shader. In contrast, to ensure proper interpolation, eigenvalue decomposition and eigenvector calculation together with image space projection need to be done in the fragment shader. Since the eigenvectors are without orientation, it is possible to have sign flips between adjacent vertices. If the interpolation takes place after

Table 1 Frames per second (fps) for different data sets with given number of triangles and numbers of tensors. The frame rates are compared to simple rendering of the geometry using Phong shading. The frame rates were obtained for an AMD Athlon(tm) 64 × 2 Dual Core Processor 3800 + (512 K L2 Cache) with an NVIDIA G80 GPU (GeForce 8,800 GTS) and 640 MB of graphics memory at a resolution of 1,024 × 768 pixels. The geometry share relates the time used by the GPU to rasterize the geometry to the overall rendering time, which contains all steps of the pipeline. The time used to render the geometry clearly dominates the rendering times and reaches up to 90% of the overall rendering time even for medium-sized geometries

Figure	Nb triangles	Nb tensors	fps	fps (Phong only)	∅ Geometry share
9	41472	63075	32	61	72%
5	58624	88803	30	60	69%
10	571776	861981	14	16	90%

the eigenvector decomposition, these sign changes can render the interpolation useless. The eigenvectors v'_{λ_1} and v'_{λ_2} need to be scaled since textures are used for transportation where each value must be in the interval [0, 1]. To simplify further data handling and storage on the GPU, we scale the eigenvectors as follows:

$$\|v\|_\infty = max\{|v_x|, |v_y|\} \tag{12}$$

$$v''_{\lambda_i} = \frac{v'_{\lambda_i}}{\|v'_{\lambda_i}\|_\infty} \text{ with } i \in \{1,2\}, \quad \text{and} \quad \|v'_{\lambda_i}\|_\infty \neq 0 \tag{13}$$

The maximum norm (L_∞-norm) ensures that one component of the eigenvector is 1 or -1 and, therefore, one avoids numerical instabilities arising when limited storage precision is available, and can use memory-efficient 8-bit textures. The special case $\|v'_{\lambda_i}\|_\infty = 0$ only appears when the surface normal and the eigenvector point in the same direction. This case needs to be handled in the shader.

2.7.2 Advection Step

During each advection iteration, the input and output texture need to be switched. This way, the advection result of the previous advection iteration can be used as input without the need to allocate and deallocate a separate texture for each iteration.

3 Results

We have introduced a method to create a fabric-like surface tensor LIC in image space, similar to the one introduced in [13]. We used ideas from [17] to transform the algorithm into image space. Our implementation, using this method, is able to reach frame rates high enough for real-time user interaction. The only bottleneck is the hardware's ability to render large and triangle-rich geometry. All further steps can be done in constant time, see Table 1.

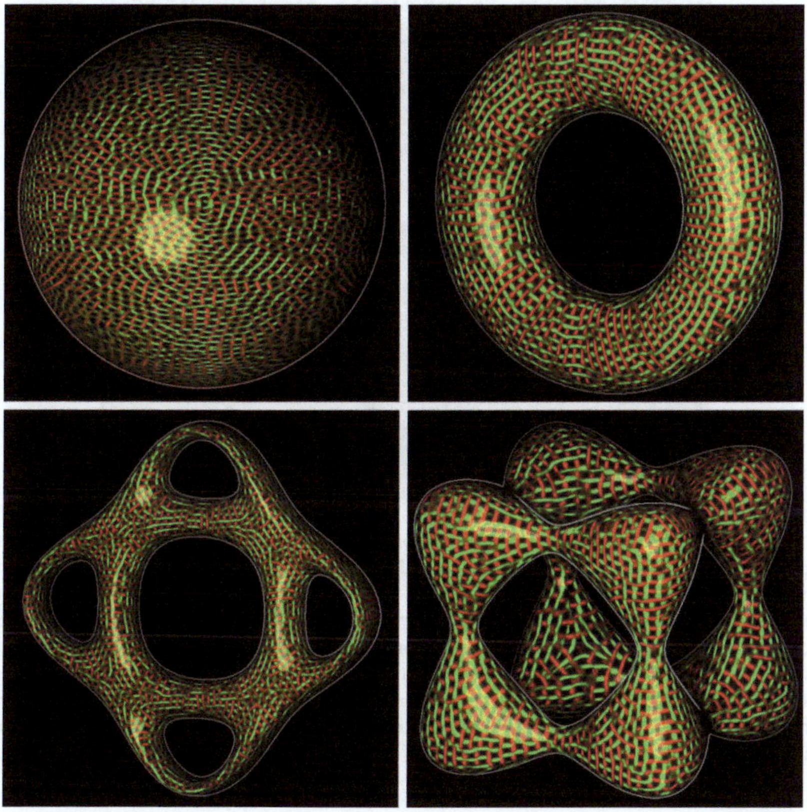

Fig. 8 Analytic test data sets. We applied our method to isosurfaces and the scalar field's Laplacian to demonstrate the suitability for complicated surfaces. Shown are the final images using our method for a sphere, torus, Tangle, and Bretzel5 data set (Eqs. 14–16)

3.1 Artificial Test Data Sets

We first applied our method to artificial test data sets that have complex topology: a torus, the Bretzel5, and the Tangle data set (cf. [16]), defined as implicit surfaces:

$$(1 - \sqrt{x^2 + y^2})(1 - \sqrt{x^2 + y^2}) + z^2 - 0.125 = 0, \tag{14}$$

$$((x^2 + .25 * y^2 - 1) * (.25 * x^2 + y^2 - 1))^2 + z^2 - 0.1 = 0, \text{ and} \tag{15}$$

$$x^4 - 5 * x^2 + y^4 - 5 * y^2 + z^4 - 5 * z^2 + 11.8 + w = 0. \tag{16}$$

We used the Laplacian on the surfaces as tensor fields. The results displayed in Fig. 8 show that neither the topology nor our artificial parameterization of the input noise texture influences the quality of the final rendering.

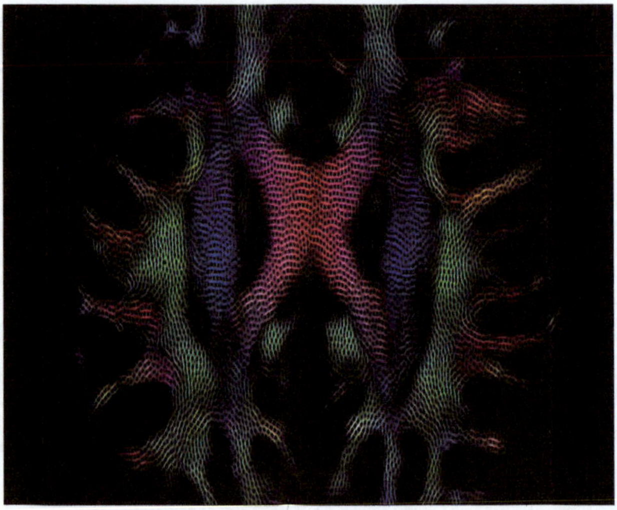

Fig. 9 An axial slice through a human brain: Corpus callosum (CC) (*red*), pyramidal tract (*blue*), and parts of the cinguli (*green* in front and behind the CC) are visible. The main direction in three-dimensional space is indicated by the RGB color map, where *red* indicates lateral (*left–right*), *green* anterior–posterior, and *blue* superior–inferior direction. The *left–right* structure of the CC can clearly be seen in its center, whereas color and pattern indicate uncertainty towards the outer parts. The same is true for the cinguli's anterior–posterior structure. As seen from the *blue color*, the pyramidal tract is almost perpendicular to the chosen plane and, therefore, secondary and ternary eigenvectors dominate the visualization. Alternatively, we could easily fade out those out-of-plane structures in cases where they distract the user

3.2 Modification for Medical Data Processing

Even though many higher-order methods have been proposed, due to scanner, time, and cost limitations, second-order tensor data is still dominant in clinical application. Medical second-order diffusion tensor data sets differ from engineering data sets because they indicate one major direction whereas the secondary and ternary directions only provide information in areas where the major direction is not well-defined, i.e., the fractional anisotropy—a measure for the tensor shape—is low. Almost spherical tensors, which indicate isotropic diffusion, occur in areas where multiple fiber bundles traverse a single voxel of the measurement or when no directional structures are present. Therefore, we modulate the color coding using additional information: In areas where one fiber direction dominates, we only display this major direction using the standard color coding for medical data sets, where x, y, and z alignment are displayed in red, green, and blue, respectively. In areas where a secondary direction in the plane exists, we display this information as well but omit the secondary color coding and display the secondary direction in gray-scale rendering mode and always below the primary direction (cf. Fig. 9). We

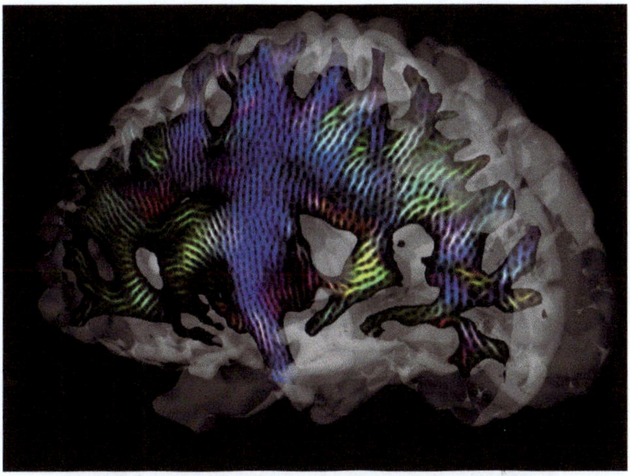

Fig. 10 Diffusion tensor data set of a human brain. We employed the method by Anwander et al. [1] to extract a surface following neural fibers and applied our method with an alternative color coding that is more suitable and can be incorporated more easily into medical visualization tools

use the method of Anwander et al. [1] to extract surfaces that are, where possible, tangential to the fiber directions. Hence, we can guarantee that the projection error introduced by our method in the surface's domain remains small. Even in areas where the fractional anisotropy is low and the color coding does no longer provide directional information, such as in some parts of the pyramidal tract in Fig. 9, the texture pattern still provides this information (Fig. 10).

3.3 *Mechanical Datasets*

Our approach is not only applicable to medical datasets, but it can also be applied to many other tensor data sets. Figures 11 and 12 show a slice in an earthquake dataset and an analytical strain tensor field. The analytical data set is the well-known single point load data set, where a single infinitesimally small point source pushes on an infinite surface. The forces and distortions inside the object are represented by stress and strain tensors, which are symmetric, second-order tensors. The earthquake data set is a simulation of a single concrete pile in solid ground excited by a measured earthquake pattern from the Kyoto earthquake (cf. Fig. 12). As shown, the material stress tensors, are defined on an irregular grid. We extracted a plane perpendicular to the pile and show the tensor information in that plane. Due to the time-dependent nature of the simulation, static images are quire complex.

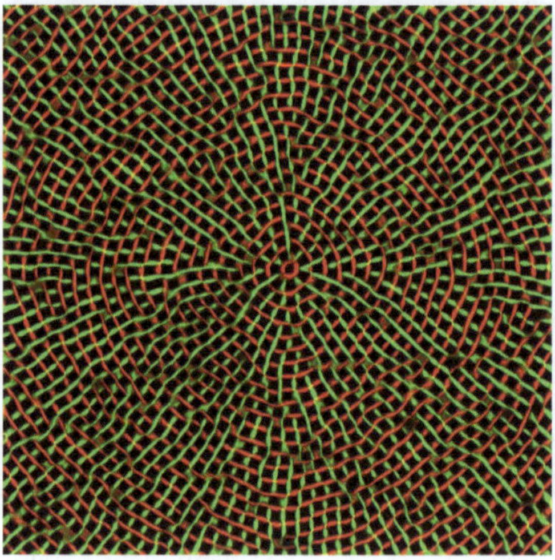

Fig. 11 A slice in the well-known single point load data set, showing the symmetric strain tensor at the surface of the slice

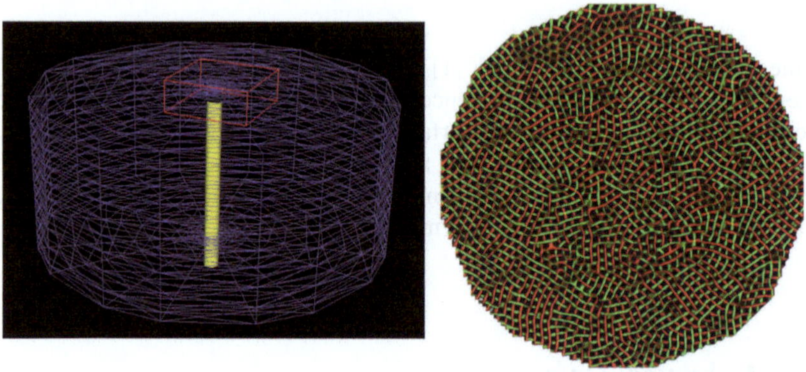

Fig. 12 A concrete pile in solid ground. *Left*: the original grid shown in purple. *Right*: a slice of the dataset showing the symmetric part of the tensor field

3.4 Performance

As indicated before, the only "bottleneck" in the visualization pipeline is the strongly geometry-dependent projection step. Since the surface needs to be rendered repeatedly in case of user interaction, the performance measures of our method consider repeated rendering of the geometry. The frame rate with geometry not being moved and, therefore, making the projection step and the edge detection step unnecessary, is considerably higher. The advection step can be done multiple times per frame. This reduces the number of frames needed until the advection is saturated. To ensure high frame-rates and smooth user-interaction, we do only one advection step per frame. To make the frame rates in the following tables comparable, user interaction is assumed and, therefore, rendering a single frame always consists of

- One projection step, including geometry rendering;
- One edge detection pass;
- Three advection iterations; and
- One output processing pass.

As seen in the previous sections, fragments not belonging to the geometry are discarded as soon as possible without using deferred shading. This also leads to performance gain in advection and output processing. In Table 1, a selection of data sets with their corresponding number of triangles and tensors are listed. The frame rates shown were obtained on an AMD Athlon(tm) 64 X2 Dual Core Processor 3800+ (512 K L2 Cache) with a NVIDIA G80 GPU (GeForce 8800 GTS) and 640 MB of graphics memory at a resolution of $1{,}024 \times 768$ pixels.

The assumption that geometry rendering with projection is the weakest component in this pipeline and that edge detection, advection, and output processing perform at a data-independent frame rate is confirmed by the frame rates shown in Table 1. It confirms that for large geometries, rendering the geometry alone is the dominant component. Since the vertex-wise calculations during projection are limited to tensor projection (Eq. 2) and noise texture transformation (Sect. 2.2.2), the most expensive calculations during projection are executed per fragment. This means that the expensive eigenvalue decomposition and eigenvector calculations are only required for fragments (pixels). To further decouple the calculation effort from the geometry's size, the depth test should be performed before performing the eigenvector decomposition. This goal can be achieved by first rendering the projected tensors to a texture, and computing the decomposition for visible fragments only. Nevertheless, this is not necessary for our current data set and screen sizes where the time required to render the geometry itself clearly dominates the time required to compute the texture pattern in image space. This can be seen in the increasing values in Table 1 with increasing size of vertices rendered.

4 Conclusions and Possible Directions for Future Research

We have presented a novel method for rendering fabric-like structures to visualize tensor fields on almost arbitrary surfaces. We have extended our method with an additional postprocessing step, which ensures crisp and clear results. With this, the tensor field structure can be grasped even better.

As our method works without generating three-dimensional textures that span the whole data set at sub-voxel resolution, it can be applied to complex data sets without introducing texture memory problems common to methods relying on three-dimensional noise textures. As major parts of the calculation are performed in image space, the performance of our algorithm is almost independent of data set size, provided that surfaces can be drawn efficiently, e.g., by using acceleration structures to draw only those parts of the geometry that intersect the view frustum or using ray tracing methods.

Whether the surface itself is the domain of the data, a surface defined on the tensor information (e.g., hyperstream surfaces), or a surface defined by other unrelated quantities (e.g., given by material boundaries in engineering data or anatomical structures in medical data) is independent from our approach. Nevertheless, the surface has to be chosen appropriately because only in-plane information is visualized. To overcome this limitation, information perpendicular to the plane could be incorporated in the color coding, but due to a proper selection of the plane that is aligned with our features of interest, this has not been necessary for our purposes.

Especially in medical visualization, higher-order tensor information is becoming increasingly important and different methods exist to visualize these tensors, including local color coding, glyphs, and integral lines. Nevertheless, an extension of our approach is one of our major aims. In brain imaging, experts agree that the maximum number of possible fiber directions is limited. Typically, a maximum of three or four directions in a single voxel are assumed (cf. Schultz et al. [21]). Whereas the number of output textures can easily be adapted, the major remaining problem is a lack of suitable decomposition algorithms on the GPU. Image space techniques, by their very nature, resample the data and, therefore, require one to use such proper interpolation schemes. In addition, maintaining orientations and assigning same fibers in higher-order data to the same texture globally is not possible today and, therefore, is a potential topic for further investigation.

Acknowledgements We thank Alfred Anwander and Thomas R. Knösche from the Max Planck Institute for Human Cognitive and Brain Sciences, Leipzig, Germany, for providing the human brain image data sets, and for fruitful discussions and comments, and Boris Jeremić, Department of Civil and Environmental Engineering, UC Davis, for providing the earthquake data set. We thank the members of the Visualization and Computer Graphics Research Group of the Institute for Data Analysis and Visualization, Department of Computer Science, UC Davis, and the members of the Abteilung für Bild- und Signalverarbeitung des Instituts für Informatik der Universität Leipzig, Germany.

This work has been supported in part by NSF grant CCF-0702817.

References

1. Anwander, A., Schurade, R., Hlawitschka, M., Scheuermann, G., Knösche, T.: White matter imaging with virtual klingler dissection. NeuroImage **47**(Suppl. 1), S105–S105 (2009). doi:10.1016/S1053-8119(09)70916-4. Organization for Human Brain Mapping 2009 Annual Meeting
2. Blinn, J.F.: Models of light reflection for computer synthesized pictures. In: SIGGRAPH '77: Proceedings of the 4th Annual Conference on Computer Graphics and Interactive Techniques, pp. 192–198. ACM, New York (1977). http://doi.acm.org/10.1145/563858.563893

3. Blinn, J.F.: Simulation of wrinkled surfaces. SIGGRAPH Comput. Graph. **12**(3), 286–292 (1978). http://doi.acm.org/10.1145/965139.507101
4. Cabral, B., Leedom, L.C.: Imaging vector fields using line integral convolution. In: SIGGRAPH '93: Proceedings of the 20th Annual Conference on Computer Graphics and Interactive Techniques, pp. 263–270. ACM, New York (1993). http://doi.acm.org/10.1145/166117.166151
5. Chu, A., Chan, W.Y., Guo, J., Pang, W.M., Heng, P.A.: Perception-aware depth cueing for illustrative vascular visualization. In: BMEI '08: Proceedings of the 2008 International Conference on BioMedical Engineering and Informatics, pp. 341–346. IEEE Computer Society, Washington, DC (2008). http://dx.doi.org/10.1109/BMEI.2008.347
6. Delmarcelle, T., Hesselink, L.: Visualization of second order tensor fields and matrix data. In: VIS '92: Proceedings of the 3rd Conference on Visualization '92, pp. 316–323. IEEE Computer Society, Los Alamitos (1992)
7. Dick, C., Georgii, J., Burgkart, R., Westermann, R.: Stress tensor field visualization for implant planning in orthopedics. IEEE Trans. Vis. Comput. Graph. **15**(6), 1399–1406 (2009). http://doi.ieeecomputersociety.org/10.1109/TVCG.2009.184
8. Enders, F., Sauber, N., Merhof, D., Hastreiter, P., Nimsky, C., Stamminger, M.: Visualization of white matter tracts with wrapped streamlines. In: Silva, C.T., Gröller, E., Rushmeier, H. (eds.) Proceedings of IEEE Visualization 2005, pp. 51–58. IEEE Computer Society, Los Alamitos (2005)
9. Grabner, M., Laramee, R.S.: Image space advection on graphics hardware. In: SCCG '05: Proceedings of the 21st Spring Conference on Computer Graphics, pp. 77–84. ACM, New York (2005). http://doi.acm.org/10.1145/1090122.1090136
10. Hasan, K.M., Basser, P.J., Parker, D.L., Alexander, A.L.: Analytical computation of the eigenvalues and eigenvectors in DT-MRI. J. Magn. Reson. **152**(1), 41–47 (2001). doi:10.1006/jmre.2001.2400
11. Hesselink, L., Levy, Y., Lavin, Y.: The topology of symmetric, second-order 3d tensor fields. IEEE Trans. Vis. Comput. Graph. **3**(1), 1–11 (1997). http://dx.doi.org/10.1109/2945.582332
12. Hlawitschka, M., Garth, C., Tricoche, X., Kindlmann, G., Scheuermann, G., Joy, K.I., Hamann, B.: Direct visualization of fiber information by coherence. International Journal of Computer Assisted Radiology and Surgery, CARS, CUARC.08 Special Issue. **5**(2), 125–131 (2010)
13. Hotz, I., Feng, L., Hagen, H., Hamann, B., Joy, K., Jeremic, B.: Physically based methods for tensor field visualization. In: VIS '04: Proceedings of the Conference on Visualization '04, pp. 123–130. IEEE Computer Society, Washington, DC (2004). http://dx.doi.org/10.1109/VIS.2004.80
14. Hotz, I., Feng, Z.X., Hamann, B., Joy, K.I.: Tensor field visualization using a fabric-like texture on arbitrary two-dimensional surfaces. In: Möller, T., Hamann, B., Russel, R.D. (eds.) Mathematical Foundations of Scientific Visualization, Computer Graphics, and Massive Data Exploration. Springer, Heidelberg (2009)
15. Iwakiri, Y., Omori, Y., Kanko, T.: Practical texture mapping on free-form surfaces. In: PG '00: Proceedings of the 8th Pacific Conference on Computer Graphics and Applications, p. 97. IEEE Computer Society, Washington, DC (2000)
16. Knoll, A., Hijazi, Y., Hansen, C., Wald, I., Hagen, H.: Interactive ray tracing of arbitrary implicits with simd interval arithmetic. In: RT '07: Proceedings of the 2007 IEEE Symposium on Interactive Ray Tracing, pp. 11–18. IEEE Computer Society, Washington, DC (2007). http://dx.doi.org/10.1109/RT.2007.4342585
17. Laramee, R.S., Jobard, B., Hauser, H.: Image space based visualization of unsteady flow on surfaces. In: VIS '03: Proceedings of the 14th IEEE Visualization 2003 (VIS'03), p. 18. IEEE Computer Society, Washington, DC (2003). http://dx.doi.org/10.1109/VISUAL.2003.1250364
18. Merhof, D., Sonntag, M., Enders, F., Nimsky, C., Hastreiter, P., Greiner, G.: Hybrid visualization for white matter tracts using triangle strips and point sprites. IEEE Trans. Vis. Comput. Graph. **12**(5), 1181–1188 (2006). http://doi.ieeecomputersociety.org/10.1109/TVCG.2006.151
19. Purnomo, B., Cohen, J.D., Kumar, S.: Seamless texture atlases. In: SGP '04: Proceedings of the 2004 Eurographics/ACM SIGGRAPH symposium on Geometry Processing, pp. 65–74. ACM, New York (2004). http://doi.acm.org/10.1145/1057432.1057441

20. Schirski, M., Kuhlen, T., Hopp, M., Adomeit, P., Pischinger, S., Bischof, C.: Virtual tubelets-efficiently visualizing large amounts of particle trajectories. Comput. Graph. **29**(1), 17–27 (2005). http://dx.doi.org/10.1016/j.cag.2004.11.004
21. Schultz, T., Seidel, H.P.: Estimating crossing fibers: a tensor decomposition approach. IEEE Trans. Vis. Comput. Grap. **14**(6), 1635–1642 (2008). http://doi.ieeecomputersociety.org/10.1109/TVCG.2008.128
22. Tricoche, X., Scheuermann, G., Hagen, H.: Tensor topology tracking: a visualization method for time-dependent 2D symmetric tensor fields. In: Eurographics 2001 Proceedings, Computer Graphics Forum 20(3), 461–470. The Eurographics Association, Saarbrücken (2001). http://dx.doi.org/10.1111/1467-8659.00539
23. Tricoche, X.: Vector and tensor field topology simplification, tracking, and visualization. Ph.D. thesis, University of Kaiserslautern (2002)
24. Turing, A.: The chemical basis of morphogenesis. Philos. Trans. R. Soc. Lond. **237**(641), 37–72 (1952)
25. Turk, G.: Generating textures on arbitrary surfaces using reaction-diffusion. In: SIGGRAPH '91: Proceedings of the 18th Annual Conference on Computer Graphics and Interactive Techniques, pp. 289–298. ACM, New York (1991). http://doi.acm.org/10.1145/122718.122749
26. Weiskopf, D., Ertl, T.: A hybrid physical/device-space approach for spatio-temporally coherent interactive texture advection on curved surfaces. In: GI '04: Proceedings of Graphics Interface 2004, pp. 263–270. Canadian Human-Computer Communications Society, School of Computer Science, University of Waterloo, Waterloo (2004)
27. van Wijk, J.J.: Image based flow visualization. In: SIGGRAPH '02: Proceedings of the 29th Annual Conference on Computer Graphics and Interactive Techniques, pp. 745–754. ACM, New York (2002). http://doi.acm.org/10.1145/566570.566646
28. van Wijk, J.J.: Image based flow visualization for curved surfaces. In: VIS '03: Proceedings of the 14th IEEE Visualization 2003 (VIS'03), p. 17. IEEE Computer Society, Washington, DC (2003). http://dx.doi.org/10.1109/VISUAL.2003.1250363
29. Zhang, S., Demiralp, C., Laidlaw, D.H.: Visualizing diffusion tensor mr images using streamtubes and streamsurfaces. IEEE Trans. Vis. Comput. Graph. **9**(4), 454–462 (2003). http://doi.ieeecomputersociety.org/10.1109/TVCG.2003.1260740
30. Zhang, E., Hays, J., Turk, G.: Interactive tensor field design and visualization on surfaces. IEEE Trans. Vis. Comput. Graph. **13**(1), 94–107 (2007). http://dx.doi.org/10.1109/TVCG.2007.16
31. Zheng, X., Pang, A.: Hyperlic. In: VIS '03: Proceedings of the 14th IEEE Visualization 2003 (VIS'03), p. 33. IEEE Computer Society, Washington, DC (2003). http://dx.doi.org/10.1109/VISUAL.2003.1250379

Beyond Topology: A Lagrangian Metaphor to Visualize the Structure of 3D Tensor Fields

Xavier Tricoche, Mario Hlawitschka, Samer Barakat, and Christoph Garth

Abstract Topology was introduced in the visualization literature some 15 years ago as a mathematical language to describe and capture the salient structures of symmetric second-order tensor fields. Yet, despite significant theoretical and algorithmic advances, this approach has failed to gain wide acceptance in visualization practice over the last decade. In fact, the very idea of a versatile visualization methodology for tensor fields that could transcend application domains has been virtually abandoned in favor of problem-specific feature definitions and visual representations. We propose to revisit the basic idea underlying topology from a different perspective. To do so, we introduce a Lagrangian metaphor that transposes to the structural analysis of eigenvector fields a perspective that is commonly used in the study of fluid flows. Indeed, one can view eigenvector fields as the local superimposition of two vector fields, from which a bidirectional flow field can be defined. This allows us to analyze the structure of a tensor field through the behavior of fictitious particles advected by this flow. Specifically, we show that the separatrices of 3D tensor field topology can in fact be captured in a fuzzy and numerically more robust setting as ridges of a trajectory coherence measure. As a result, we propose an alternative structure characterization strategy for the visual analysis of practical 3D tensor fields, which we demonstrate on several synthetic and computational datasets.

X. Tricoche (✉) · S. Barakat
Purdue University, Lafayette, IN, USA
e-mail: xmt@purdue.edu; sbarakat@purdue.edu

M. Hlawitschka · C. Garth
University of California at Davis, Davis, CA, USA
e-mail: hlawitschka@ucdavis.edu; cgarth@ucdavis.edu

D.H. Laidlaw and A. Vilanova (eds.), *New Developments in the Visualization and Processing of Tensor Fields*, Mathematics and Visualization,
DOI 10.1007/978-3-642-27343-8_5,

1 Introduction and Motivation

Tensor fields are ubiquitous in the theory of continuum mechanics. They offer an elegant mathematical language to describe the forces acting upon solids and fluids. Their analysis is therefore needed in application disciplines ranging from structural mechanics and fluid dynamics to geophysics, earthquake research, materials engineering, and aeronautics. The theoretical and practical importance of tensor fields has led to a dedicated research effort in the scientific visualization community aimed at devising analysis tools that allow scientists and engineers to make sense of the corresponding datasets. Yet, the task is challenging owing to the size, dimensionality, and many degrees of freedom of the data.

To address this difficulty, a general approach in the visualization literature consists in extracting salient structures from the data in a pre-processing stage. The information obtained through this computation is then used to facilitate the visual inspection of large and complex datasets. Specifically, it allows subsequent data depictions to focus on remarkable geometric descriptors, thus avoiding visual clutter while improving the interactivity of the visualization. Topology in particular provides a theoretical framework within which the notions of structure and saliency can be articulated in a principled way. Following the introduction of this formalism in vector field visualization, topology was extended to tensor fields over 15 years ago and a complete algorithmic framework is now available for the extraction of the so-called topological skeleton in three-dimensional datasets [18].

One could therefore assume that a general solution has been found to the visual analysis of 3D tensor fields. Unfortunately, a rapid glance at the recent literature reveals unambiguously that topology has fallen short of offering a globally valid approach for this problem and it has not been adopted by visualization practitioners in the investigation of their tensor data. The shortcomings of the topological approach in 3D concern its significant algorithmic complexity and its lack of numerical robustness. The latter aspect is particularly problematic since it essentially disqualifies this method from being applied to any measured or simulated numerical dataset. Instead, the characterization of important structures in application datasets has been mainly driven by domain-specific feature of interest that lack generality and are typically defined in an ad-hoc manner. A prime example of this trend concerns the large body of work dedicated to diffusion tensor imaging (DTI) data, where anatomical structures such as fiber bundles are the natural focus of both analysis and visual representation.

We propose in this paper to revisit the basic idea underlying topology from a different perspective. Specifically, we introduce a Lagrangian metaphor that transposes to the structural analysis of eigenvector fields a mathematical theory that has recently gained popularity in the fluid dynamics community. Building upon the strong theoretical connections that exist between vector and eigenvector fields, we show that the topology of 3D tensor fields can be characterized through extremal manifolds of a trajectory coherence measure obtained by processing eigenvector fields. This approach significantly improves upon the topological method however in that it yields a fuzzy and numerically more robust characterization that is well suited for practical datasets. We demonstrate our technique and compare the extracted structures to topology in a benchmark analytical datasets and in a computational fluid dynamics simulation. Our results document the potential of this general strategy for the visual analysis of symmetric 3D tensor fields across engineering and scientific applications.

The remainder of this paper is organized as follows. We review previous work in tensor field visualization with an emphasis on the topological framework in Sect. 2. The theoretical foundations of our approach, which span dynamical systems, differential geometry, and computer vision, are summarized in Sect. 3. The proposed model of structure is described in Sect. 4 along with some algorithmic considerations. Finally, results are shown in Sect. 5 and we point out promising avenues for future research in Sect. 6.

2 Related Work

2.1 Topological Methods

The topological framework was first applied to the visualization of second-order tensor field by Delmarcelle and Hesselink [4]. Leveraging ideas introduced previously for the topology-based visualization of vector fields [9, 12], these authors proposed to display a planar tensor field through the topological structure of its two orthogonal eigenvector fields. As discussed in their work, the lack of orientation of eigenvector fields leads to singularities that are not seen in regular vector fields. Indeed, those *degenerate points* correspond to locations where the tensor field becomes isotropic, i.e. where both eigenvalues are equal and the eigenvectors are undefined. Yet, this seminal work shows that a similar synthetic representation is obtained in the tensor setting through topological analysis: degenerate points are connected in graph structure through curves called *separatrices* that are everywhere tangent to an eigenvector field. Refer to Fig. 1.

The three-dimensional case was first considered in a subsequent paper by Hesselink et al. [13]. Interestingly, their discussion was primarily focused on the types of degenerate *points* that can occur in this setting. As such it did not explicitly mention that the most typical singularities in 3D are lines and not isolated points. In fact, this basic property was first pointed out in the work of Zheng and Pang [34] who also proposed the first algorithm for the extraction of these line features. In a nutshell, their method consists in computing the intersection of these lines with the faces of a voxel grid, by solving a set of seven cubic equations. This method was later improved by allowing for the continuous tracking of intersection points across the voxel interior [35]. Additionally, a geometric formulation was proposed as an alternative to the system of equations [35]. Most recently, Schultz et al. discussed three-dimensional tensor field topology in the context of DT-MRI data [27].

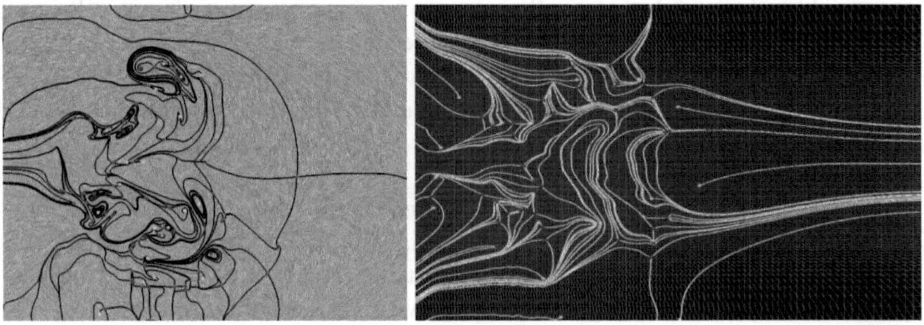

Fig. 1 Topological graph. Singularities correspond to the nodes of the graph, while separatrices form the edges. *Left*: vector field topology of a turbulent flow. *Right*: topology of a rate of strain symmetric tensor field

Following a systematic approach, their work demonstrates the shortcomings of this mathematical framework in the structural analysis of the typically noisy images acquired in practice. As an alternative, they proposed an approach where structure is defined with respect to a stochastic assessment of the connectivity along integral curves.

2.2 Ridges and Valleys

The detection of creases, in other words ridges and valleys, in scalar images is a topic of traditional interest in a variety of disciplines, most prominently in image processing and computer vision [17]. Among the multiple definitions proposed in the literature, the one introduced by Eberly et al. is widely used in practice [5]. In essence, this definition generalizes the intuitive height-based definition of ridges and valleys [3] to d-dimensional manifolds embedded in n-dimensional image space [6].

From an algorithmic standpoint, several methods have been proposed that permit the extraction of these manifolds from numerical data. Many of them apply a principle similar to Marching Cubes [19], effectively interpreting creases as 0-level sets of the dot product between the gradient of the considered scalar image and one or several eigenvectors of its hessian matrix. The lack of intrinsic orientation of those eigenvectors requires the use of heuristics to provide them with an arbitrary but locally consistent orientation. Some authors match sets of eigenvectors across the faces of a voxel [21] while others determine a local reference by computing the average orientation of the eigenvector field over a face [30]. A scale-space approach is discussed in [7]. Peikert and Roth introduced the notion of *Parallel Vector Operator* [22] as a computation primitive in flow visualization and they showed that it could be used to find the intersection of ridge and valley lines with the faces of a computational mesh [23]. Computationally, the method can be implemented in a variety of ways, including isocontour intersection, iterative numerical search, and through the solution of an eigensystem.

It is interesting to observe that several applications of this general methodology to Scientific Visualization problems have been presented in recent years. Sahner et al. extract a skeleton of vortices in three-dimensional flows as valley lines of a Galilean invariant (invariant under changes of inertial reference frame) called λ_2 [25]. Their algorithm combines ideas developed by Eberly with a *Feature Flow Field* approach [31]. In a work most closely related to ours, Kindlmann et al. extract ridge and valley surfaces of the *Fractional Anisotropy* (FA) in DTI volumes using a modified version of *Marching Cubes*. In particular, their scheme uses smooth reconstruction kernels and an orientation tracking scheme along edges to assign a coherent orientation to an eigenvector field on a voxel face. In addition, Sadlo and Peikert applied the scheme proposed by Furst and Pizer [8] to extract *Lagrangian Coherent Structures* from transient flows as ridge and valley surfaces of a scalar measure of particle coherence [24].

3 Theory

We review in this section the two major models proposed to date in the visualization literature to identify salient structures in tensor fields, namely topology and creases, and underscore their connections. In doing so, we explicitly restrict our considerations to techniques applicable to a broad range of applications and as such do not assume a specific physical interpretation for the tensor field. We then briefly introduce the notion of Lagrangian coherent structures, which has recently attracted significant attention in the fluid dynamics community and stems from the theory of dynamical systems. Finally, we describe how this conceptual framework can be extended to apply to tensor field, a generalization that we justify by the mathematical link that exists between vector and line fields.

3.1 *Tensor Field Topology*

A three-dimensional second-order symmetric tensor (simply called *tensor* hereafter) is fully represented by its three real eigenvalues (tensor *shape*) and an associated set of mutually orthogonal eigenvectors (tensor *orientation*). For a tensor field, the ordering of the three eigenvalues $\lambda_1 \geq \lambda_2 \geq \lambda_3$ thus defines *major*, *medium*, and *minor* eigenvector fields. Because such fields carry neither norm nor intrinsic orientation, they form *line fields*. In each eigenvector field, one can define curves that are everywhere tangent to the field. These curves are generally referred to as *hyperstreamlines* in the visualization literature [4].

One can characterize the topology of an eigenvector field in terms of the connectivity established by its hyperstreamlines. In other words, topology segments the domain into regions where hyperstreamlines share the same end points. This formalism is directly related to the topological framework used to study vector fields, where it characterizes regions of similar asymptotic behavior of the corresponding flow [26]. Note that the three eigenvector fields associated with a tensor field are mutually orthogonal and their topologies are closely related.

In the tensor setting, singularities of the topology corresponds to locations where the directional information of an eigenvector field is degenerate, which occurs when two or more eigenvalues are equal. Three degenerate configurations are possible in 3D, namely $\lambda_1 = \lambda_2 > \lambda_3$ (*planar anisotropy*), $\lambda_1 > \lambda_2 = \lambda_3$ (*cylindrical anisotropy*), and $\lambda_1 = \lambda_2 = \lambda_3$ (*spherical isotropy*). While the latter case is in fact numerically instable and typically absent from practical datasets, the first two degeneracies are stable features of the tensor topology. In their recent work Zheng and Pang have shown that these features are in general lines [34, 35].

A major drawback of the topological approach lies in its lack of robustness. Indeed, the structures identified by a topological analysis are very sensitive to noise and therefore essentially meaningless in the context of measured data such as Diffusion Tensor Imaging (DTI), where low signal-to-noise are typical in clinical practice [27]. This result echoes our observation that alternative structure definitions are needed to address the visual analysis needs of a variety of problems. We illustrate this point with our results on FA in DTI in Sect. 5.

3.2 *Crease Manifolds in Tensor Fields*

The ridges and valleys (collectively, creases) of a scalar field f can be defined in terms of the gradient $\mathbf{g} = \nabla f$ and Hessian $\mathbf{H}$ of the field [6]. In other words, creases are the manifolds along which f is at a local extremum, when constrained to the line or plane defined by one or two eigenvectors of the Hessian. A function is at extrema where its gradient is orthogonal to the constraint surface [20], thus ridges and valleys are where the gradient $\mathbf{g}$ is orthogonal to one or two of the unit-length eigenvectors $\{\mathbf{e}_1, \mathbf{e}_2, \mathbf{e}_3\}$ (with corresponding eigenvalues $\lambda_1 \geq \lambda_2 \geq \lambda_3$) of the Hessian $\mathbf{H}$:

	Surface	Line
Ridge	$\mathbf{g} \cdot \mathbf{e}_3 = 0$	$\mathbf{g} \cdot \mathbf{e}_2 = \mathbf{g} \cdot \mathbf{e}_3 = 0$
	$\lambda_3 < 0$	$\lambda_3,\ \lambda_2 < 0$
Valley	$\mathbf{g} \cdot \mathbf{e}_1 = 0$	$\mathbf{g} \cdot \mathbf{e}_1 = \mathbf{g} \cdot \mathbf{e}_2 = 0$
	$\lambda_1 > 0$	$\lambda_1,\ \lambda_2 > 0$

Observe that the sign and the magnitude of the eigenvalue(s) determine the crease strength. In particular, $|\lambda_1|$ (resp. $|\lambda_2|$) measure the feature strength of a valley line (resp. surface), while $|\lambda_3|$ (resp. $|\lambda_2|$) measure the feature strength of a ridge line (resp. surface) [6].

A link between creases and tensor fields can be established through the study of scalar invariants. Invariants of second-order three-dimensional tensors can be intuitively understood as measurements of tensor *shape*, which is independent of tensor *orientation*. As such, they are defined in terms of the tensor's eigenvalues. In particular, an invariant called *mode* [2] provides a conceptual link between creases and tensor field topology [33] while the ridge manifolds of FA have been shown to delineate the one- and two-dimensional core structures of major fiber bundles in the brain white matter, and valley surfaces of FA constitute boundaries between adjacent fiber bundles with distinct orientations [15, 16, 28, 33].

We show in the following that creases can be applied to the direction information of a tensor field to reveal salient manifolds that relate to the topological skeleton.

3.3 Lagrangian Coherent Structures

As a preamble to the Lagrangian definition of structure for tensor fields that we discuss in the next section, we briefly introduce in the following the notion of Lagrangian coherent structures in vector fields. The conceptual link between these two structure types is established in Sect. 4.1 through the interpretation of an eigenvector field as a bidirectional flow.

A vector field $\mathbf{v}$ can be associated with a *dynamical system* through following equations.

$$\begin{cases} \dot{\mathbf{x}}(t,\, t_0,\, \mathbf{x_0}) = \mathbf{v}(t,\, \mathbf{x}(t,\, t_0,\, \mathbf{x_0})) \\ \mathbf{x}(t_0,\, t_0,\, \mathbf{x_0}) = \mathbf{x_0}, \end{cases}$$

where the dot designates derivation with respect to the time variable t, $\mathbf{x_0}$ is the *initial condition*. The trajectory $\mathbf{x}(\cdot,\, t_0,\, \mathbf{x_0}) \;:\; t \;\mapsto\; \mathbf{x}(t,\, t_0,\, \mathbf{x_0})$ is obtained by integrating the system. The map $\mathbf{x_t} \;:=\; \mathbf{x}(t,\, t_0, \cdot)$ is called *flow map*: $\mathbf{x_t}(\mathbf{x_0})$ corresponds to the position reached at time t by a particle released at $\mathbf{x_0}$ at time t_0.

The coherence of particle trajectories can be quantified through the *finite-time Lyapunov exponent* (FTLE) [11]. Specifically, stable and unstable *Lagrangian coherent structures* (LCS) are characterized as ridge manifolds of the FTLE field. With previous notation, one considers the flow map $\mathbf{x}_T$ which maps a position $\mathbf{x_0}$ occupied by a particle at initial time t_0 to the position reached by this particle at time $T \;=\; t_0 + \tau$, where τ is finite. The spatial variations of this flow map around a given position $\mathbf{x_0}$ are locally determined by its spatial gradient, the Jacobian matrix $J_{\mathbf{x}}(t, t_0, \mathbf{x_0}) \;:=\; \nabla_{\mathbf{x_0}}\mathbf{x}(t, t_0, \mathbf{x_0})$ at $\mathbf{x_0}$. This gradient can be used to determine the maximal dispersion after time τ of particles in a neighborhood of $\mathbf{x_0}$ at time t_0 as a function of the direction $\mathbf{d_{t_0}}$ along which we move away from $\mathbf{x_0}$: $\mathbf{d_t} \;=\; J_{\mathbf{x}}(t, t_0, \mathbf{x_0})\;\mathbf{d_{t_0}}$. Maximizing the norm $|\mathbf{d_t}|$ over all possible unit directions $\mathbf{d_{t_0}}$ corresponds to computing the spectral norm of $J_{\mathbf{x}}(t, t_0, \mathbf{x_0})$ (i.e., the square root of the maximum eigenvalue of $J^T J$). Therefore, maximizing the dispersion of particles around $\mathbf{x_0}$ at t_0 over the space of possible directions around $\mathbf{x_0}$ is equivalent to evaluating

$$\sigma_\tau(t_0, \mathbf{x_0}) := \sqrt{\lambda \text{max}(J_{\mathbf{x}}(t, t_0, \mathbf{x_0})^T J_{\mathbf{x}}(t, t_0, \mathbf{x_0}))}. \quad (1)$$

Linearization and normalization by advection time τ yields following expression for the finite-time Lyapunov exponent:

$$\lambda(t, t_0, \mathbf{x_0}) = \frac{1}{|\tau|} \log \sqrt{\lambda \max(J_{\mathbf{x}}(t, t_0, \mathbf{x_0})^T J_{\mathbf{x}}(t, t_0, \mathbf{x_0}))}. \tag{2}$$

This rate can be evaluated for both forward and backward advection (positive or negative τ). Large values of λ for forward (resp. backward) advection correspond to unstable (resp. stable) manifolds with repelling (resp. attracting) impact on nearby particles.

It is key to note that the separatrices of the topology belong to the hyperbolic manifolds that are characterized as LCS. Therefore the structures that are identified in numerical datasets using the standard topological method can typically be characterized in the LCS framework. The LCS method is also more robust to noise and uncertainty since it defines structures as the ridge surfaces of a continuously varying measure. Hence, the LCS approach provides a conceptual framework that elegantly generalizes the topological method while overcoming some of its most basic limitations in visual data analysis.

4 A Lagrangian Model of Structure in Tensor Fields

4.1 An Extension of LCS to Tensor Fields

As previously defined, LCS and FTLE are notions that pertain only to vector fields. Yet, visualization research has successfully exploited the connections between vector and eigenvector fields. A profound mathematical link exists between vector fields and so-called *line fields* (in other words a field associated each point with a line direction [29]), of which eigenvector fields are a particular example. This connection can be intuitively understood by considering the projection that associates a vector field defined over a *twofold covering space* and a line field. *Branching points* in the covering result in (topological) singularities. Refer to Fig. 2.

It was shown in previous work [26, 32] that it provides a high-level theoretical justification for the transposition of the topological (and other vector field visualization approaches) to the study of eigenvector fields. In the light of the parallel drawn in Sect. 3.3 between LCS and vector field topology, this fundamental connection permits an extension of the notion of LCS to tensor fields.

To provide a more formal motivation for this generalization, it is necessary to consider the definition of the separatrices in the topological skeleton of a tensor field. In the 3D case, these separatrices are two-dimensional manifolds that originate along the 1D singularities. More specifically, these manifolds form in the vicinity of the singularity the boundary of so-called hyperbolic sectors [32, 37].

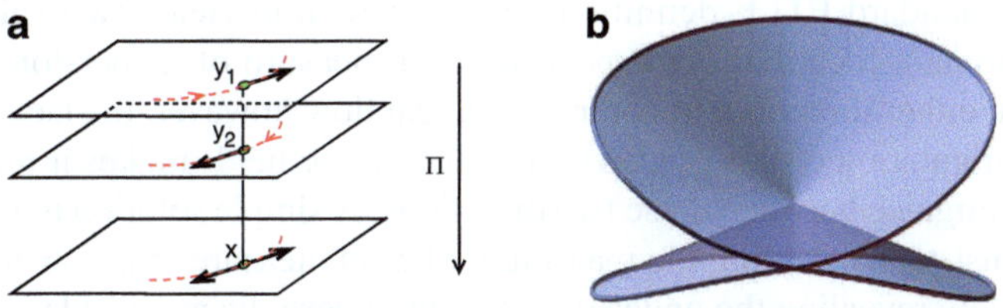

Fig. 2 Covering spaces provide a theoretical connection in the form of a *projection* between a vector field defined over a self intersecting two-manifold embedded in a 3D *ambient space* and a 2D eigenvector field. (**a**) Twofold covering. (**b**) Branched covering

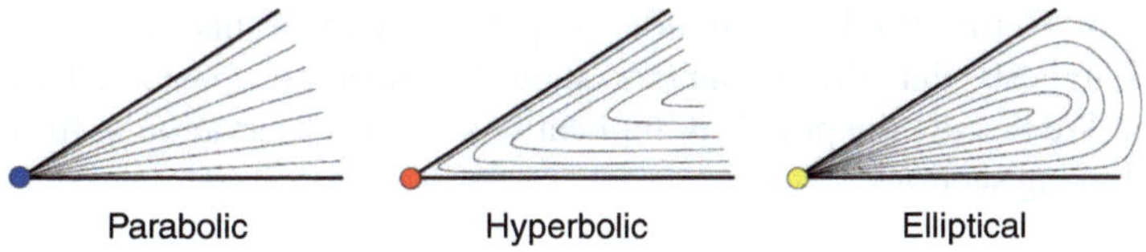

Fig. 3 Sector types in the vicinity of a singularity correspond to different patterns form by integral curves. *Left*: parabolic type. *Center*: hyperbolic type. *Right*: elliptic type. These patterns are the only possible ones

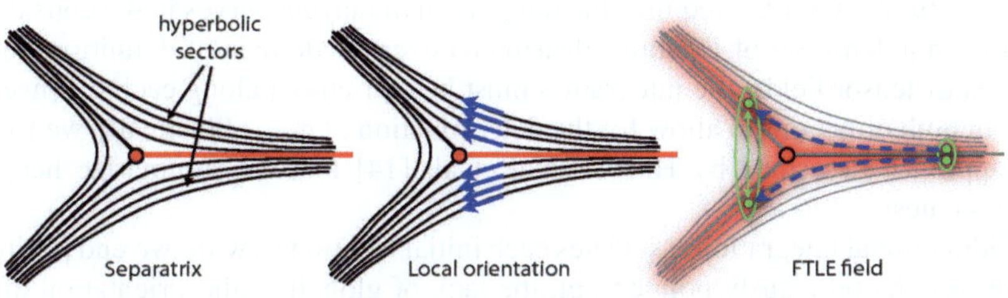

Fig. 4 FTLE computed in a locally defined and normalized vector field that is everywhere tangent with the underlying eigenvector field

Refer to Fig. 3 for an illustration of the possible sector types. It is the dispersion of the hyperstreamlines in the vicinity of separating manifolds that enables their characterization as ridges of a measure similar to FTLE.

Eigenvector fields do not possess an orientation and the presence of singularities in the topology clearly makes a globally consistent orientation of hyperstreamlines impossible in general. However, such an orientation can be assigned locally to yield a partial vector field and associated flow. The magnitude of this vector field is meaningless and can be considered normalized. Observe that the construction we just described is in fact the one that is implicitly taking place when hyperstreamlines are being constructed through numerical integration: a vector field is locally fitted to the underlying line field to advance the integration. With that setup in place, we can now define a finite-time Lyapunov exponent computed in this locally valid vector field. Refer to Fig. 4 for an illustration of this procedure.

Like the standard FTLE definition, this construction yields two values σ_1 and σ_2 at any domain location, corresponding to a measure of dispersion rate of the local flow in either direction. In contrast to a real flow however, the lack of globally consistent orientation of this piecewise defined vector fields makes it impossible to globally distinguish between these two directions. A simple solution to this problem however consists in selecting the maximum of both measures $\sigma_{max} = \max(\sigma_1, \sigma_2)$, thus effectively revealing the underlying salient hyperbolic manifolds.

The second fundamental difference between a vector and an eigenvector field from the point of view of this structure characterization concerns the eigenvectors' lack of intrinsic norm. The local vector field mentioned previously can be assumed to be normalized. This, in turn, means that the integration time τ present in the definition of FTLE (Eq. 1) amounts to a spatial length in this context. This new meaning suggests that this parameter should in fact be considered as a scale parameter. We discuss the practical implications of this observation in further detail in the following sections.

4.2 Computation

The computation of LCS requires the integration of tangent curves in the considered field from a dense set of locations distributed over the domain of definition. In the context of tensor fields, the integration must be carried out along each eigenvector field in both directions to allow for the determination of σ_{max}. Practically we follow the approach described by Hlawitschka et al. [14] that we summarize here for completeness.

Bidirectional integration associates each initial location $\mathbf{x}_0$ with two end positions $\mathbf{x}_1$ and $\mathbf{x}_2$. As previously pointed out, the lack of globally valid orientation of the eigenvector field implies that the respective order of these positions is arbitrary. Hence to compute the Jacobian of the flow map in each direction $\mathbf{J}_\mathbf{x}(t, t_0, \mathbf{x}_{1,2})$, we record at each point the vector chosen locally to play the role of forward direction. This vector is then used in a subsequent step to determine what indices should be used to compute the two Jacobian values through central differences.

The integration length (denoted t by analogy with the vector case) is a spatial scale parameter that must be selected carefully to reveal interesting structures. Excessive values not only lead increase the complexity of the characterized structures (by compounding the impact on multiple manifolds on individual trajectories), they also lead to issues associated with the boundaries of the domain. Indeed, trajectories whose requested length cannot be reached within the domain cause normalization issues in the computation of $\sigma_{1,2}$. Our solution to this problem consists in computing σ_{max} across a range of integration lengths in order to identify a posteriori the most relevant length. These discrete length samples can also be used to form a scale space in which a continuous analysis could be performed. Though we did not explore this avenue depth in the present work, we illustrate in the following section the incidence of this parameter on the resulting structures.

Once the FTLE fields have been computed, the next stage consists in extracting the ridges that form the salient manifolds of the tensor field's structure. We are using to that effect the method recently proposed by Barakat and Tricoche [1], in which the ridge extraction is formulated as a ray casting problem in a view-dependent setting. This method offers indeed the significant benefit of running interactively on the GPU, thus allowing us to test the implications of various parameters used to filter the ridges.

5 Results

To show the relationship between the separatrices of tensor field topology and the LCS computed in eigenvector fields we first consider the double point load dataset that has been studied in previous work [34–37] since it provides a basis for comparison. Specifically, we applied the method described by Zheng et al. [36] to extract separating surfaces along the degenerate lines of the topology. These degenerate lines are shown as red curves in the images below. Unfortunately, this method turned out to be numerically challenging in the context of this particular dataset, as shown next.

We start by looking at the structures associated with the major eigenvector field, see Fig. 5. It can be seen that the separatrices of the topology were only incompletely characterized, owing to the near degeneracy of the tensor field in the entire region surrounding the upper part of the P-type degenerate lines. As expected, the ridges of the FTLE field in contrast prove much more robust and properly capture the symmetric geometry of the separatrices. A close-up (Fig. 5, bottom row) sheds some additional light on the issues associated with the topology. It can indeed be observed that the separating surfaces are starting along inconsistent directions. Again, the LCS do not suffer from this shortcoming. The topology associated with the minor eigenvector field (linear type degenerate lines) is shown in Fig. 6.

As mentioned previously, the integration length used in the construction of the FTLE field is a degree of freedom of the analysis that can be tuned to control the geometric complexity of the structures. We document the impact of this parameter on the resulting visualization in Fig. 7 where the topology of the minor eigenvector field can be seen increase monotonically with the integration length.

While topology has been shown in previous work [27] to yield fragile and therefore unreliable structures in the context of noisy numerical datasets, our proposed approach is fundamentally more robust and enables the analysis of challenging engineering datasets. To document the performance of our method in such demanding scenarios, we considered two CFD simulations exhibiting turbulent flows in canonical configurations. The first dataset correspond to a single time step out of a transient simulation of a flow past a protruding cone, leading to the formation of downstream vortices. In the second dataset fast and slow fluid flow layers mix after passing a thin plate. The shear induced by the differing velocities causes strong turbulence. In these flow phenomena, the considered symmetric tensor

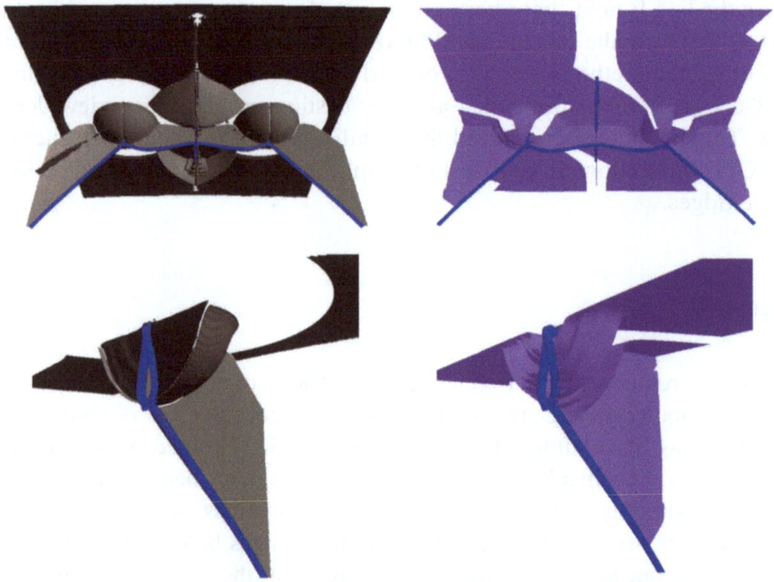

Fig. 5 Comparison between the separating surfaces associated with planar type (*blue*) degenerate lines and LCS computed in the major eigenvector field of the double point load dataset. *Left*: Ridges of FTLE. *Right*: separating surfaces and their corresponding degenerate lines

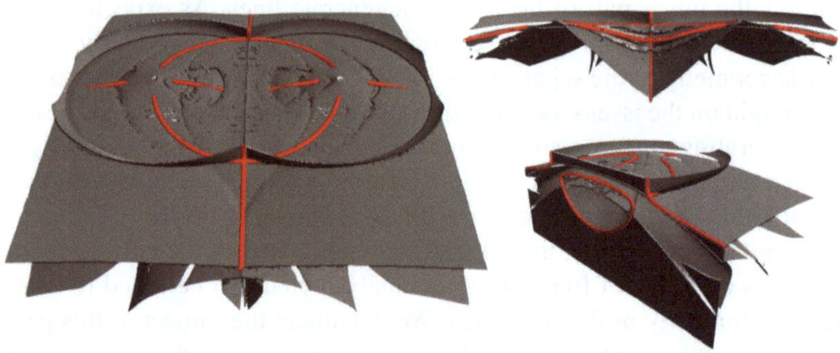

Fig. 6 Separatrices of the planar-type degenerate lines of the double point load dataset extracted as ridges of the minor eigenvector field

field is the rate of strain, which is known to be closely related to major features of interest such as vortices and flow separation in fluid dynamics problems [10].

The surfaces characterized in those datasets form boundaries between regions of different strain behaviors, whereby each region is associated with a locally uniform pattern of a principal strain direction. The interpretation of the role of these regions and associated boundaries in the behavior of the flow in turn depends on additional parameters such as the relative magnitude of the eigenvalues, i.e. the tensor

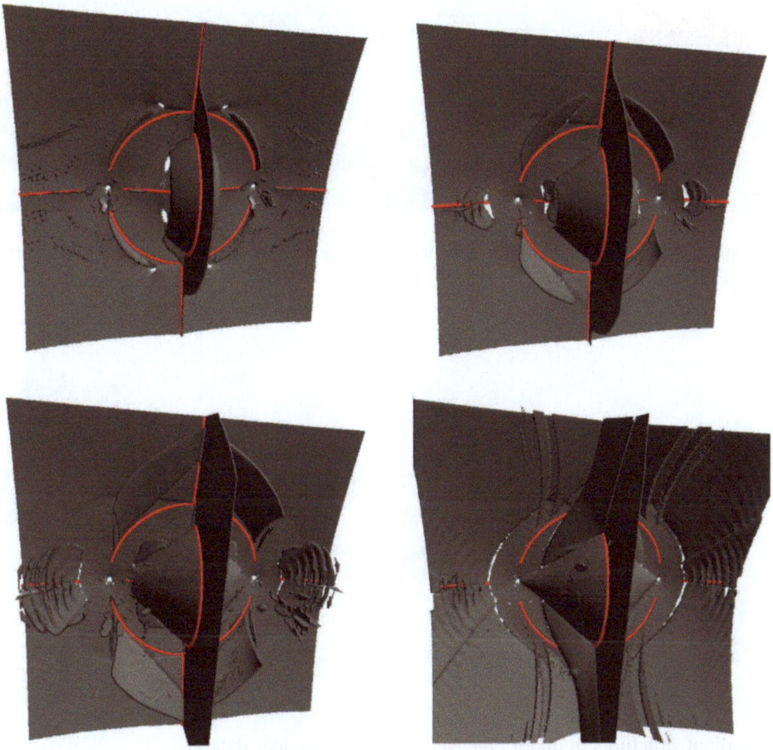

Fig. 7 Evolution of the extracted manifolds under increasing integration length in the minor eigenvector field of the double point load dataset

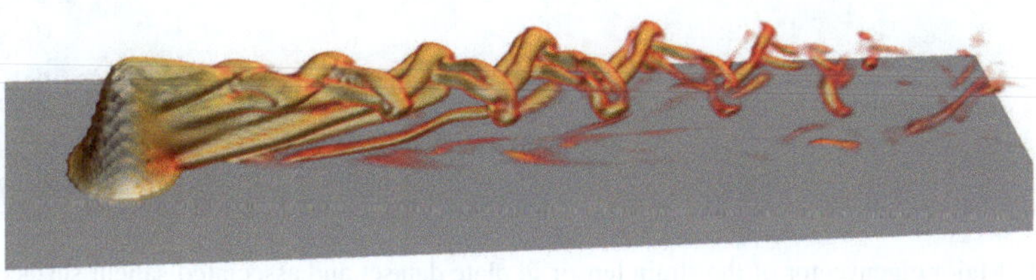

Fig. 8 Vortex shedding through a cone obstacle

field anisotropy. Note that while these quantitative considerations are basically orthogonal to the structure of the eigenvector fields, they offer a complementary perspective that is key to a physical analysis of the considered phenomenon.

We start by considering the simpler of the two datasets, in which a protuberant cone causes vortex shedding. An illustration of the resulting vortices is presented in Fig. 8.

To reveal the relationship between the salient manifolds in the tensor FTLE field and the patterns of coherent orientation of the strain principal direction, we show

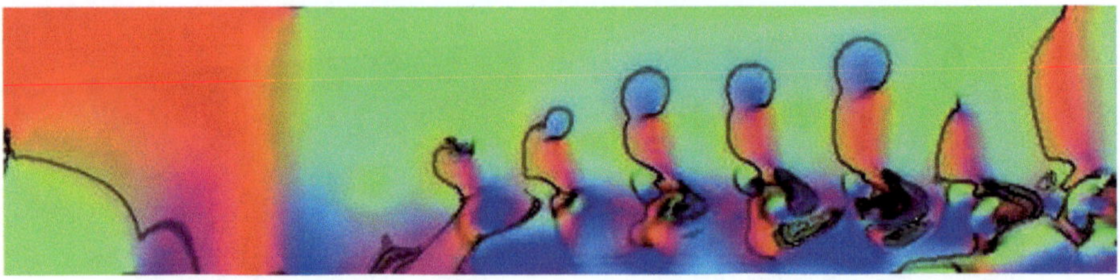

Fig. 9 Major eigenvector of the strain tensor in shedder dataset combined with salient surfaces of tensor FTLE

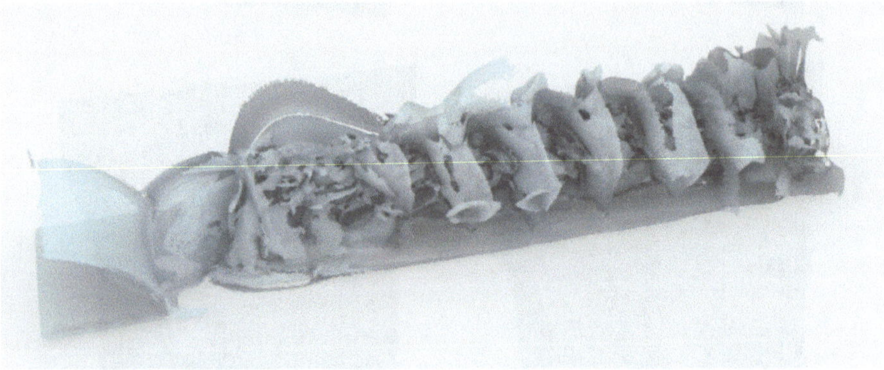

Fig. 10 Salient structures in major eigenvector field of shedder dataset

Fig. 11 Major eigenvector of the strain tensor in plate dataset and associated salient surfaces

in Fig. 9 color coding of the eigenvector orientation (using the standard symmetric RGB encoding) combined with the geometry of those manifolds. It can be seen that these surfaces (shown in a 2D slice) properly delineate regions of different behaviors, corresponding to different colors. A 3D view of the surfaces is shown in Fig. 10.

The plate dataset considered hereafter exhibits significantly more complicated structures due to higher turbulence. This leads to convoluted patterns of the major eigenvector field of the strain tensor that are visible in Fig. 11. Here again the salient surfaces obtained through Lagrangian processing successfully highlight the boundaries of significantly different regions and their impact on the flow.

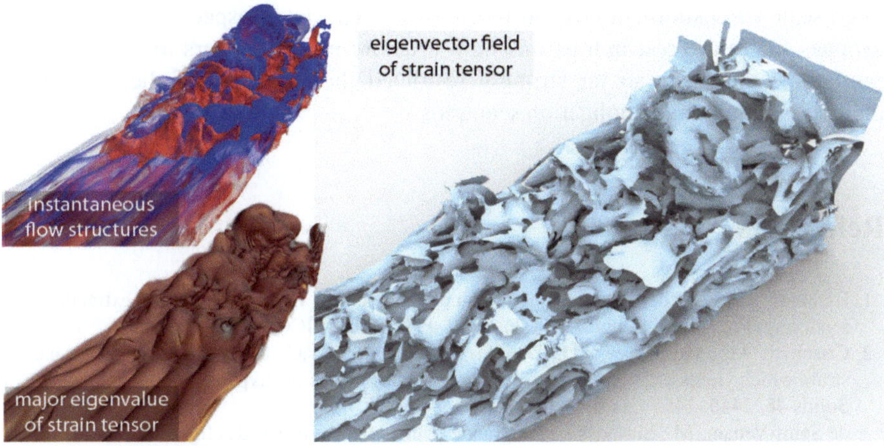

Fig. 12 Plate dataset. The *top left* image shows a overview of the instantaneous flow structures formed by the interaction of a shearing flow with a plate obstacle. The resulting turbulence induces the effective mixing of slow and fast moving layers. The *bottom left* image corresponds to the major eigenvalue of the strain tensor field. The *right* image shows the ridge surfaces extracted from the FTLE field computed in the strain tensor field

A detailed 3D view of this dataset is proposed in Fig. 12. The patterns of the flow itself (top left), as characterized through the standard LCS approach computed in the velocity vector field at fixed time, shows the typical turbulent patterns that are expected in this case. The major eigenvalue of the strain tensor (lower left) takes on high values that are directly correlated with the location of these structures. Hence we use this field as a mask to spatially confine the Lagrangian computation of salient structures. The result is shown on the right hand side in Fig. 12. It can be seen that very complicated geometric structures emerge from this analysis. Further investigation would be necessary to determine their role in the organization of the flow. As stated previously, such a study requires to take into account the influence of the eigenvalues and their interplay with the geometry of the eigenvector field.

6 Conclusion and Future Work

We have presented a generalization to tensor fields of Lagrangian coherent structures, a dynamical systems' concept applied so far to the analysis of vector fields. Our proposed approach is built upon a Lagrangian metaphor for eigenvector fields that finds its theoretical justification in the connection that exists between vector fields and line fields. After reviewing the state of the art of structure-based tensor field visualization techniques, we have shown that LCS improve upon the results achieved by the topological method in a standard benchmark synthetic dataset. More importantly, our results document the ability of a LCS-based analysis to reveal salient structures in highly complex 3D tensor fields, such as those associated with large-scale CFD simulations of turbulent flows. This latter aspect opens promising avenues for future research as this new structure model appears to overcome the limitations that have so far strongly restricted the relevance of the topological method in demanding application scenarios.

References

1. Barakat, S., Tricoche, X.: An image-based approach to interactive crease extraction and rendering. Procedia Comput. Sci. **1**(1), 1709–1718 (2010). ICCS 2010
2. Criscione, J.C., Humphrey, J.D., Douglas, A.S., Hunter, W.C.: An invariant basis for natural strain which yields orthogonal stress response terms in isotropic hyperelasticity. J. Mech. Phys. Solids **48**, 2445–2465 (2000)
3. de Saint-Venant, M.: Surfaces à plus grande pente constituées sur des lignes courbes. Bulletin de la Société Philomathématique de Paris 24–30 (1852)
4. Delmarcelle, T., Hesselink, L.: The topology of symmetric, second-order tensor fields. In: Proceedings of IEEE Visualization 1994, pp. 140–147. IEEE Computer Society, Los Alamitos (1994)
5. Eberly, D.: Ridges in Image and Data Analysis. Kluwer, Boston (1996)
6. Eberly, D., Gardner, R., Morse, B., Pizer, S.: Ridges for image analysis. J. Math. Imaging Vis. **4**, 351–371 (1994)
7. Furst, J.D., Pizer, S.M., Eberly, D.H.: Marching cores: a method for extracting cores from 3d medical images. In: Proceedings of IEEE Workshop on Mathematical Methods in Biomedical Image Analysis, pp. 124–130. IEEE Computer Society, Los Alamitos (1996)
8. Furst, J.D., Pizer, S.M.: Marching ridges. In: Proceedings of 2001 IASTED International Conference on Signal and Image Processing. IASTED/ACTA Press, Anaheim/Calgary (2001)
9. Globus, A., Levit, C., Lasinski, T.: A tool for visualizing the topology if three-dimensional vector fields. In: IEEE Visualization Proceedings, pp. 33–40. IEEE Computer Society, Los Alamitos, CA, October 1991
10. Haimes, R.: Using residence time for the extraction of recirculation regions. In: AIAA Paper 99–3291. Norfolk, VA (1999)
11. Haller, G.: Lagrangian structures and the rate of strain in a partition of two-dimensional turbulence. Phys. Fluids **13**(11), 3365–3385 (2001)
12. Helman J., Hesselink, L.: Surface representation of two and three-dimensional fluid flow topology. In: Proceedings of IEEE Visualization '90 Conference, pp. 6–13. IEEE Computer Society, Los Alamitos, CA (1990)
13. Hesselink, L., Levy, Y., Lavin, Y.: The topology of symmetric, second-order 3D tensor fields. IEEE Trans. Vis. Comput. Graph. **3**(1), 1–11 (1997)
14. Hlawitschka, M., Garth, C., Tricoche, X., Kindlmann, G., Scheuermann, G., Joy, K., Hamann, B.: Direct visualization of fiber information by coherence. Int. J. Comput. Assist. Radiol. Surg. **5**(2), 125ff (2010)
15. Kindlmann, G., Tricoche, X., Westin, C.-F.: Delineating white matter structure in diffusion tensor MRI with anisotropy creases. Med. Image Anal. **11**(5), 492–502 (2007)
16. Kindlmann, G., San Jose Estepar, R., Smith, S.M., Westin, C.-F.: Sampling and visualizing creases with scale-space particles. IEEE Trans. Vis. Comput. Graph. **15**(6), 1415–1424 (2009)
17. Koenderink, J.J., van Doorn, A.J.: Local features of smooth shapes: ridges and courses. In: Vemuri, B.C. (ed.) Proceedings of Geometric Methods in Computer Vision II, B.C. Vemuri (ed.), vol. 2031, pp. 2–13 (1993)
18. Laramee, R.S., Hauser, H., Zhao, L., Post, F.H.: Topology-based flow visualization, the state of the art. In: Hege, H.-C., Polthier, K., Scheuermann, G. (eds.) Topology-Based Methods in Visualization II, Mathematics and Visualization, pp. 1–19. Springer, Berlin (2007)

19. Lorensen, W.E., Cline, H.E.: Marching cubes: a high resolution 3d surface construction algorithm. Comput. Graph. **21**(4), 163–169 (1987)
20. Marsden, J.E., Tromba, A.J.: Vector Calculus. W.H. Freeman, New York (1996)
21. Morse, B.S.: Computation of object cores from grey-level images. PhD thesis, University of North Carolina at Chapel Hill, Chapel Hill (1994)
22. Peikert, R., Roth, M.: The Parallel Vectors operator – a vector field visualization primitive. In: IEEE Visualization Proceedings '00, pp. 263–270. IEEE Computer Society, Los Alamitos, CA (2000)
23. Roth, M.: Automatic extraction of vortex core lines and other line-type features for scientific visualization. PhD thesis, ETH Zürich (2000)
24. Sadlo, F., Peikert, R.: Efficient visualization of lagrangian coherent structures by filtered amr ridge extraction. IEEE Trans. Vis. Comput. Graph. **13**(6), 1456–1463 (2007)
25. Sahner, J., Weinkauf, T., Hege, H.-C.: Galilean invariant extraction and iconic representation of vortex core lines. In: Brodlie, K.J.K., Duke, D. (eds.) Proceedings of Eurographics/IEEE VGTC Symposium on Visualization (EuroVis '05), pp. 151–160. Eurographics Association, Leeds, UK, June 2005
26. Scheuermann, G., Tricoche, X.: Topological methods in flow visualization. In: Johnson, C., Hansen, C. (eds.) Visualization Handbook, pp. 341–356. Academic Press, Copyright Elsevier Butterworth-Heinemann, Oxford, UK (2004)
27. Schultz, T., Theisel, H., Seidel, H.-P.: Topological visualization of brain diffusion MRI data. IEEE Trans. Vis. Comput. Graph. **13**(6), 1496–1503 (2007)
28. Schultz, T., Theisel, H., Seidel, H.-P.: Crease surfaces: from theory to extraction and application to diffusion tensor MRI. IEEE Trans. Vis. Comput. Graph. (2009). RapidPost, accepted 16 April 2009. doi:10.1109/TVCG.2009.44
29. Spivak, M.: A Comprehensive Introduction to Differential Geometry, vol. III. Publish or Perish Incs., Berkeley (1979)
30. Stetten, G.D.: Medial-node models to identify and measure objects in real-time 3-d echocardiography. IEEE Trans. Med. Imaging **18**(10), 1025–1034 (1999)
31. Theisel, H., Seidel, H.-P.: Feature flow fields. In: Proceedings of Joint Eurographics – IEEE TCVG Symposium on Visualization (VisSym '03), pp. 141–148. Eurographics Association Aire-la-Ville, Switzerland (2003)
32. Tricoche, X., Zheng, X., Pang, A.: Visualizing the topology of symmetric, second-order, time-varying two-dimensional tensor fields. In: Weickert, J., Hagen, H. (eds.) Visualization and Processing of Tensor Fields, pp. 225–240. Springer, Berlin (2006)
33. Tricoche, X., Kindlmann, G., Westin, C.-F.: Invariant crease lines for topological and structural analysis of tensor fields. IEEE Trans. Vis. Comput. Graph. **14**(6), 1627–1634 (2008)
34. Zheng, X., Pang, A.: Topological lines in 3D tensor fields. In: VIS '04: Proceedings of the Conference on Visualization '04, Washington, DC, USA, pp. 313–320. IEEE Computer Society, Los Alamitos (2004)
35. Zheng, X., Parlett, B., Pang, A.: Topological lines in 3D tensor fields and discriminant hessian factorization. IEEE Trans. Vis. Comput. Graph. **11**(4), 395–407 (2005)
36. Zheng, X., Parlett, B., Pang, A.: Topological structures of 3d tensor fields. In: Proceedings of IEEE Visualization, pp. 551–558. IEEE Computer Society, Los Alamitos, CA (2005)
37. Zheng, X., Tricoche, X., Pang, A.: Degenerate 3d tensors. In: Visualization and Processing of Tensor Fields, pp. 241–256. Springer, Berlin (2006)

Tensor Field Design: Algorithms and Applications

Eugene Zhang

Abstract Tensor field design has found increasing applications in computer graphics, geometry processing, and scientific visualization. In this chapter, we review recent advances in tensor field design and discuss possible future research directions.

1 Introduction

In this survey we review recent advances in tensor field design, which refers to creating a tensor field based on user and application-specific constraints. Tensor field design has many applications in computer graphics, geometry processing, and scientific visualization. A tensor field can be used to guide the orientations of many objects in their respective applications, such as brush strokes and hatches in non-photorealistic rendering [13–15], texture and geometry patterns in pattern synthesis [23, 40, 44], mesh elements in geometry remeshing [1, 24, 27, 33], and road network in street modeling [5]. Different tensor fields can lead to different visual effects in these applications. Tensor field design can also be used to modify an input tensor field, such as the curvature tensor, in order to remove errors or reduce geometric complexity in the input field, as well as to add synthetic features in order to achieve some application-specific goals [51]. For visualization, tensor field design can be used to create data sets with specific geometric and topological characteristics in order to test the efficiency of certain visualization methods [4, 41, 42, 51].

There are some fundamental challenges in tensor field design. First, tensors of different *orders* have drastically different properties, which in turn require different design approaches (Sect. 2). Second, the requirements of tensor field design are

E. Zhang (✉)
School of Electrical Engineering and Computer Science, 2111 Kelley Engineering Center, Oregon State University, Corvallis, OR 97331, USA
e-mail: zhange@eecs.oregonstate.edu

D.H. Laidlaw and A. Vilanova (eds.), *New Developments in the Visualization and Processing of Tensor Fields*, Mathematics and Visualization,
DOI 10.1007/978-3-642-27343-8_6, © Springer-Verlag Berlin Heidelberg 2012

highly application-dependent. Consequently, a design system needs to be tailored with respect to application-specific goals. Third, tensor field design requires efficient tensor field analysis. However, past research has largely focused on first- and second-order tensors, and relatively few methods support higher-order tensor field analysis. Fourth, in many applications tensor field design is performed on surfaces. The geometric and topological features in the surface often make design more challenging. Finally, a tensor field design system needs to provide interactive design and feedback capabilities. Consequently, design operations need to be implemented efficiently, and proper visualization techniques should be supplied.

The remainder of this survey is organized as follows. We will discuss applications of tensor field design in Sect. 2. Theory and algorithms for the design of first-order, second-order, and higher-order tensor fields are described in Sects. 3, 4, and 5, respectively. Adapting these techniques to manifold surfaces is discussed in Sect. 6. Interactive higher-order tensor field visualization is reviewed in Sect. 7. Finally, we discuss future research directions in tensor field design in Sect. 8.

2 Applications of Tensor Field Design

In this survey we focus on tensor fields defined on two-dimensional manifolds, as most applications of tensor field design involve surfaces [4, 5, 11, 28, 34, 35, 50, 51].

The behaviors of a tensor greatly depend on its order N, which is the number of indices for the entries in the tensor. A first-order tensor $V = (v_i)$ and second-order tensor $M = (m_{ij})$ is a vector and a matrix, respectively. When $N > 2$, a tensor $T = (t_{i_1,i_2,\ldots,i_N})$ is referred to as a *higher-order* tensor. Existing applications of tensor field design require that the tensors be *symmetric*, i.e., given a tensor $T = t_{i_1,i_2,\ldots,i_N}$, we have $t_I = t_{I'}$ if I' is a permutation of I. Under this definition, any vector is automatically, although trivially, symmetric. When $N = 2$, this definition corresponds to the set of symmetric matrices. For convenience, we will mean symmetric tensors when mentioning tensors in the remainder of the chapter.

A tensor field is a continuous tensor-valued function defined over some domain. In this section we will review applications of tensor field design based on the order of the tensor field.

2.1 Vector Field Design

A first-order tensor is a vector. There are a number of motivating applications for vector field design.

In example-based texture synthesis, an input 2D texture is used to decorate a surface [32, 40, 44] (Fig. 1 (middle)). When the texture is *anisotropic*, a vector field is needed to orient the texture on the surface. Different vector fields can lead to

Fig. 1 A vector field is needed to guide texture and geometry synthesis on surfaces. The singularities in the vector field (*left*) can lead to the breakup of the synthesis pattern (*middle*). A leaf pattern is placed on the Bunny surface following a vector field (not shown) (The right image is a courtesy of [23], ©2011 IEEE)

different visual effects. Moreover, the quality of the synthesis is affected by the properties of the vector field. For example, the singularities in the vector field (Fig. 1 (left): colored dots) can lead to the breakup of texture patterns (Fig. 1 (middle)). In addition, the lack of smoothness in the field can make synthesis more difficult. It is therefore important to control the topology of the vector field such as the number and location of the singularities as well as the smoothness of the field. Moreover, more natural looking result can be obtained if the vector field aligns with the sharp edges in the underlying surface [48]. Vector field design can also guide geometry pattern synthesis [2, 23, 53] with similar requirements (Fig. 1 (right)).

Another use of vector field design is to guide particle motions in fluid simulation [6, 36]. In this application the vector field modifies either the velocity of the particles or the external force. Vector field design can also be used to drive crowd simulation [31, 38] in which a potential field (scalar field) representing the goal and obstacles helps the crowd moving quickly and smoothly in the environment. The gradient of the potential field, which is a vector field itself, is used as a force that constantly modifies the velocity of the individuals in the crowd. In both applications the vector fields of interest are time-varying. Consequently, we will defer the discussion of them pending future development in time-varying vector field design.

Vector field design has been used to generate data sets to test the efficiencies of vector field visualization techniques [4, 41, 42]. In these applications, it is crucial to have control over the scale and complexity of the features in the field. For example, Chen et al. [4] develop techniques to generate periodic orbits in order to test their technique for periodic orbit extraction and vector field simplification.

The gradient of a scalar field is a curl-free vector field. Consequently, scalar field design can be considered as a special instance of vector field design. Applications of scalar field design include stable surface feature extraction [10], topological noise removal [47], topological shape matching [16], surface parameterization [26, 49], and remeshing [8, 9]. These applications often make use of the Reeb graph or

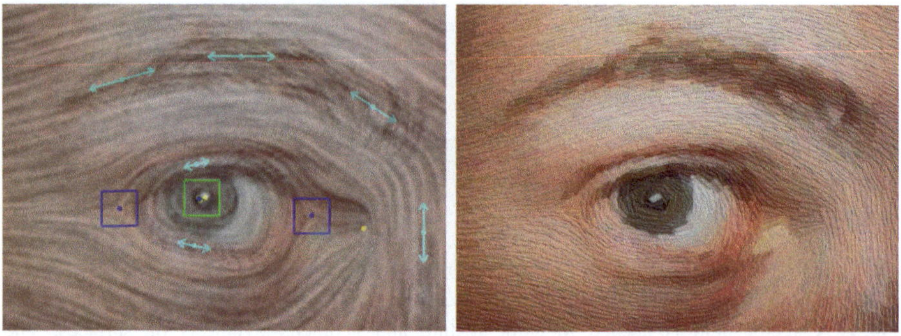

Fig. 2 This figure demonstrates how painterly rendering can benefit from tensor field design [51]: (*left*) an input image of a human eye and a user-designed tensor field, and (*right*) a van Gogh style rendering in which brush strokes follow the tensor field. The *arrows* and *boxes* indicate user-specified design constraints, and the *colored dots* show the singularities in the field. The painterly image was produced by the algorithm of Hays and Essa [13] (Image courtesy of [51], ©2007 IEEE)

Morse-Smale complex derived from the scalar field or its gradient vector field. The quality of Reeb graph and Morse-Smale complex greatly impacts subsequent applications.

2.2 Second-Order Tensor Field Design

Some original motivating applications of vector field design such as painterly rendering and pen-and-ink sketching of surfaces [50], have been shown to be better addressed as second-order or higher-order tensor field design [28, 51].

Unlike vector fields, second-order tensor fields can represent bidirectional objects such as the edges in an image, principal curvature directions of a manifold surface, and street networks. This is due to the fact that if a vector v is an eigenvector of a second-order tensor M, so is $-v$.

Zhang et al. [51] advocate the use of second-order tensor field design to align brush strokes along the edges in an image in painterly rendering [13, 14] (Fig. 2) as well as hatches with the curvature tensor in pen-and-ink sketching of surfaces [51]. In these applications, tensor field design can be used to create different visual effects, add synthetic features to the rendered result, and reduce errors of the numerical estimates of the curvature tensor. The latter is also useful in quadrangular geometry remeshing [1,24], in which excessive umbilical points can lead to meshing difficulties.

Tensor field design has also found applications in the modeling of street networks [5], in which highways as well as major and minor roads are designed to follow the major and minor eigenvectors of a tensor field. The tensor field can be created from scratch or based on a terrain map, and it may or may not respect the

Fig. 3 Tensor field design for procedural street modeling of part of Chicago: (**a**) a designed tensor field respecting topographical water maps and (**b**) a road network generated from hierarchical, evenly-spaced hyperstreamlines following the major and minor eigenvectors of this tensor field (Image courtesy of Guoning Chen)

natural features in the terrain, such as lake shores (Fig. 3). Moreover, the tensor field can be discontinuous.

2.3 *Higher-Order Tensor Field Design*

For pen-and-ink sketching, Hertzmann and Zorin [15] treat the hatches in the cross-hatch regions as a set of crosses rather than two families of mutually perpendicular lines. There are two benefits to this. First, smoothing a cross field typically generates a smoother directional layout than smoothing two line fields. Second, the singularities of a line field can sometimes better model sharp corners in a surface.

In a parallel development, Ray et al. [33] propose the concept of *periodic global parameterization*, which they use for quadrangular remeshing. Instead of tracing streamlines according to the major and minor eigenvectors of the curvature tensor as done in [1], a global parameterization is computed on the mesh surface such that the parameter lines (levelsets in the parameter space) can have a rotation of a multiple of $\frac{\pi}{2}$ when crossing an edge in the mesh. Since a quad element is invariant under a fourfold rotational symmetry, the parameter lines naturally form a quadrangulation of the mesh. Similar to pen-and-ink sketching, allowing the fourfold rotational symmetry greatly increases the quality of resulting quad meshes and can better model corner points in the mesh.

Nieser et al. [27] show the benefit of designing a sixfold rotational symmetry field for triangular remeshing and surface tiling with regular texture and geometry patterns. Palacios and Zhang point out that N-fold rotational symmetries can be

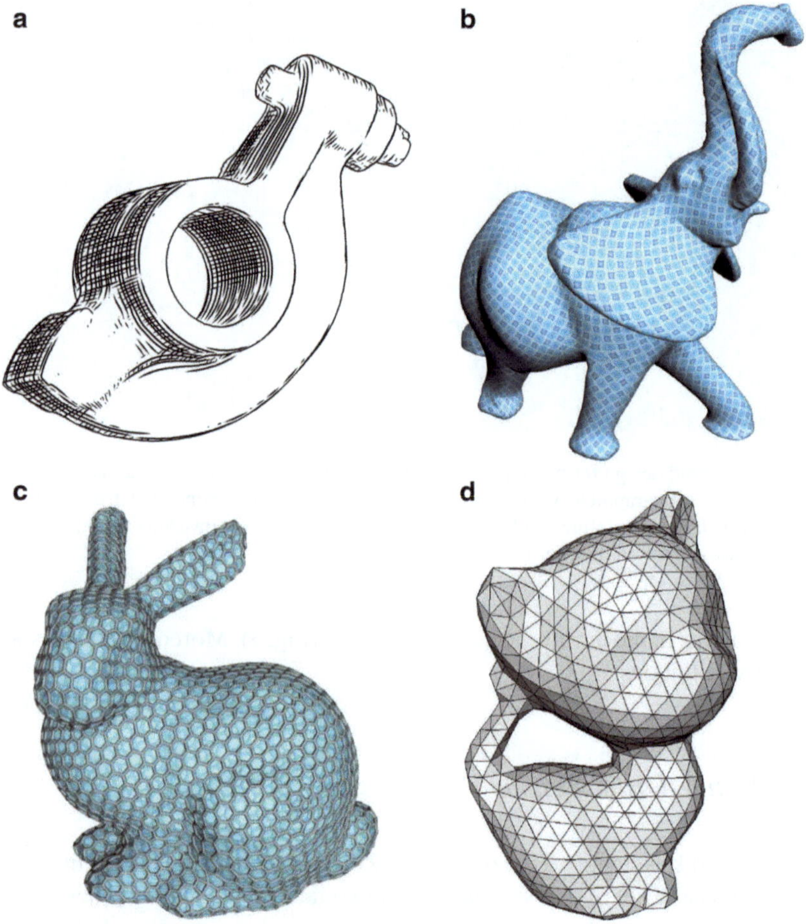

Fig. 4 Higher-order tensors have been used to present N-way rotational symmetries (N-RoSy [28]), with applications in pen-and-ink sketching (**a**), regular pattern synthesis (**b**), architectural modeling (**c**), and geometry remeshing (**d**) (Image (**a**) and (**c**) are courtesy of [29], ©2011 IEEE)

represented as symmetric N-th order tensor fields. Consequently, four-fold and six-fold rotational symmetries can be treated as fourth and sixth order tensors.

In all of these applications, tensor field design can be used to create any arbitrary tensor field or to remove artifacts from an intrinsic tensor field derived from the surface. See Fig. 4 for some example applications of higher-order tensor field design.

Table 1 Applications of tensor field design and their respective design requirements. A design capability is desirable when it is *not* required

Application	Tensor field order	Field smoothness	Feature alignment	Topological control	Editing from existing fields
Texture synthesis	First-	Required	Desirable	Required	Desirable
Geometry synthesis	First-	Required	Desirable	Required	Desirable
Painterly rendering	Second-	Required	Desirable	Required	Required
Street modeling	Second-	Required	Required	Desirable	Required
Pen-and-ink sketching	Fourth-	Required	Required	Required	Required
Quadrangular remeshing	Fourth-	Required	Required	Required	Required
Surface parameterization	Fourth- and sixth-	Required	Required	Required	Required
Regular surface tiling	Fourth- and sixth-	Required	Desirable	Required	Required
Triangular remeshing	Sixth-	Required	Required	Required	Required
Scientific visualization	Any	Required	Required	Required	Desirable

2.4 Requirements

A tensor field can be designed either from scratch or by modifying an existing tensor field. The field must satisfy the user requirements, which can be a combination of geometric constraints, such as field smoothness and surface feature alignment, as well as topological ones, such as the control over the number, type, and location of the singularities in the field. These requirements are the result of the application. For instance, recall that the singularities in a vector field can lead to the breakup of texture patterns in texture synthesis. Also, in remeshing applications it is important that the higher-order tensor field is aligned with sharp features in the surface. Table 1 lists the applications of tensor field design as well as for each application its requirements and the order of tensors used. Notice that these requirements often conflict with each other, and when this happens the desirable properties will be given a lower priority.

3 First-Order Tensor Field Design

In this section we will describe past work in first-order tensor field design, or vector field design which will be the term used in the remainder of this survey.

A number of requirements exist for vector field design:

Fig. 5 An initial vector field can be created using *singular elements* (*left*, *colored boxes*), and *regular elements* (*middle*, *colored arrows*). The centers of the colored boxes are the locations of the desired singularities. Notice that there can be unspecified singularities in these examples

1. Any vector field can be generated.
2. The design process should be interactive and intuitive to a user.
3. Geometric and topological aspects of a vector field can be controlled, possibly based on the characteristics of the underlying manifold.

Given these desired properties, various design techniques have been proposed. The input to these techniques is a set of user design primitives (or design elements), which include *regular elements* and *singular elements*. A regular element consists of a point $\mathbf{p}_0$ on the surface and a tangent vector $v(\mathbf{p}_0)$, while a singular element includes a point $\mathbf{p}_0$, a tangent vector $v(\mathbf{p}_0) = 0$, and a matrix $M(\mathbf{p}_0)$ which is the desired Jacobian at $\mathbf{p}_0$. Singular elements can be used to generate singularity patterns such as sources, sinks, saddles, swirls, and centers. Figure 5 shows regular elements (colored arrows) and singular elements (colored boxes) with some example vector fields. Note that singularities can also be generated with a set of regular elements. However, doing so often leads to difficulty in fine control over the exact patterns near the singularities.

A regular design primitive can be specified with two points: the first point gives the center location of the primitive, and a second point whose displacement with respect to the first point provides the vector value of the regular primitive. For a singular design primitive, one needs to specify the center location of the primitive and the desirable Jacobian. The location of the design primitive as well as the Jacobian can be modified through a graphical user interface.

It is often more convenient for the user to draw a desired streamline, which could be a periodic orbit. Vector field design systems then sample the streamline and create a dense set of regular elements [4, 5, 11, 35] (Fig. 6).

Given the constraints, there have been two classes of methods that can generate a vector field respecting the constraints.

In the first class, each design primitive is used to generate a basis vector field $v_i(\mathbf{p})$ and summed. Zhang et al. [50] provide the formulas for the basis vector fields for regular and singular elements. Chen et al. [4] propose additional formulas for basis vector fields in order to create periodic orbits (Fig. 7). While

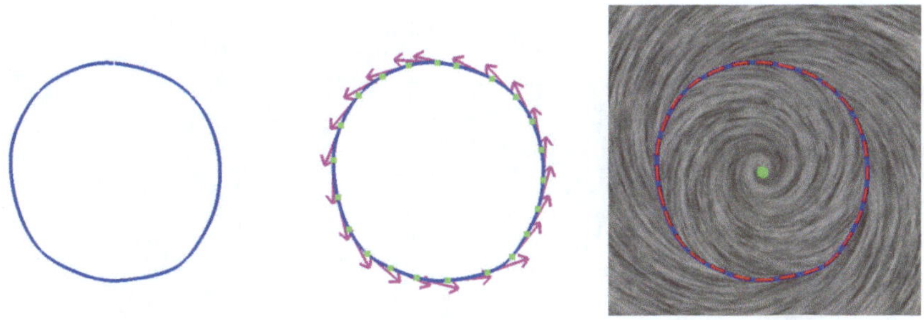

Fig. 6 Given an oriented loop (*left*), a sequence of sample points are created (*middle*: *dots*) where tangent vectors are evaluated (*middle*: *arrows*). A vector field that contains a periodic orbit (*right*: *red dashed lines*) is then generated based on these vector values (Image courtesy of [4], ©2007 IEEE)

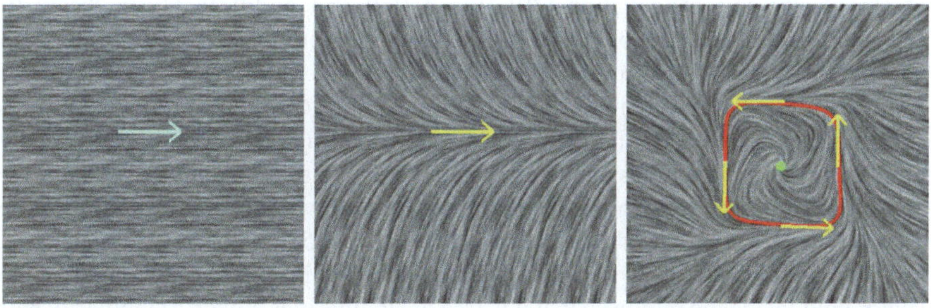

Fig. 7 This figure compares the basis vector field corresponding to a regular element (*left*) and an attachment element (*middle*). The periodic orbit in the right was created by using four attachment elements (Image courtesy of [4], ©2007 IEEE)

this technique is efficient for vector field design on the plane, its complexity is linearly dependent on the number of design elements. Moreover, when performing vector field design on surfaces, a surface parameterization is needed for each design primitive. Consequently, this method is rather ineffective for surfaces. In the second class, the design primitives are converted into the boundary conditions of an energy minimization problem whose solution gives rise to the resulting vector field. This method, while faster, often lacks control over the exact patterns near singular constraints.

Topological control is of paramount importance in various graphics applications vector field design. Specifically, the number, location, and type of singularities are important quality measures for a vector field. To control these aspects, two classes of methods have been proposed. In the first class, vector field smoothing is performed either globally or locally [1,37]. This method, while efficient in reducing the geometric and topological complexity, does not provide the fine control over the number and location of singularities. The second technique is based on the notions

Fig. 8 Two consecutive topological editing operations are performed on a vector field shown in the *left*: (1) one of the center singularity is cancelled with a nearby saddle, and (2) the remaining saddle is moved

of singularity pair cancellation [39] (Fig. 8 (left and middle)). Given the well-known Poincaré index theory [39], a singularity cannot be removed from the field without other topological implications. The atomic operation is to cancel a pair of singularities with opposite indexes. Tricoche et al. [39] cancel a pair of singularities connected by an orbit by finding a narrow region around this orbit and iteratively modifying the vector values in the region. This process is computationally expensive and often ineffective due to the region being too narrow. Zhang et al. [50] propose a different algorithm, based on Conley index theory [25]. According to this theory, given an isolating block with a trivial Conley index, it is possible to modify the vector field inside the region so that it is free of singularities. Given a singularity pair that consists of a repeller and an attractor, Zhang et al. [50] find an isolating block for the pair by computing the intersection between two regions, one reachable from the repeller in the singularity pair following the vector field, and the other reachable from the attractor in the pair following the reversed flow. Local smoothing is then performed inside the isolating block.

To control the location of singularities, Zhang et al. [50] also propose a new operation: singularity movement. Given the current and desirable locations of a singularity, the system finds an isolating block that contains both locations. Then a small region surrounding the new location will be specified such that it forces a singularity inside. The boundary of the two regions bounds a ring-shaped region that has a trivial Conley index. Results from Conley index theory state that it is possible to construct a vector field in such a region so that it is free of singularities. Again, Zhang et al. use Laplacian smoothing inside the ring. The aggregated effect of this computation is a moved singularity (Fig. 8 (middle and right)).

Chen et al. [4] extend the idea of singularity pair cancellation to pair cancellations involving periodic orbits. They have identified six direct cancellation scenarios and seven indirect scenarios. Besides the aforementioned singularity pair cancellation, the other five direct cancellation scenarios are shown in Fig. 9.

It is also desirable to use the features in the surface, such as ridges and valleys, to guide the vector field. Xu et al. [48] handle this by iteratively selecting a vector at each feature to increase the smoothness in the field. However, such an approach

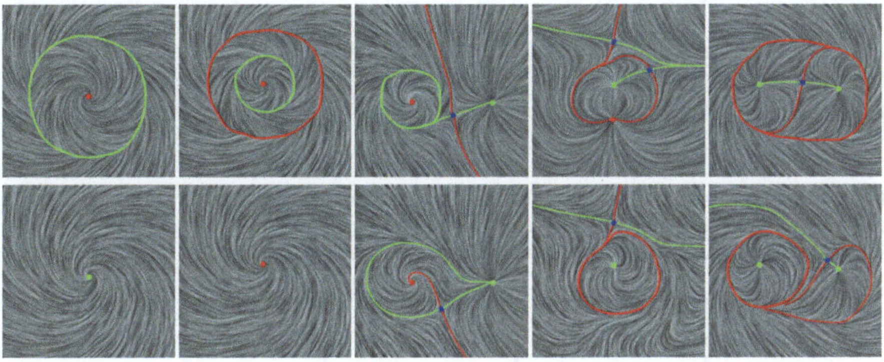

Fig. 9 Five direct cancellation scenarios that involve periodic orbits (Image courtesy of [4], ©2007 IEEE)

cannot adequately solve the problem as it attempts to use a vector field to model bidirectional objects (sharp edges in a surface). A more effective approach is to make use of second-order tensor field design, which we describe next.

4 Second-Order Tensor Field Design

Second-order tensors, or matrices, possess an interesting property: if a vector v is an eigenvector of a matrix M, so is $-v$. This property has the following significance.

1. A second-order tensor can represent line features, such as silhouettes, contours, ridges and valleys, as well as principal curvature directions.
2. A second-order tensor field has very different geometric and topological properties than vector fields.

In fact, as summarized in [51], a second-order tensor field possesses more complicated structures than a vector field in the following sense. Given a continuous vector field V, one can construct a continuous second-order tensor field T such that $V(\mathbf{p})$ is an eigenvector of $T(\mathbf{p})$ except when $V(\mathbf{p}) = 0$. Consequently, the hyperstreamline plot of T will resemble the streamline plot of V. Moreover, any first-order singularity in V is a second-order singularity of T. In contrast, given a continuous second-order tensor field T, one in general cannot produce a continuous vector field whose streamline plot matches the hyperstreamline plot of T. In particular, no first-order singularity in T (Fig. 10: wedge, trisector) can be represented by any singularity in V.

Because of the fundamental differences between vectors and second-order tensors, the design of the latter has practical uses in applications that deal with lines, or bidirectional objects. For example, to correct errors in numerical estimates of the curvature tensor of a surface or edges in an image, vector field design is

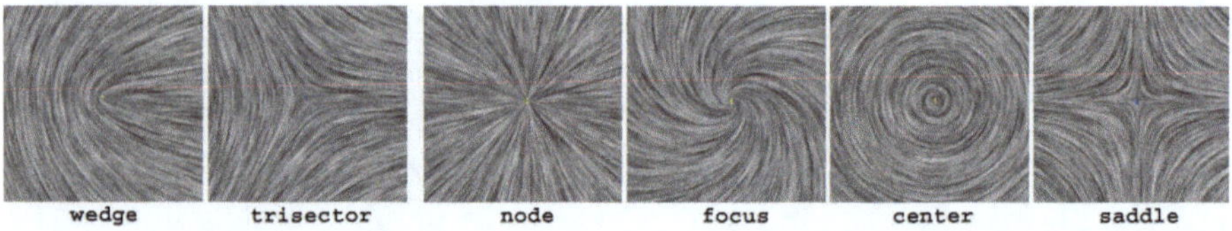

Fig. 10 Some canonical first- and second-order singularities in a second-order tensor field. Notice that the second-order points (node, focus, center, and saddle) are visually similar to first-order singularities in a vector field (Image courtesy of [51], ©2007 IEEE)

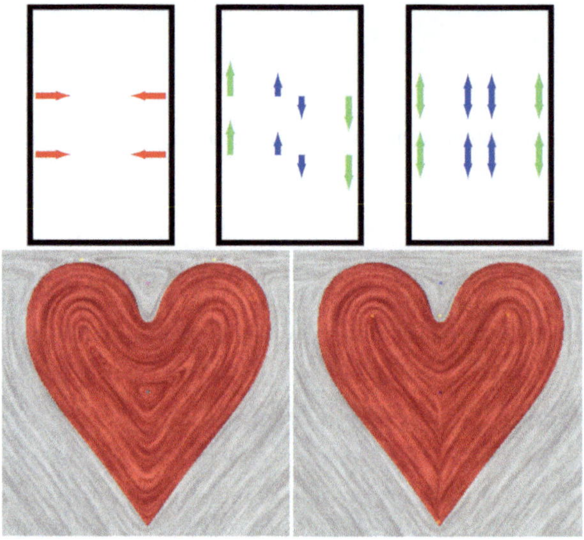

Fig. 11 This figure illustrates the difference between vector-based image edge fields (VIEF) and tensor-based image edge fields (TIEF). For the rectangle (*top row*), the image gradient vector field along the walls points to the other side (*left*, *red arrows*). This causes VIEF to point in opposite directions, and extrapolating values from the wall to the interior of the rectangle cause singularities (*middle*, *green* and *blue arrows*). TIEF does not suffer from this problem due to the sign ambiguity of directions (*right*). For the heart, TIEF (*right*) is smoother than VIEF (*left*) in the interior region due to the richer vocabulary in tensor fields than vector fields. Inside the heart region, VIEF can only have an elongated center as the singularity. On the other hand, TIEF contains a trisector and two wedges, which are more natural here (Image courtesy of [51], ©2007 IEEE)

inadequate as converting the tensor field to a vector field will lead to additional, artificial singularities and the loss of umbilical points such as wedges and trisectors. Second-order tensor field design, on the other hand, will not have these problems. See Figs. 11 and 12 for examples and justification. Here, a vector-based edge field (VIEF) is obtained by rotating the image gradient vector field by $\frac{\pi}{2}$ counterclockwise everywhere in the domain. In contrast, a tensor-based edge field (TIEF) satisfies that its minor eigenvector field is always collinear with the image gradient vector field. We refer to [51] for more detailed discussion on the topic.

Fig. 12 Comparison between the vector-based image edge field (VIEF, *left*) and the tensor-based image edge field (TIEF, *right*) for painterly rendering of an image of a duck. Notice that TIEF is smoother than VIEF (*top row*), and their impact on the painterly results are clearly visible near the beak of the duck (Image courtesy of [51], ©2007 IEEE)

The actual design of second-order tensor fields is surprisingly similar to that of vector fields. This is because, the kind of second-order symmetric tensors we are interested in, can be converted into a vector field that has the same set of singularities. More specifically, any symmetric, traceless second-order tensor has the following form:

$$\begin{pmatrix} a & b \\ b & -a \end{pmatrix} = \rho \begin{pmatrix} \cos\theta & \sin\theta \\ \sin\theta & -\cos\theta \end{pmatrix} \tag{1}$$

The major eigenvectors of this tensor are $\pm\lambda \begin{pmatrix} \cos\frac{\theta}{2} \\ \sin\frac{\theta}{2} \end{pmatrix}$ $(\lambda > 0)$. The following map τ from the set of symmetric, traceless second-order tensors to vectors

$$\tau(\begin{pmatrix} a & b \\ b & -a \end{pmatrix}) = \begin{pmatrix} a \\ b \end{pmatrix} \tag{2}$$

is bijective and continuous. Consequently, the set of singularities in a second-order tensor field T has a one-to-one correspondence to the set of singularities in the vector field $\tau(T)$ (see Fig. 13). Moreover, any algebraic operations applied to the tensors, which are needed by operations such as field interpolation, basis field blending, constrained optimization, smoothing, field arithmetic, and topological editing, can be performed in terms of the corresponding vectors. Consequently, the design of second-order tensor fields, can be converted into vector field design.

There are several items that this conversion cannot handle, such as tensor field analysis and visualization. Fortunately, tensor field analysis for 2D manifolds have

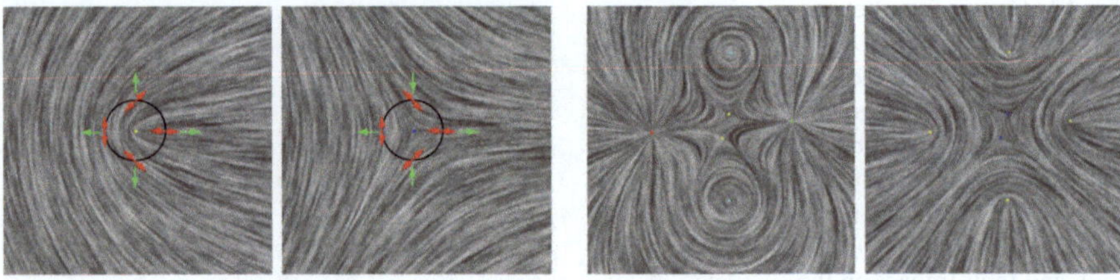

Fig. 13 Tensor-to-vector conversion are applied to two canonical tensor fields (*left* and *middle-left*). In the *left*, after doubling the angle between the major eigenvector field (indicated by *red double arrows*) and the x-axis and then converting it into a vector field (*green arrows*), a wedge is turned in a source. In the *middle-left*, a trisector is transformed into a saddle. In the third, also more sophisticated example, a tensor field (*middle-right*) is converted into a vector field (*right*). Notice this conversion does not change the number and location of the singularities (Image courtesy of [51], © 2007 IEEE)

been thoroughly inspected in [7]. Zhang et al. [51] adapt the image-based flow visualization technique from vector field visualization to tensor field visualization. The detail of this will be covered in Sect. 7.

Chen et al. [5] apply second-order tensor field design to street modeling. In this case, evenly-spaced hyperstreamlines following a second-order tensor field is used to generate a street network, which include highways, major roads, or minor roads. Moreover, there can be discontinuity in the field, as local roads do not always match along major roads. Furthermore, at the street intersection the angle between the two roads is not always $\frac{\pi}{2}$. In fact, the user can adjust this angle by designing a scalar field.

5 Higher-Order Tensor Field Design

In various applications it has been pointed out that N-way rotational symmetries (N-RoSy) play an important role, such as 4-RoSy's for quadrangular remeshing [33] and pen-and-ink sketching [15] as well as 6-RoSy's for triangular and hexagonal remeshing [28, 35]. Furthermore, 4- and 6-RoSy's can be used to tile 3D surfaces with wallpaper patterns with fourfold or sixfold rotational symmetries [21, 28, 35] (Fig. 4b and c).

Given a RoSy field, there are a number of issues to address before it can be designed. The following is a list of such questions:

1. How to interpolate two N-RoSy's, defined at two points in the domain?
2. How to smooth an N-RoSy field?
3. What is the topology of an N-RoSy field?
4. How to exact the topology of an N-RoSy field?

To address these questions it is crucial to find a mathematical representation for N-RoSy fields, and the following realization provides the answer [28]:

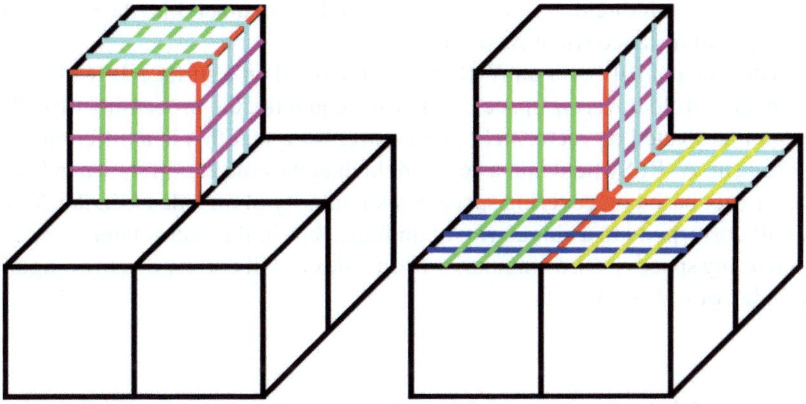

Fig. 14 The separatrices of an N-RoSy field are the boundaries of the hyperbolic sectors in the vicinity of a singularity. These sectors overlap. Here we show the cases for a positively indexed singularity (*left*) and a negatively indexed singularity (*right*)

Given an N-RoSy

$$s = \{\begin{pmatrix} R\cos(\theta + \frac{2k\pi}{N}) \\ R\sin(\theta + \frac{2k\pi}{N}) \end{pmatrix} \mid 0 \le k \le N-1\} \tag{3}$$

it can be represented by the following N-th order symmetric tensor:

$$t_{i_1 \dots i_N} = \begin{cases} R\cos(N\theta) & \text{if } i_1 + i_2 + .. + i_N \equiv 0 \mod 4 \\ R\sin(N\theta) & \text{if } i_1 + i_2 + .. + i_N \equiv 1 \mod 4 \\ -R\cos(N\theta) & \text{if } i_1 + i_2 + .. + i_N \equiv 2 \mod 4 \\ -R\sin(N\theta) & \text{if } i_1 + i_2 + .. + i_N \equiv 3 \mod 4 \end{cases} \tag{4}$$

in the sense that the N directions in s are exactly the set of the local maxima of $t_{i_1 \dots i_N}$ on the set of unit vectors. In fact, this correspondence is unique. Note that the set of N-RoSy's is a proper subset of N-th order symmetric tensors when $N > 1$. Palacios and Zhang [28] further point out that such tensors have only two independent variables, namely, $t_{00\dots0}$ and $t_{10\dots0}$, which enables the following bijective map between the set of such tensors to the set of two-dimensional vectors:

$$\tau(t_{i_1 \dots i_N}) = \begin{pmatrix} t_{00\dots0} \\ t_{10\dots0} \end{pmatrix} \tag{5}$$

This correspondence allows us to perform algebraic operations on N-RoSy's such as addition, subtraction, and scalar multiplication based on vector algebra. Consequently, we can perform linear interpolation and smoothing on RoSy fields. Moreover, we are now able to define and extract singularities in an N-RoSy field as well as to trace separatrices (Fig. 14). Finally, this representation allows one to convert higher-order field design to vector field design, much like the second-order case (a special instance when $N = 2$).

Kälberer et al. [19] cover an N-RoSy field defined on a surface M with a vector field on an N-fold covering space of M. Consequently, the processing of N-RoSy fields can be treated as vector field processing. Note that this is a different way of converting an N-RoSy field into a vector field than the conversion shown in Eq. 5. In the latter (not based on covering spaces), a singularity of an index k in the N-RoSy field will correspond to a singularity of an index kN in the vector field. In contrast, with covering spaces [19], a singularity of an index k will correspond to a singularity of an index of $kN - (N - 1)$.

6 Tensor Field Design on Manifold Surfaces

Many applications of tensor field design require that the underlying domain be a mesh surface. The geometry and topology of the underlying surface can have a profound impact on the design process as well as the quality of the resulting tensor field.

First, due to a lack of a global surface parameterization for an arbitrary surface, it is no longer a trivial task to build a basis tensor field from a design primitive. Surface geometry often makes it more difficult to generate a high-quality basis field, especially around the center of the design primitive. To address this, Zhang et al. [50] compute a geodesic polar map [45] for each design primitive. To construct a geodesic polar map with respect to a point $\mathbf{p}_0$ on the surface, one can first use a fast marching technique such as that of Kimmel and Sethian [20] to compute the radial component ρ for each vertex. The gradient of ρ, a vector field, can then be used to trace each vertex in the mesh back to center of design primitive. The direction in which this trajectory approaches the center location gives the angular component θ (an example is shown in Fig. 15). While this method leads to better control of the field behavior near the center of design primitives, it requires the construction of a geodesic polar map for each design primitive and thus can make interactivity prohibitively expensive.

Second, design with constrained optimization, while faster than using basis fields, faces from some serious difficulties. One of these challenges is in which domain is the constrained optimization performed. Chen et al. [4] solves the Laplacian equation in 3D, i.e., treating the vector field as a 3D vector field, before projecting the result onto the tangent plane at each vertex. While this method is efficient in matching the constraints, it often generates additional singularities and large divergent and convergent regions. Palacios and Zhang [28] use the notion of *parallel transport* from classical differential geometry, which treats the tensor field as a tangent bundle. When solving the Laplacian at a vertex, tangent tensors from neighboring vertices are parallel transported onto the vertex's tangent plane. This leads to more desirable results. A similar method has been used in generating

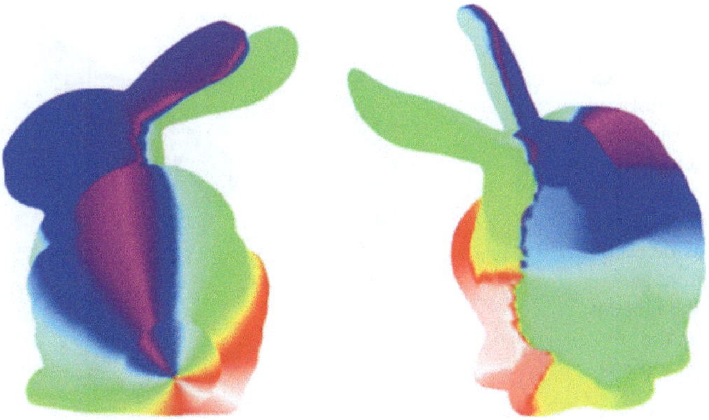

Fig. 15 A visualization of the angular component of the geodesic polar map on the bunny when the origin is in the middle of the tail. Zhang et al. [50] use such polar maps to compute a tangent vector field on the surface based on design elements. Notice the discontinuity in the polar map away from the origin (*right*). This can lead to excessive singularities in the resulting vector field that need to be removed with additional design elements

the hatching field in non-photorealistic rendering [15] and the direction field for quadrangular remeshing [33].

Third, there are often excessive singularities in the design field, and they often appear in undesirable locations on the surface. To deal with this problem and to achieve efficient control, Fisher et al. [11] use discrete differential forms [17] to represent the vector field. With this framework, vector field design is converted to differential form design, which tends to produce highly smooth results. Fisher et al. [11] apply their technique to interactive texture synthesis in which the user can specify the location and type of singularities as well as certain streamlines. These streamlines are then converted into directional constraints. Ray et al. [34] use a face-based vector field representation. In this representation, singularities are defined in terms of the turning angles of the vector field across the edges in the mesh, thus enabling efficient control over the number and location of the singularities. This work is further advanced so that the turning angles across edges can be automatically computed based on user specifications [35]. Lai et al. [21] also address the problem of singularity control with the concept of *holonomy*, which measures the total turning angle excess for each vertex. By generating a *flat metric*, they allow the generation of a vector field whose number of singularities reaches the theoretical lower bound.

When streamline tracing is important for the application, such as pen-and-ink sketching, additional steps are needed to *reconstruct* an everywhere continuous tensor field based on the tensor values defined at the vertices or triangles. For vertex-based vector field representations, Zhang et al. [50] introduce a non-linear interpolation scheme that is based on the notion of *parallel transport* from differential geometry (Fig. 16). This interpolation scheme enables the adaptation of vector field design and editing from the plane onto surfaces (Figs. 5 and 8).

Fig. 16 The non-linear interpolation scheme of Zhang et al. [50] ensures vector field continuity on a triangular mesh surface with non-zero discrete Gaussian curvatures. Notice that the vector field is not linearly interpolated inside each face of the octahedron, as evidenced by the tangent directions along the streamlines (*red curves*)

The interpolation scheme is later extended to second-order tensor fields [51] and higher-order tensor fields [28]. When representing vector fields with differential forms, Fisher et al. [11] explicitly construct a continuous vector field using edge-based subdivision schemes [43].

7 Visualization

High-quality and interactive visualization of tensor fields is critical to the user experiences during tensor field design. There have been abundant techniques for the first-order case, i.e., vector field visualization. Texture-based vector field visualization techniques [3,41,46] are most desirable due to the high frame rate and high quality that they offer. Such techniques have been extended to vector fields on surfaces [22,42] and second-order tensor fields [51,52].

For higher-order tensor fields, there are N directions at each point except the singularities. While it is possible to show only one of the N directions, such a visualization often contains visual artifacts due to the discontinuity introduced when selecting one direction from N directions. Moreover, it can be difficult to understand the structure of the tensor field, especially around singularities (Fig. 17, left). Showing all directions at each point addresses these problems (Fig. 17, right).

One way to accomplish this is to draw streamlines following the N directions in the field. However, due to the existence of singularities, an N-RoSy field cannot always be converted into a vector field without visual artifacts. Palacios and Zhang [29] develop a technique that addresses this difficulty. Basically, an N-RoSy field is converted into $M(N)$ vector fields where $M(N) = 2N$ when N is odd and $M(N) = N$ when N is even. These fields are designed in a way that the first $\frac{M(N)}{2}$ fields will capture all possible directions in the N-RoSy field at each point. These vector fields, when visualized using vector field visualization methods, will contain visual artifacts at exactly the same locations. The second $\frac{M(N)}{2}$ fields are designed to also provide a full coverage of N-RoSy directions and have visual artifacts in the same positions, which are complementary to those from the first half of the fields. Combining the visualization of these $M(N)$ fields with an appropriate confidence map as the weights leads to a high-quality visualization framework that is also interactive when N is small enough, such as $N = 4$ and $N = 6$. Figure 18 illustrates this framework with an example N-RoSy field.

Averaging many images can lead to the loss of contrast. Palacios and Zhang handle this by recognizing that LIC images follow Gauss distributions in terms of color intensity. They then borrow results from probability theory to find the exact amount of loss of intensity during image composition and compensate it by scaling the original LIC images [29] (Fig. 19).

8 Conclusion

In this survey, we review applications, requirements, and existing approaches to tensor field design. There are a number of future research directions.

First, while N-RoSy field design has been largely solved, other types of symmetries exist, such as reflections and gliding reflections. These symmetries, along with rotational symmetries, lead to the 17 members of the *wallpaper groups*. It is not clear how or whether it is possible to represent these additional types of symmetries using tensors. Without a proper mathematical representation, however, it is challenging to study the algebraic, geometric, and topological properties of these symmetry fields.

Second, past research has focused on perfect rotational symmetries, i.e., there are a set of N angularly evenly-spaced directions everywhere in the domain. Real world symmetries are often imperfect and/or inhomogeneous, leading to near-regular symmetries as well as mixed symmetries. Properly modeling these types of symmetries has practical applications in graphics and geometry processing. Given the link between tensors and perfect rotational symmetries, we are hopeful that near-regular and mixed symmetries can also be modeled by tensor fields.

Third, past work in tensor field design has focused on 2D, time-independent, symmetric tensors with the exception of some work in shape deformation [12], hair modeling [30], and painterly rendering of videos [18]. Extending this to 3D, to asymmetric tensors, and to time-varying tensor fields pose new challenges as we need the ability to perform topological analysis and visualization of these objects.

It is our belief that the graphics and geometry communities have just scratched the surface of the problem, and as more research effort is invested tensor field design can be established as a field of study just like shape modeling and flow visualization.

Acknowledgements I wish to thank Greg Turk, Konstantin Mischaikow, Guoning Chen, Jonathan Palacios, Robert S. Laramee, James Hays, Pawel Pilarczyk, Yu-Kun Lai, Xianfeng Gu, Bruno Lévy, Greg Esch, Peter Wonka, Pascal Müller, Matthias Nieser, and Konrad Polthier for their valuable discussions. The images shown in this survey are also courtesy of many of them. The reviewers' constructive comments have made the survey stronger. Finally, I wish to thank my family for their support that they have provided during my attending the Dagstuhl seminar on Tensor Field Processing in 2009 as well as afterwards when I wrote the survey. The survey is

Fig. 17 Visualizing only one of the directions of an N-RoSy field (*left*) not only results in a visual discontinuity but also does not allow a user to see the distinct patterns of the field around features like singularities. Rendering all N directions at each point resolves this issue (*right*) (Image courtesy of [29], ©2011 IEEE)

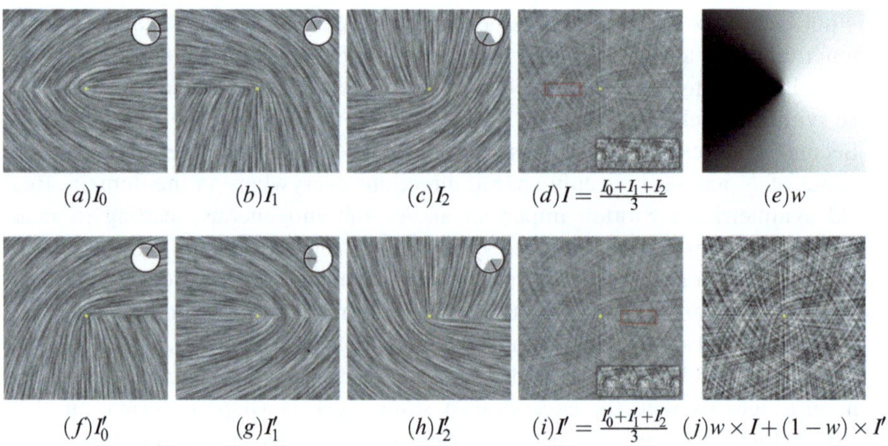

Fig. 18 A demonstration of the visualization technique of Palacios and Zhang [29]. In (**a**)–(**c**), the LIC algorithm [3] is applied to V_0, V_1, and V_2 (the guiding angle for each is shown in the *upper right corner*) to obtain I_0, I_1 and I_2, respectively. Notice that while (**a**)–(**c**) provide a complete coverage of the streamlines passing through any regular point in the domain, they have the same regions of discontinuity (*left* X-axis). By blending them uniformly, one obtains I (**d**), a visualization of S with visual artifacts in the same place (a close-up of the artifact, highlighted in *red*, is seen inset with the contrast enhanced; note the curving patterns in a region that should be regular). To remedy the problem, one can apply the LIC algorithm to V_0', V_1', and V_2', generating the images I_0' (**f**), I_1' (**g**), and I_2' (**h**), and blend them uniformly to obtain I' (**i**). The visual artifacts in I' appear on the *right* side (again, a close-up is inset) of the X-axis. By blending I and I' using the weight map w (**e**), one obtains the final image in (**j**) in which the artifacts due to field discontinuities are no longer visible (Image courtesy of [29], ©2011 IEEE)

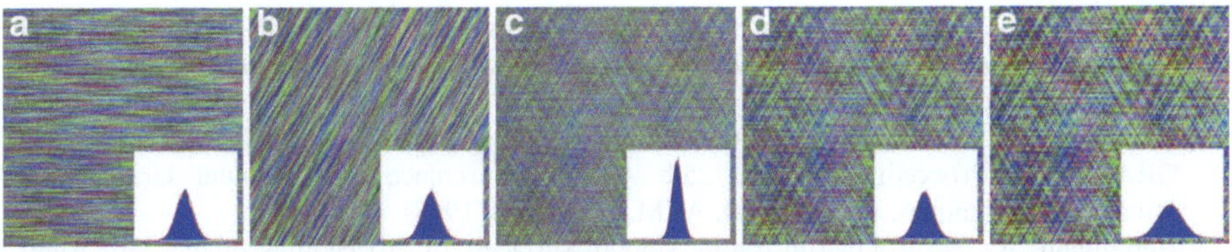

Fig. 19 This figure illustrates the contrast correction and enhancement technique for a constant 6-RoSy field [29]. When combining three LIC images (only two shown in (**a**) and (**b**)) uniformly to generate a visualization for the 6-RoSy field (**c**), the image appears to be washed out due to a lower variance than the original images (**a** and **b**). With results from probability theory, the problem can be corrected (**d**). The resulting image can be further enhanced (**e**) (Image courtesy of [29], ©2011 IEEE)

partially funded by National Science Foundation grants IIS-0546881, CCF-0830808, and IIS-0917308.

References

1. Alliez, P., Cohen-Steiner, D., Devillers, O., Lévy, B., Desbrun, M.: Anisotropic polygonal remeshing. ACM Trans. Graph. (SIGGRAPH 2003) **22**(3), 485–493 (2003)
2. Bhat, P., Ingram, S., Turk, G.: Geometric texture synthesis by example. In: SGP '04: Proceedings of the 2004 Eurographics/ACM SIGGRAPH Symposium on Geometry Processing, pp. 41–44. ACM, New York (2004)
3. Cabral, B., Leedom, L.C.: Imaging vector fields using line integral convolution. In: SIGGRAPH '93: Proceedings of the 20th Annual Conference on Computer Graphics and Interactive Techniques, pp. 263–270. ACM, New York (1993)
4. Chen, G., Mischaikow, K., Laramee, R.S., Pilarczyk, P., Zhang, E.: Vector field editing and periodic orbit extraction using morse decomposition. IEEE Trans. Vis. Comput. Graph. **13**(4), 769–785 (2007)
5. Chen, G., Esch, G., Wonka, P., Müller, P., Zhang, E.: Interactive procedural street modeling. ACM Trans. Graph. **27**(3), 103:1–103:10 (2008)
6. Chenney, S.: Flow tiles. In: SCA '04: Proceedings of the 2004 ACM SIGGRAPH/Eurographics Symposium on Computer Animation, pp. 233–242. Eurographics Association, Aire-la-Ville (2004)
7. Delmarcelle, T., Hesselink, L.: The topology of symmetric, second-order tensor fields. In Proceedings of the Conference on Visualization '94, VIS '94, Washington, DC, pp. 140–147. IEEE Computer Society Press, Los Alamitos (1994)
8. Dong, S., Kircher, S., Garland, M.: Harmonic functions for quadrilateral remeshing of arbitrary manifolds. Comput. Aided Geom. Des. **22**(5), 392–423 (2005)
9. Dong, S., Bremer, P.T., Garland, M., Pascucci, V., Hart, J.C.: Spectral surface quadrangulation. ACM Trans. Graph. **25**(3), 1057–1066 (2006)
10. Edelsbrunner, H., Letscher, D., Zomorodian, A.: Topological persistence and simplification. In: FOCS '00: Proceedings of the 41st Annual Symposium on Foundations of Computer Science, p. 454. IEEE Computer Society, Washington, DC (2000)
11. Fisher, M., Schröder, P., Desbrun, M., Hoppe, H.: Design of tangent vector fields. In: SIGGRAPH '07: ACM SIGGRAPH 2007 Papers, p. 56. ACM, New York (2007)
12. von Funck, W., Theisel, H., Seidel, H.P.: Vector field based shape deformations. In: SIGGRAPH '06: ACM SIGGRAPH 2006 Papers, pp. 1118–1125. ACM, New York (2006)

13. Hays, J., Essa, I.: Image and video based painterly animation. In: NPAR '04: Proceedings of the 3rd International Symposium on Non-photorealistic Animation and Rendering, pp. 113–120. ACM, New York (2004)
14. Hertzmann, A.: Painterly rendering with curved brush strokes of multiple sizes. In: SIGGRAPH '98: Proceedings of the 25th Annual Conference on Computer Graphics and Interactive Techniques, pp. 453–460. ACM, New York (1998)
15. Hertzmann, A., Zorin, D.: Illustrating smooth surfaces. In Proceedings of the 27th Annual Conference on Computer Graphics and Interactive Techniques, SIGGRAPH '00, pp. 517–526. ACM Press/Addison-Wesley Publishing Co., New York (2000). doi: 10.1145/344779.345074
16. Hilaga, M., Shinagawa, Y., Kohmura, T., Kunii, T.L.: Topology matching for fully automatic similarity estimation of 3d shapes. In: SIGGRAPH '01: Proceedings of the 28th Annual Conference on Computer Graphics and Interactive Techniques, pp. 203–212. ACM, New York (2001)
17. Hirani, A.N.: Discrete exterior calculus. Ph.D. thesis, California Institute of Technology, Pasadena (2003). Adviser-Marsden, Jerrold E
18. Kagaya, M., Brendel, W., Deng, Q., Kesterson, T., Todorovic, S., Neill, P.J., Zhang, E.: Video painting with space-time-varying style parameters. IEEE Trans. Vis. Comput. Graph. **17**, 74–87 (2011)
19. Kälberer, F., Nieser, M., Polthier, K.: Quadcover – surface parameterization using branched coverings. Comput. Graph. Forum **26**(3), 375–384 (2007)
20. Kimmel, R., Sethian, J.A.: Computing geodesic paths on manifolds. Proc. Natl. Acad. Sci. **95**(15), 8431–8435 (1998)
21. Lai, Y.K., Jin, M., Xie, X., He, Y., Palacios, J., Zhang, E., Hu, S.M., Gu, X.: Metric-driven rosy field design and remeshing. IEEE Trans. Vis. Comput. Graph. **16**, 95–108 (2010)
22. Laramee, R.S., Jobard, B., Hauser, H.: Image space based visualization of unsteady flow on surfaces. In Proceedings of the 14th IEEE Visualization 2003 (VIS '03), p. 18. IEEE Computer Society, Washington, DC (2003). doi: 10.1109/VISUAL.2003.1250364
23. Li, Y., Bao, F., Zhang, E., Kobayashi, Y., Wonka, P.: Geometry synthesis on surfaces using field-guided shape grammars. IEEE Trans. Vis. Comput. Graph. **17**, 231–243 (2011)
24. Marinov, M., Kobbelt, L.: Direct anisotropic quad-dominant remeshing. In Proceedings of the Computer Graphics and Applications, 12th Pacific Conference, PG '04, pp. 207–216. IEEE Computer Society, Washington, DC (2004)
25. Mischaikow, K., Mrozek, M.: Conley index. In: Handbook of Dynamical Systems, vol. 2, pp. 393–460. North-Holland, Amsterdam (2002)
26. Ni, X., Garland, M., Hart, J.C.: Fair morse functions for extracting the topological structure of a surface mesh. In: SIGGRAPH '04: ACM SIGGRAPH 2004 Papers, pp. 613–622. ACM, New York (2004)
27. Nieser, M., Palacios, J., Polthier, K., Zhang, E.: Hexagonal global parameterization of arbitrary surfaces. IEEE Trans. Vis. Comput. Graph. **18**(6), 865–878 (2012). doi: 10.1109/TVCG.2011.118
28. Palacios, J., Zhang, E.: Rotational symmetry field design on surfaces. ACM Trans. Graph. **26**(3), 55:1–55:10 (2007)
29. Palacios, J., Zhang, E.: Interactive visualization of rotational symmetry fields on surfaces. IEEE Trans. Vis. Comput. Graph. **17**(7), 947–955 (2011)
30. Paris, S., Chang, W., Kozhushnyan, O.I., Jarosz, W., Matusik, W., Zwicker, M., Durand, F.: Hair photobooth: geometric and photometric acquisition of real hairstyles. In: SIGGRAPH '08: ACM SIGGRAPH 2008 Papers, pp. 1–9. ACM, New York (2008)
31. Patil, S., van den Berg, J., Curtis, S., Lin, M.C., Manocha, D.: Directing crowd simulations using navigation fields. IEEE Trans. Vis. Comp. Graph. **17**(2), 244–254 (2011). doi: 10.1109/TVCG.2010.33
32. Praun, E., Finkelstein, A., Hoppe, H.: Lapped textures. In: SIGGRAPH '00: Proceedings of the 27th Annual Conference on Computer Graphics and Interactive Techniques, pp. 465–470. ACM/Addison-Wesley, New York (2000)

33. Ray, N., Li, W.C., Lévy, B., Sheffer, A., Alliez, P.: Periodic global parameterization. ACM Trans. Graph. **25**(4), 1460–1485 (2006)
34. Ray, N., Vallet, B., Li, W.C., Lévy, B.: N-symmetry direction field design. ACM Trans. Graph. **27**(2), 1–13 (2008)
35. Ray, N., Vallet, B., Alonso, L., Levy, B.: Geometry-aware direction field processing. ACM Trans. Graph. **29**(1), 1–11 (2009)
36. Stam, J.: Flows on surfaces of arbitrary topology. ACM Trans. Graph. (SIGGRAPH 2003) **22**(3), 724–731 (2003)
37. Tong, Y., Lombeyda, S., Hirani, A.N., Desbrun, M.: Discrete multiscale vector field decomposition. ACM Trans. Graph. **22**(3), 445–452 (2003)
38. Treuille, A., Cooper, S., Popović, Z.: Continuum crowds. In: SIGGRAPH '06: ACM SIGGRAPH 2006 Papers, pp. 1160–1168. ACM, New York (2006)
39. Tricoche, X., Scheuermann, G., Hagen, H.: Continuous topology simplification of planar vector fields. In: VIS '01: Proceedings of the Conference on Visualization '01, pp. 159–166. IEEE Computer Society, Washington, DC (2001)
40. Turk, G.: Texture synthesis on surfaces. In Proceedings of the 28th Annual Conference on Computer Graphics and Interactive Techniques, SIGGRAPH '01, pp. 347–354. ACM, New York, NY (2001). doi: 10.1145/383259.383297
41. van Wijk, J.J.: Image based flow visualization. ACM Trans. Graph. (SIGGRAPH 2002) **21**(3), 745–754 (2002)
42. van Wijk, J.J.: Image based flow visualization for curved surfaces. In Proceedings of the 14th IEEE Visualization 2003 (VIS'03), p. 17. IEEE Computer Society, Washington, DC (2003). doi: 10.1109/VISUAL.2003.1250363
43. Wang, K., Weiwei, Tong, Y., Desbrun, M., Schröder, P.: Edge subdivision schemes and the construction of smooth vector fields. In: SIGGRAPH '06: ACM SIGGRAPH 2006 Papers, pp. 1041–1048. ACM, New York (2006)
44. Wei, L.Y., Levoy, M.: Texture synthesis over arbitrary manifold surfaces. In: SIGGRAPH '01: Proceedings of the 28th Annual Conference on Computer Graphics and Interactive Techniques, pp. 355–360. ACM, New York (2001)
45. Welch, W., Witkin, A.: Free-form shape design using triangulated surfaces. In: SIGGRAPH '94: Proceedings of the 21st Annual Conference on Computer Graphics and Interactive Techniques, pp. 247–256. ACM, New York (1994)
46. van Wijk, J.J.: Spot noise texture synthesis for data visualization. In: SIGGRAPH '91: Proceedings of the 18th Annual Conference on Computer Graphics and Interactive Techniques, pp. 309–318. ACM, New York (1991)
47. Wood, Z., Hoppe, H., Desbrun, M., Schröder, P.: Removing excess topology from isosurfaces. ACM Trans. Graph. **23**(2), 190–208 (2004)
48. Xu, K., Cohne-Or, D., Ju, T., Liu, L., Zhang, H., Zhou, S., Xiong, Y.: Feature-aligned shape texturing. ACM Trans. Graph. (Proceedings SIGGRAPH Asia 2009) **28**(5), Article 108 (2009)
49. Zhang, E., Mischaikow, K., Turk, G.: Feature-based surface parameterization and texture mapping. ACM Trans. Graph. **24**, 1–27 (2005)
50. Zhang, E., Mischaikow, K., Turk, G.: Vector field design on surfaces. ACM Trans. Graph. **25**(4), 1294–1326 (2006)
51. Zhang, E., Hays, J., Turk, G.: Interactive tensor field design and visualization on surfaces. IEEE Trans. Vis. Comput. Graph. **13**(1), 94–107 (2007)
52. Zheng, X., Pang, A.: HyperLIC. In Proceedings of the 14th IEEE Visualization 2003 (VIS'03), p. 33. IEEE Computer Society, Washington, DC (2003)
53. Zhou, K., Huang, X., Wang, X., Tong, Y., Desbrun, M., Guo, B., Shum, H.Y.: Mesh quilting for geometric texture synthesis. ACM Trans. Graph. **25**(3), 690–697 (2006)

Part III
Applications of Tensor-Field Analysis and Visualization

Interactive Exploration of Stress Tensors Used in Computational Turbulent Combustion

Adrian Maries, Md. Abedul Haque, S. Levent Yilmaz, Mehdi B. Nik, and G. Elisabeta Marai

Abstract Simulation and modeling of turbulent flow, and of turbulent reacting flow in particular, involves solving for and analyzing time-dependent and spatially dense tensor quantities, such as turbulent stress tensors. The interactive visual exploration of these tensor quantities can effectively steer the computational modeling of combustion systems. In this chapter, we discuss the challenges in dense symmetric-tensor visualization applied to turbulent combustion calculation, and analyze the feasibility of using several established tensor visualization techniques in the context of exploring space-time relationships in computationally-simulated combustion tensor data. To tackle the pervasive problems of occlusion and clutter, we propose a solution combining techniques from information and scientific visualization. Specifically, the proposed solution combines a detailed 3D inspection view based on volume rendering with glyph-based representations—used as 2D probes—while leveraging interactive filtering and flow salience cues to clarify the structure of the tensor datasets. Side-by-side views of multiple timesteps facilitate the analysis of time-space relationships. The resulting prototype enables an analysis style based on the overview first, zoom and filter, then details on demand paradigm originally proposed in information visualization. The result is a visual analysis tool to be

A. Maries (✉) · Md. Abedul Haque · G.E. Marai
Department of Computer Science, University of Pittsburgh, Pittsburgh, PA 15260, USA
e-mail: amaries@cs.pitt.edu; mdabedul@cs.pitt.edu; marai@cs.pitt.edu

S.L. Yilmaz
Center for Simulation and Modeling, University of Pittsburgh, Pittsburgh, PA 15260, USA
e-mail: slyilmaz@pitt.edu

M.B. Nik
Department of Mechanical Engineering and Materials Science, University of Pittsburgh, Pittsburgh, PA 15260, USA
e-mail: smb51@pitt.edu

D.H. Laidlaw and A. Vilanova (eds.), *New Developments in the Visualization and Processing of Tensor Fields*, Mathematics and Visualization,
DOI 10.1007/978-3-642-27343-8_7,

utilized in debugging, benchmarking, and verification of models and solutions in turbulent combustion. We demonstrate this analysis tool on three example configurations and report feedback from combustion researchers.

1 Introduction

Research into optimization of power generation systems for advanced energy and emissions performance has become increasingly important in the last two decades largely due to alarming environmental concerns and stringent regulations of NO_x and SO_x emissions, and other greenhouse gases such as CO_2 [8]. Microturbine generation systems are on the forefront of this research due to the promise of high-efficiency, ultra-clean systems that can be used to produce electrical energy as well as thermal energy in co-generation applications. There is a continuing need to develop and implement advanced computational tools for modeling and prediction of turbulent combustion for a wide range of mixing, fuel compositions and flow configurations [11, 28]. In fact, reliable and flexible computational modeling is considered a key element in achieving the objectives of modern gas-turbine design [22].

Tensor quantities are quite common in turbulence modeling, in particular in the three principal approaches to computational turbulent combustion [9]: Direct numerical simulation (DNS), Reynolds-averaged Navier-Stokes (RANS) and Large eddy simulation (LES). For example, in LES a spatially filtered form of the Navier-Stokes equations is considered, whereby only the flow features (eddies) that are larger than a characteristic filter size are resolved without approximation. Certain *subfilter* (or *subgrid*) quantities appear unclosed and need to be modeled. An important agent of such quantities is the non-linear convection term, which is defined as the *subgrid scale* (SGS) *stress tensor* [25] and the SGS scalar fluxes (for flows involving scalar transport) [9]. In eddy-diffusivity type LES models, SGS tensors are correlated to resolved quantities, such as the resolved strain tensor [5,21].

Visually identifying the characteristics of such tensor quantities in finer details can bring significant insights into the computational modeling process. For example, DNS solutions are commonly used in verification of LES and RANS models, where the cases have typically been limited to simple canonical configurations [15]. In recent years, however, progress in DNS modeling has lead to DNS being applied to more realistic configurations with finer resolution [16, 29]. Visual exploration of tensor quantities in such complex configurations would ultimately steer the model verification and development process. Questions posed during analysis may be of the form, what artifacts do LES introduce in the simulation of a complex configuration, compared to the canonical DNS solution? How do these artifacts evolve over time? What are the regions and magnitude of error? In the long run, the visualization of tensor quantities could help turbulent combustion researchers identify regions of interest in the flow in order to design simulation schemes which are both computationally affordable and sufficiently accurate.

In this chapter we investigate the challenges associated with the exploratory visualization of tensor quantities in turbulent combustion simulations. We then propose, implement and evaluate an interactive prototype for exploring turbulent combustion tensor data that addresses some of these challenges. To the best of our knowledge, this is the first successful interactive visualization of time-varying stress tensors in the context of high-density turbulent combustion data. The result is a visual analysis tool developed through tight collaboration with researchers in computational combustion. We evaluate the tool on several computational-combustion datasets of particular interest, and show the importance of the proposed approach for debugging the numerical simulation of complex configurations. Last but not least, we contribute a discussion of lessons learned, current limitations, and future directions of research as motivated by the driving computational application.

2 Tensors in Turbulent Combustion

2.1 Turbulent Combustion Modeling

A tensor is an extension of the concept of a scalar and a vector to higher orders. Scalars and vector are 0-th and 1-st order tensors, respectively. In general, a k-th order tensor can be represented by a k-dimensional array, e.g. a second order tensor is a 2D array (a matrix). A stress vector is the force acting on a given unit surface, and a stress tensor is defined as the components of stress vectors acting on each coordinate surface; thus it can be described by a symmetric 2-nd order tensor. The diagonal components of the stress tensor represent normal forces, i.e. compression and tension, and the non-diagonal components represent the shearing forces. The eigenvalues of the stress tensor are its principal stresses and the associated eigenvectors are the principal directions of the stress tensor. Since the stress tensor is symmetric, its eigenvalues are real. Strain tensor is a related quantity, the components of which are the components of the rate of deformation in each coordinate direction (i.e. derivatives of velocity). For the fluid flow motion, there are different so called constitutive models that relate strain to stress. In this work, we examine a Newtonian fluid, where the stress-strain relationship is linear.

A sufficiently accurate, flexible and reliable model can be used for an *in silico* combustor rig test as a much cheaper alternative to the real life rig tests employed in combustor design and optimization. In order to achieve such a model, the methodology should be well tested and proven with lab-scale configurations. Figure 1 shows an example configuration for a lab Bunsen burner, and example results from a corresponding computational model and simulation, showing the gaseous distribution at a specific time-frame.

Multiple numerical approaches exist for generating such computational models of combustion, most notably DNS, RANS and LES [9]. DNS consists of solving the Navier-Stokes transport equations of fluid flow and resolving all of the scales of

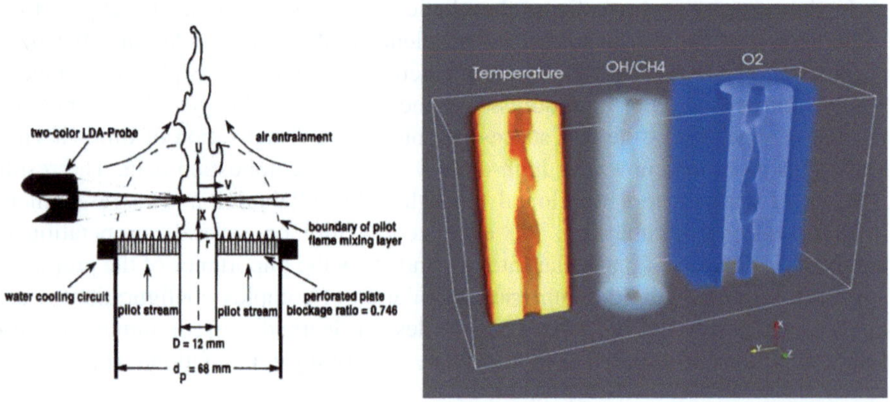

Fig. 1 The experimental setup which introduces the Bunsen burner experiment (*left*)[4], and example results from a corresponding computational LES model and simulation, showing the gaseous distribution at a specific time-frame (*right*) [39]

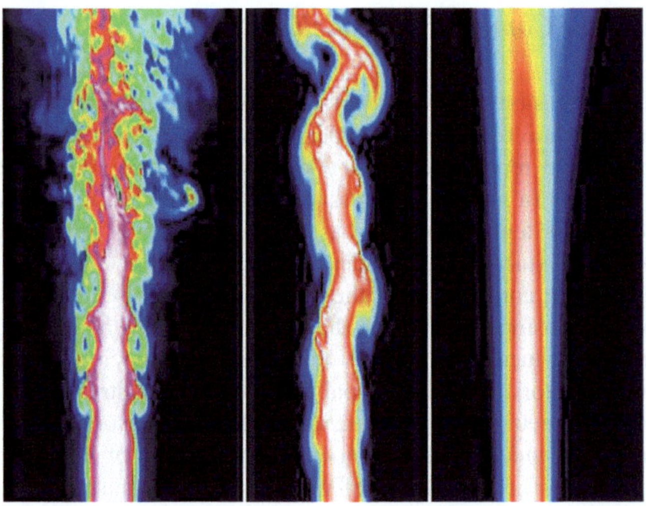

Fig. 2 DNS (*left*), LES (*middle*) and RANS (*right*) predictions of a turbulent jet. LES requires less computational effort than DNS, while delivering more detail than the inexpensive RANS

motion. In RANS, the transport equations are time-averaged based on knowledge of the properties of flow turbulence, and the mean transport equations are solved. In LES, the spatially filtered equations of motion are considered, the solution of which portrays the large scale motion.

DNS, RANS and LES have complementary strengths. Figure 2 demonstrates the characteristically different results that can be obtained with these models. DNS provides a very accurate, model-free representation of the unsteady evolution of turbulent flows. However, applications are largely restricted by the computational power required by DNS and can be carried out only in limited and highly idealized cases [12,29,32]. RANS is significantly less expensive than DNS, and, as such, it is the more popular approach for engineering applications [24]. Finally, LES provides a balance between computational cost and the level of fidelity in the results. The LES methodology has been the subject of much modern research and is becoming increasingly more popular in combustion research [10].

All models begin by describing the compressible reacting flow via a set of partial differential equations (PDEs) that represent the conservation of mass, momentum, and energy. These PDEs are a fully coupled set of multi-dimensional non-linear equations and can be posed in a variety of forms depending on the flow conditions (compressibility, scale, flow regime, etc.). Details are beyond the scope of this chapter, and the reader is referred to many excellent books available on the subject [20, 23, 33]. In this chapter, we exemplify the visualization of stress/strain tensors, and therefore restrict the presentation to the pertinent subset of this PDEs, namely the momentum transport equation.

The velocity stress/strain tensor fields are manifested in the transport of fluid momentum, which is a vector quantity governed by the following conservation equation:

$$\frac{\partial \rho u_i}{\partial t} + \frac{\partial \rho u_i u_j}{\partial x_j} = -\frac{\partial p}{\partial x_i} + \frac{\partial \tau_{ij}}{\partial x_j} \quad \text{for } i = 1, 2, 3 \tag{1}$$

where the Cartesian index notation is employed in which the index $i = 1, 2, 3$ represents spatial directions along the x, y, and z Cartesian coordinates, respectively; and the repeated index j implies summation over the coordinates. t is time, ρ is the fluiddensity, $\mathbf{u} \equiv [u_1, u_2, u_3]$ is the Eulerian fluid velocity, p is the pressure, and τ is the stress tensor defined as:

$$\tau_{ij} = \mu S_{ij} \tag{2}$$

where μ is the dynamic viscosity coefficient (a fluid-dependent parameter) and S is the velocity strain tensor defined as:

$$S_{ij} = \frac{1}{2}\left(\frac{\partial u_i}{\partial x_j} + \frac{\partial u_j}{\partial x_i}\right) \tag{3}$$

From the computational modeling perspective, another tensor quantity of interest is the so-called turbulent stress tensor. In DNS of turbulent flow the conservation PDEs are solved in their exact form (as shown above for the momentum equation). On the other hand, in LES an alternative form is derived via the following spatial filtering operation:

$$\langle Q \rangle (\mathbf{x}, t) = \int_{-\infty}^{+\infty} Q\left(\mathbf{x}', t\right) G\left(\mathbf{x} - \mathbf{x}'\right) \mathrm{d}\mathbf{x}' \tag{4}$$

where $Q(\mathbf{x}, t)$ is any turbulent quantity such as velocity or density. The filtering operation applied to the momentum equation, Eq. (1), gives:

$$\frac{\partial \langle \rho \rangle \langle u_i \rangle_L}{\partial t} + \frac{\partial \langle \rho \rangle \langle u_i \rangle_L \langle u_j \rangle_L}{\partial x_i} = -\frac{\partial \langle p \rangle}{\partial x_j} + \frac{\partial \langle \tau_{ij} \rangle}{\partial x_i} - \frac{\partial T_{ij}}{\partial x_i} \tag{5}$$

where the notation $\langle\ \rangle_L$ indicates the density weighted filter:

$$\langle Q \rangle_L = \frac{\langle \rho Q \rangle}{\langle \rho \rangle}, \tag{6}$$

Equation (5) greatly simplifies the filtered form of the PDEs and is commonly employed in LES and RANS of turbulent flow, instead of solving Eq. (1) directly. The important term in Eq. (5) is the turbulent stress tensor T which is defined as:

$$T_{ij} = \langle \rho \rangle \left(\langle u_i u_j \rangle_L - \langle u_i \rangle_L \langle u_j \rangle_L \right) \tag{7}$$

In LES the combustion flow solution is sought for the filtered terms *directly* by solving the simplified, filtered form of the equations (Eq. 5). However, a closure problem originates from the non-linearity in the second term on the left-hand-side of Eq. (1) and is manifested as the unknown term $\langle u_i u_j \rangle_L$ in Eq. (5). **There is no equation in the filtered set of PDEs that corresponds to this term! Indeed, the whole literature on modeling of turbulent flow is providing closure models for T in one way or another.**[1] The same is true in RANS where ensemble averaging replaces filtering. For these reasons, the stress/strain tensors and the turbulent stress tensor are of particular interest to computational modelers.

2.2 Challenges

While many tensor visualization techniques have been proposed and implemented in various systems, in particular in medical imaging and civil engineering, the visualization of turbulent combustion tensors presents several key challenges. First, because combustion datasets are the result of computational simulations often performed at high resolutions, they tend to be particularly dense. Such high densities lead naturally to clutter and occlusion problems when visualizing the data, as well as to slow interaction when visually exploring the data. Second, because researchers are particularly interested in the 3D structure of the flow and the possible ways to numerically decompose and simulate this flow, it is important to show the 3D context of the data. At the same time, because the researchers are typically trained to visualize their data using 2D exploration tools such as TecPlot or ParaView, it is important to still give them access to similar 2D projection tools when exploring

[1]In reacting flow there are additional and generally more significant closure problems associated with chemical source terms in the mass and energy conservations equations, however this is beyond the scope of this text. The reader is referred to the literature [9, 12].

the data. Furthermore, care should be exercised when applying tensor visualization techniques from other fields (such as diffusion tensor imaging) to combustion data; because of the high density of simulation data and the range of values these data take, many existing tensor representations do not have an intuitive equivalent in combustion flow. Finally, many exploratory tasks in this context are visual comparison tasks, where the user aims to compare multiple steps of a specific numerical simulation, or different numerical simulations. To this end, it is important to provide a quick overall sense of the tensor field structure, while also allowing the user to explore particular regions of interest in more detail.

3 Related Work

Several methods have been proposed for the visualization of tensor datasets. They include eigenvector color maps, glyphs, streamlines, volume rendering and volume deformation. Most of them are used to visually represent Diffusion Tensor Magnetic Resonance Imaging (DT-MRI) data, while some have been used in mechanical engineering to display different types of quantities, such as stress or fluid flow.

The most basic method uses eigenvector color mapping. It assigns an (R,G,B) color according to the (X,Y,Z) components of the principal eigenvector and a saturation level that depends on the magnitude of the anisotropy metric [41]. Colormaps are commonly used by mechanical engineers for component-by-component visualization of vector and tensor quantities. The limitation of colormaps is that they cannot show directional information and can only display one type of information at any one time, e.g., tensor component or eigenvalue.

Another approach utilizes glyphs, which are 3D icons whose shapes, colors, textures and locations correspond in some way to the properties of the data. There are many ways these mappings can be made. Generally, the shape indicates the directions of the eigenvectors at particular voxels, while colors can point to the value of the anisotropy at those voxels [41]. Two examples of glyphs are composite glyphs [38] and superquadrics [16], both of them used in brain imaging. In the engineering field, a type of glyphs called stress hedgehogs have been used to visualize mechanical stress [13]. The disadvantage of glyphs is that, for dense datasets such as the ones generated through simulation, they lead to clutter and occlusions.

Streamlines (sometimes called hyper-streamlines) and streamtubes are often used for tracing white-matter fibers in DTI-MRI datasets [40]. They follow the direction of the main eigenvector through the volume and can be grouped using clustering algorithms into clusters that have anatomical correspondents [30, 44]. In addition to visualizing white matter fibers, streamtubes have been used to simulate blood flow through an artery in order to detect the effect of the flow on the walls of the artery [37]. In the mechanical engineering field, hyper-streamlines have been used to visually represent mechanical stress [13]. Slavin et al. use streamtubes to visualize topological defects in nematic liquid crystals, which are revealed by discontinuities in the orientation order of molecules in the liquid [31]. Unlike glyphs, which can show directional information only at certain points in the volume, streamlines can display directional information continuously in the volume. Nevertheless, streamlines are also prone to clutter and occlusions.

Volume rendering is a method of mapping points in the volume to the screen using a transfer function which controls how transparent or opaque each point is. Previously-used methods include barycentric opacity maps, hue-balls and lit-tensors [19]. Bhalerao and Westin use a textured mapped approach called tensor splatting [1]. Volume rendering of stress magnitudes, combined with tracing of short line segments to show stress direction, can help with designing and positioning of implants in hip joint replacement planning [7]. The volume rendering approach holds promise for dense datasets, although it can primarily show scalar information.

The volume deformation method is generally used to visualize mechanical stress. It allows the viewer to infer the characteristics of the tensor field by observing the effect it has on an object. Zheng et al. present two techniques: normal vector deformation and anisotropic deformation. The former is best used at indicating the direction of the tensor field while the latter can show the compressing and shearing properties of the tensor field [43]. Because of the fact that this method requires an object onto which the deformation has to be applied, it is in general not applicable to visualizing combustion flow.

Most of the work in the field tensor visualization of has been focused on symmetric tensor fields. There has recently been some interest in visualizing asymmetric tensor fields. One example is the use of tensor lines to analyze the gradient of the velocity vector field, an asymmetric tensor field [42].

Because of the complexity of a 3D tensor field, each tensor potentially having six components (three diagonal and three non-diagonal elements), three eigenvalues and three eigenvectors to visualize, tensor visualization techniques often have occlusion and cluttering problems. In order to fix these problems, researchers often reduce the dimensionality and/or use interaction [3, 7, 14]. Both Jianu et al. and Chen et al. provide linked views of 3D and 2D representations of brain imaging data as well as several types of interaction with the 3D and 2D models, such as selection and coloring of white-matter fiber clusters. Dick et al. adopt a focus+context technique, where the user can move a circular region over the background visualization in order to view it in more detail. Sherbondy et al. use interaction in an application that uses dynamic queries to display neural pathways between volumes of interest [26].

Interactive filtering has been used more extensively outside the tensor visualization field, in particular in information visualization [6]. In general, the idea of bridging information visualization and scientific visualization is gaining momentum [26], and several examples have been published in recent years [7, 14]. Nevertheless, the challenges in this line of research relate to the specific forms scientific data takes.

4 Methods

To the best of our knowledge, this study is the first exploratory study of tensor visualization techniques in the context of turbulent combustion flow. To address the challenges outlined in Sect. 2.2, we propose a prototype for interactive visualization of combustion tensor quantities. It combines glyph-based representations—used as a 2D-projection exploratory tool—with real-time volume rendering—used as a 3D-context visual anchor, and with velocity streamlines—serving as flow salience cues. To further address the problem of clutter and occlusions we implement interactive filtering techniques, allowing the user to focus on and compare specific regions of interest of the tensor field. We begin by describing the data.

4.1 Datasets

We have employed results of three simulations in this study. The first dataset is a laboratory-scale Bunsen burner flame. The second is a canonical test problem employed in turbulent reacting flow research, namely the temporal mixing layer configuration. The third and most intriguing dataset is the result of LES simulation of a turbulent jet configuration known as the Sandia-D experiment.

The Bunsen burner is a centimeter-scale fuel jet surrounded coaxially by a hot pilot stream of burnt gas and impinging onto quiescent air. It is a high Reynolds number (24,000) configuration with a wide range of spatial scales, DNS of which is computationally unreasonable. The state of the art in DNS is the two-dimensional Bunsen burner configuration (slot-burner) with two-orders of magnitude lower Reynolds number than what is considered here. Therefore, only LES of this flame is available. The data employed in this work for the Bunsen burner are the filtered turbulent velocity vector and the turbulent stress tensor fields taken at a snapshot in time and discretized over a uniform Cartesian grid of size 101 in each direction (1M grid points).

The temporal mixing layer, on the other hand, is a simple configuration where two streams of fuel and oxidizer flow over and against each other. The flow speeds are adjusted for a low Reynolds number yielding a narrow range of length scales, and this configuration can be easily tackled with DNS and then used as a benchmark. The data for the temporal mixing layer is similarly at a snapshot in time and at the full DNS resolution over a grid of size 193 grid points in two Cartesian directions and 194 in the other (approx. 8M grid points).

The Sandia-D dataset is very similar to the Bunsen burner, in that it is also a centimeter-scale jet configuration with a fuel jet at the center (methane-air mixture for this dataset) surrounded coaxially by a slower speed hot pilot flame of burnt premixture of acetylene, hydrogen, and air. The difference is that the pilot flame is further surrounded by a co-flowing hot air stream. It has a slightly lower Reynolds number (22,400) than the Bunsen burner dataset, but the DNS is also computationally unreasonable. The LES simulation provides solutions of the turbulent fields of species as well as the velocity field as a function of 3D space and time. The temporal resolution is dependent on the numerical specifications and is in the level of micro-seconds in the scales of this flame. The data used in this experiment is comprised of the filtered turbulent velocity vector and the turbulent stress tensor fields taken at a snapshot in time and discretized over a uniform Cartesian grid of size 200 in streamwise direction and 160 in each of cross stream directions (approx. 5M grid points).

4.2 Glyph Representation

Representations that combine the different tensor components into a single image were of immediate interest to our collaborators in combustion research. In particular, previous studies of the Bunsen burner dataset routinely used component-by-component colormap representations (ParaView) to identify discontinuities in the tensor field, and we speculated that a combined representation of the tensor components would prove useful for this task.

As a first iteration, we pursued Westin's model of displaying the tensor as a composite glyph [38], which consists of a rod, a disc and a sphere. The eigenvectors and eigenvalues of the velocity strain tensor are calculated for each evenly spaced grid point, then mapped to the composite glyph. The direction of the main eigenvalue is deemed more important than the direction of the second and third eigenvalues, thus the rod component points in the direction of the main eigenvector. All that can be said about the second eigenvector is that is in the plane of the disc, while the third component, the sphere, does not retain the direction of the third eigenvector at all. The eigenvalues are sorted in descending order and then assigned to the rod, disc and sphere component of the glyph. Since, for many of the tensors in the Bunsen burner dataset, the three eigenvalues are very close in value, the length of the rod is set to three times the size of the main eigenvalue, while the radii of the disc and sphere are equal to the second and, respectively, third eigenvalues. This is done so that the direction of the main eigenvector can be distinguished more easily. Color is used to distinguish the three components of the glyph: green for the rod, magenta for the disc and red for the sphere. Figure 3 (left) shows a slice through the tensor field, each tensor being represented as a composite glyph. As anticipated, mapping the glyphs to the 3D flow lead to clutter and occlusions, even when the glyph field was subsampled by a factor of 25.

Following feedback from our collaborator, we simplified the glyph by removing the disc and sphere components of the composite glyph and only displaying the direction of the main eigenvector as well as the value of the main eigenvalue. For this purpose, the rod component was replaced with a line. Figure 3 (right) displays the same part of the tensor field as Fig. 3 (left), except it uses the piecewise linear representation of the tensors main eigenvector. While the simplified, linear result was easier to interpret, in particular in 2D cross-sections of the field, 3D views of the

Fig. 3 Glyph (*left*) and piecewise-linear (*right*) visualization of the tensor field. In the glyph-based representation color identifies the various glyph components (*green* for the rod, *magenta* for the disc and *red* for the sphere.) To reduce clutter, the number of glyphs was subsampled by displaying only every fifth element along each direction, resulting in a dataset which was 1/25 of the original size. The subsampling factor has been empirically determined

representation were still illegible due to clutter, and failed to deliver a sense of the 3D flow. The line glyphs were still kept in the final version of the prototype, the only difference being that an arrow was added to point in the direction of the main eigenvector. In the degenerate case, when the difference between the first two eigenvalues is less than 0.001 (heuristically determined) the corresponding glyph is displayed as a gray sphere, shown in Figs. 5 and 7.

4.3 *Volume Rendering and Streamlines*

To provide 3D-context to tensor quantities while reducing clutter and occlusions, we integrated volume rendering into the tensor field visualization. Volume rendering can be a very effective way of visualizing 3D volumes, although it requires mapping a color transfer function to a scalar value. The scalar quantity we turned to for the volume rendering is *divergence*, which can be calculated as the trace of the strain tensor in Eq. (3). Under conditions of no mass source, divergence represents the change of density in time. The divergence reveals the 3D structure of the tensor field and effectively shows how the density of the fluid changes in different regions. Figure 4 shows two example images of divergence for the temporal mixing layer flow.

Our custom implementation of volume rendering uses ray-casting. The opacity transfer function for volume rendering was designed interactively so that high-divergence regions had higher opacity and low-divergence regions had low opacity. This mapping highlighted the most interesting regions of the flow. The color transfer function was designed following the same principle; Fig. 4 shows the color transfer

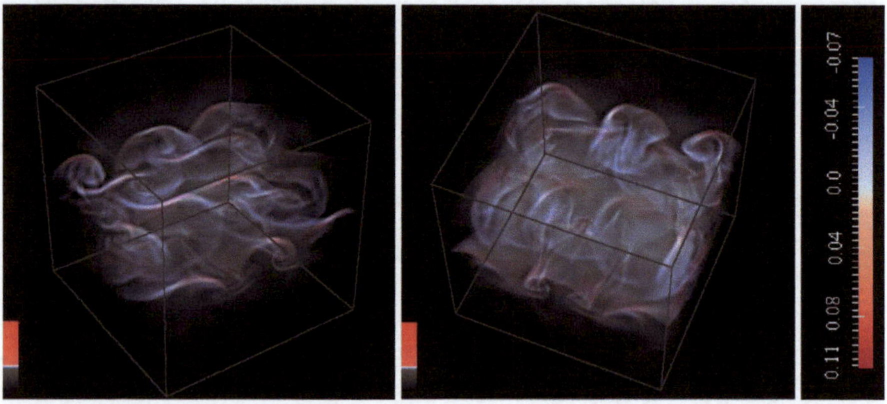

Fig. 4 Volume rendering of *divergence* of the temporal mixing layer dataset (*left* and *middle*), and color transfer function (*right*) used for volume rendering. Regions with saturated *blue* or *red* indicate higher magnitudes of divergence, either positive or negative. This rendering shows very clearly the 3D nature of the flow which is difficult to extract through other visualization methods

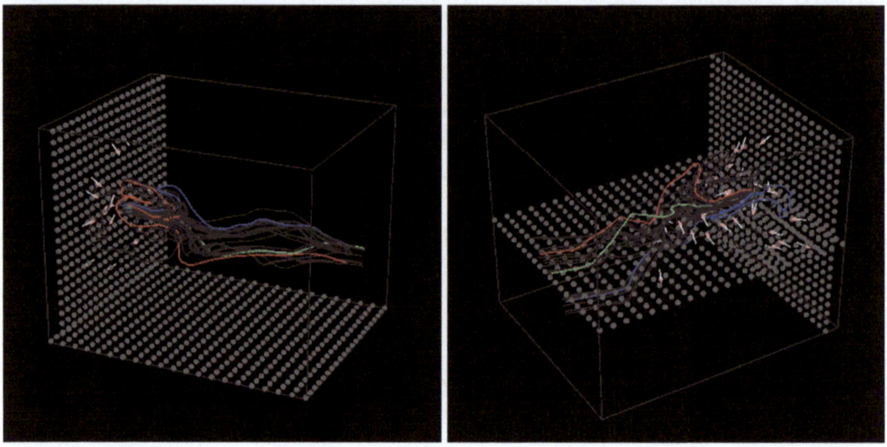

Fig. 5 Streamline and glyph visualization for the Bunsen burner dataset. The velocity field is unidirectional at the *bottom* of the volume (*left*). At the center of the volume, the velocity field is more varied (*right*). The streamlines emphasize the flow features in the central part of the simulation

function for the temporal mixing layer dataset. The user can customize transparency and color to better explore the underlying 3D stress tensor field.

To further emphasize the flow-context of the tensor data, the volume rendering was augmented with velocity streamlines—a technique borrowed from flow visualization. To create streamlines we calculated Runge-Kutta 4 integral paths through the velocity field. The starting points for the streamlines are a "cube" of evenly spaced grid points. Figure 5 shows a combined visualization of glyphs and streamlines for the Bunsen burner dataset.

The visualization tool was implemented in C/C++ with OpenGL for rendering. Custom code was used for ray-casting.

4.4 Interactive Filtering

Early feedback from combustion researchers indicated that the glyphs, volume renderings and streamlines complement each other nicely in terms of the information displayed. However, the sheer volume of the combined information is overwhelming. To tackle this challenge, we followed a strategy based on Shneiderman's info-vis mantra [2] and Tufte's principles [34–36]. Following Tufte's principles, the information content of the image is maximized by combining volume rendering, streamlines and the glyph-based representations. To this combined representation we add zooming and filtering via interaction. Following further Shneiderman's mantra, the volume rendering gives an overview of the flow and serves as a visual anchor, while the streamlines enable filtering of interesting regions, and glyph representations function as details on demand.

The visualization tool has two modes of operation, *explore mode* and *filter mode*. In the explore mode, a user can manipulate the scene and zoom in and out through mouse or keyboard interaction. Rendering the full resolution volume image during rotation or zooming slows down significantly the interaction. Therefore, during rotation or zooming, we use low-resolution volume rendering to maintain interactivity of the program: fewer rays are used for ray-casting than in the full-resolution rendering, and the remaining pixels are interpolated. We also reduce the sampling rate along a ray through the volume. The low resolution rendering produces an intensity image, while the full resolution rendering produces a color image. These modifications make the volume rendering fast and help to maintain interactive rates. The intensity volume-image during interaction helps to give an idea of how the volume is changing so that the user does not lose context between the start of the interactive step and its end.

In the filter mode, the user can focus on a sub-region of the flow by highlighting and comparing streamlines of interest. The filtering operation follows Tufte's principles and an approach originally used by the New York Times to explore stock market trends [6]. By default, all streamlines are muted gray, while mouse interaction allows the user to highlight and contrast two or more representative streamlines. Streamline seed-points can also be interactively dragged to new locations in the volume. After selecting a set of streamlines, the user can switch back to the explore mode and rotate/zoom the scene to better gauge the characteristics of the streamlines.

Figure 6 shows how the filtering operation can help declutter the visualization. In both modes, a user can control whether streamlines, volumes or the glyph-representations will be rendered or not. To further reduce clutter, the glyph representations are mapped to axis-oriented cutting planes, which are also controlled by the user.

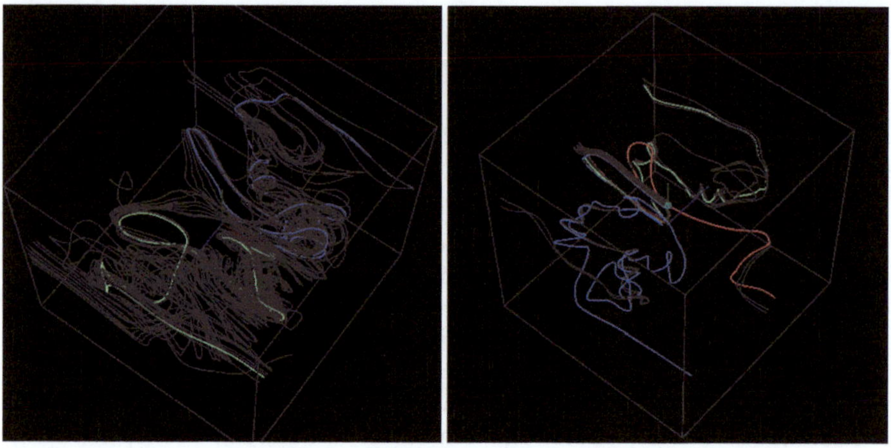

Fig. 6 Streamlines through the velocity field of the temporal mixing layer dataset. Twenty seven streamlines are shown (*left*); even such small numbers clutter the image. Filtering streamlines interactively (*right*) helps reduce the clutter and enables the user to compare families of streamlines.

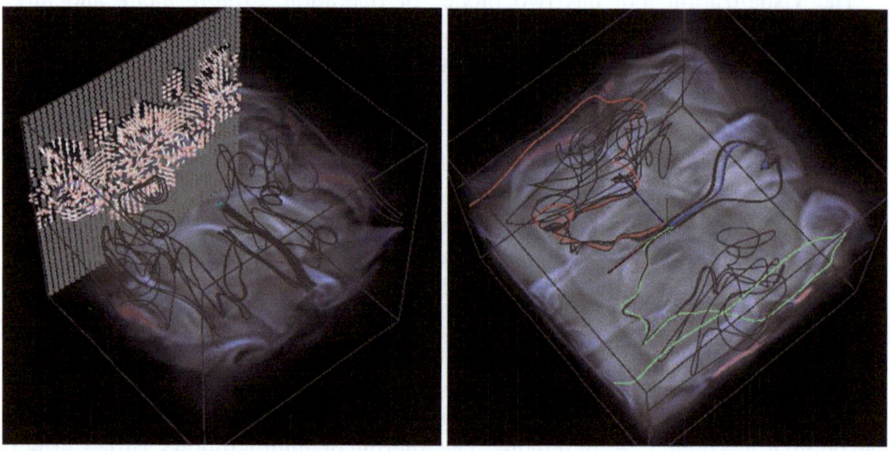

Fig. 7 Exploratory visualization tool for analyzing stress tensors used in computational turbulent combustion. The tool combines 3D inspection views based on volume renderings of divergence (shown in *purple*) with glyph-based representations (*light gray* and *pink* cutting planes), while leveraging interactive filtering of velocity streamlines (*dark gray* and color) and side-by-side views (not shown here) to clarify the structure of the tensor datasets. These snapshots highlight asymmetries in the combustion flow for a mixing layer configuration

Side-by-side views of multiple time-steps further facilitate the analysis of time-space relationships. While animated 3D views have certain advantages over side-by-side views (in particular with respect to display real-estate), the information visualization literature [27] and our experience indicate that side-by-side views are more effective at capturing time relationships.

5 Results and Discussion

We evaluate our analysis tool on the three datasets described in Sect. 2.2: a snapshot of the Bunsen burner simulation, a snapshot of the Temporal Mixing Layer simulation, and a three-step time sequence of the Sandia-D experiment. Two of the authors are computational combustion researchers who used the system and provided the following feedback.

Using the integrated exploratory system on the Bunsen burner dataset, it was noticed that the tensor field was remarkably smooth and homogeneous in the outer co-flow and the inner pilot regions. The tool was proven useful even for this elementary level of analysis, and will be further employed for more detailed LES datasets. For the mixing layer configuration, similar observations were immediately visible where the "mushroom" pattern around the shear layer at the mid-zone is distinguished well from the zero-divergence outer zones (Fig. 7). Interestingly, the analysis of the Sandia-D dataset provided a surprise—a rippling artifact surrounding the central jet—which can be seen in Fig. 8 and which showcases the advantages of the tool as a means of debugging numerical simulations. Shown in Fig. 8 are snapshots of the trace of strain tensor field at two instances in time, at 0.7 and 1.8th of a residence time, respectively; one residence time is equal to the total streamwise length divided by mean jet velocity, i.e., roughly the time it takes a fluid particle to traverse the whole length of the domain. The field shown is for the trace of the strain tensor. The regions at and near the jet core portray non-zero trace, and are indicative of the high spatial and temporal gradient in the gas-mixture density in these regions, in other words, of the active reaction zones with high compressibility. The far field trace on the other hand is much closer to zero and suggests (as expected from theory) close-to-zero compressibility of uniform and low density flow in these regions. This is especially clear in rightmost snapshot where the effects of high-frequency pressure waves are lesser compared to earlier in the simulation. This non-physical ripple effect could be attributed to numerical artifacts of the employed discretization scheme in the LES simulation, and is pronounced only in the incompressible regions, as clearly shown in the volume rendered tensor field. Overall, the researchers found the system as an exploratory tool "very good" and "cool", and also commented on the potential explanatory power of the system to be utilized in comparisons of simulations against experimental data.

With respect to the various components of the visualization scheme, the glyph representation for the Bunsen burner dataset was considered somewhat useful, and only in 2D cross-sections. In 2D probes the flow patterns were "not sufficiently expressed", while in 3D the information was "hard to read" due to clutter and occlusions, as the glyphs became "intermingled and cluttered". Because the shown turbulent stress tensor field was mostly uniform throughout the domain, ellipsoid or line-style representations did not fare better; tensors did not "intuitively have direction". When exploring the more anisotropic mixing layer dataset, 2D planes of glyphs were considered useful, although only when combined with either streamlines or volume rendering to provide 3D context. Perhaps due to researchers'

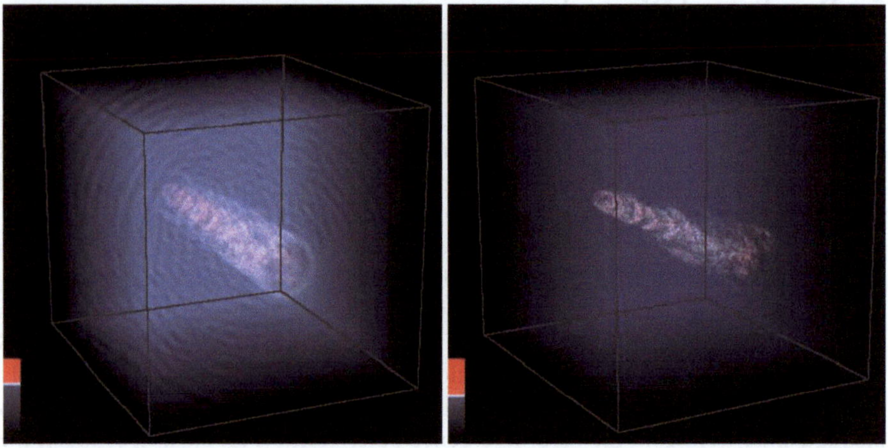

Fig. 8 Volume rendering of *divergence* of the third dataset (Sandia-D). Two timestamps of a reactive flow simulation are shown here, increasing in time from *left* to *right*. The rippling effect disappears as the simulation goes further in time. In this example, the user customized interactively the transfer function (color and opacity) to emphasize the central flow

familiarity with 2D tools such as TecPlot, the 2D content as exposed by the glyphs was still considered important for these datasets.

The volume rendering combined with the interactive streamlines generated the most excitement. The researchers were particularly impressed by the ability to interact with the streamlines, by selecting and comparing individual streamlines and following their progress through the divergence volume. In particular, the researchers noted with the mixing layer dataset that the tangled, asymmetric streamlines in the mid-plane illustrate well the turbulent shear layer behavior where opposing streams of fuel and oxidizer meet. Both researchers made remarks about the tool's ability to focus on the interesting region of the volume (e.g., "[compared to this, in other tools] interactive selection is a beast"). The researchers also commented on the resolution and interactivity of the volume rendering, which was eight times more dense and, in their estimate, ten times faster than ParaView, the visual tool they had often used for volume rendering.

Overall, the integrated prototype compared favorably against existing visualization software that offers similar, though less interactive, visualization features, such as ParaView, VisIt, Ensight and Tecplot. The essential advantages of using our utility for turbulent tensor visualization were its performance and its being tailored to this specific application. The interactive rendering rates, the real-time selection of seed points for the streamline data, and the overall easy flow of interaction were major points repeatedly emphasized by the combustion scientists.

It is important to emphasize that the crux of the study in this paper is not the utility itself but rather the application of the aforementioned visualization algorithms in the context of turbulent reacting flow. Currently, to the best of our collaborators' knowledge, commonly employed visualization strategies for serious analysis in this field is limited to 2D cross sectional contours, or rather general 3D techniques which do not necessarily highlight the significant features required, in particular in the context of visual comparison tasks.

Much of the researcher feedback was directed towards potential extensions of the tool, as well as future application to other datasets. In particular, interest was expressed with respect to loading, exploring and comparing multiple datasets. A suggestion for future work, which came from both evaluators, was the implementation of a "field calculator"—an interface to allow the on-the-fly calculation of various scalar, vector and tensor quantities of interest. The "calculator" would include operators to add scalar fields or find a derivative or gradient for a certain field before displaying it.

In terms of limitations, all datasets we analyzed are canonical datasets of relative modest size, by computational combustion standards. If the dimensions of the volume were increased significantly, the current volume rendering visualization would require sub-sampling to stay interactive; more performant volume renderers would be necessary. Furthermore, in our approach we explored only two glyph representations. Alternative representations [17] may help clarify whether glyph size and shape have a significant influence on the visualization. Also, since the glyphs are evenly spaced throughout the volume, there is no guarantee that they capture the important features of the volume. One possible direction for future work may be the use of glyph packing [18]. In terms of assisting combustion researchers in designing simulation schemes which are both computationally affordable and sufficiently accurate, techniques for segmentation of combustion tensor fields, perhaps interactively, hold particular promise.

One step further in terms of future research, exploratory visualization of massive combustion tensor datasets poses additional particularly significant challenges. Combustion simulations use thousands of CPUs to generate snapshots with millions of grid points; copying the data on a server for visualization is simply not feasible. Novel, memory-efficient exploratory visualization techniques will be necessary for such datasets.

6 Conclusion

In this chapter, we examined the challenges associated with tensor-field visualization in the context of turbulent combustion calculations and we proposed a prototype tool that can be used to visually explore combustion datasets. Our approach leverages interactive filtering and flow salience cues to clarify the structure of the tensor datasets, while effectively addressing the problems of occlusion and clutter. Side-by-side views of multiple timesteps facilitate the analysis of time-space relationships. The resulting framework enables an analysis style based on the overview first, zoom and filter, then details on demand paradigm originally proposed in information visualization. The result is a visual analysis tool to be utilized in debugging, benchmarking, and verification of models and solutions in turbulent combustion. We evaluated the proposed tool on three examples of turbulent reacting flow. Feedback from combustion researchers indicates that the tool is very useful in the exploration of turbulent combustion simulations, and emphasizes the urgent need of the field for visual analysis tools. However, significant research is still required to arrive at standard visualization in this application domain.

Acknowledgements Acknowledgements to the Pitt Visualization Research group and to NSF-IIS-0952720.

References

1. Bhalerao, A., Westin, C.-F.: Tensor splats: visualising tensor fields by texture mapped volume rendering. In: Proceedings of Sixth International Conference on Medical Image Computing and Computer-Assisted Intervention (MICCAI '03), Montréal, pp. 294–901 (2003)
2. Card, S.K., Mackinlay, J.D., Shneiderman, B.: Readings in Information Visualization: Using Vision to Think. Morgan Kaufmann, San Francisco (1999)
3. Chen, W., Ding, Z., Zhang, S., MacKay-Brandt, A., Correia, S., Qu, H., Crow, J.A., Tate, D.F., Yan, Z., Peng, Q.: A novel interface for interactive exploration of DTI fibers. IEEE Trans. Vis. Comput. Graph. **15**(6), 1433–1440 (2009)
4. Chen, Y., Peters, N., Schneemann, G.A., Wruck, N., Renz, U., Mansour, M.S.: The detailed flame structure of highly stretched turbulent premixed methane-air flames. Combust. Flame **107**, 223–226 (1996)
5. Ciofalo, M.: Large Eddy Simulation: a ritical survey of models and applications. In: Advances in Heat Transfer, vol. 25, pp. 321–419. Academic, New York (1994)
6. Cox, A., Xaquin, G.V., Leonhardt, D.: How this bear market compares. http://www.nytimes.com/interactive/2008/10/11/business/20081011_BEAR_MARKETS.html (2008), last accessed July 4th 2012
7. Dick, C., Georgii, J., Burgkart, R., Westermann, R.: Stress tensor field visualization for implant planning in orthopedics. IEEE Trans. Vis. Comput. Graph. **15**(6), 1399–1406 (2009)
8. Dunn-Rankin, D. (ed.): Lean Combustion: Technology and Control. Academic, New York (2008)
9. Fox, R.O.: Computational models for turbulent reacting flows. Cambridge University Press, Cambridge (2003)
10. Givi, P.: Filtered density function for subgrid scale modeling of turbulent combustion. AIAA J. **44**(1), 16–23 (2006)
11. Hack, R.L., McDonell, V.G.: Impact of ethane, propane, and diluent content in natural gas on the performance of a commercial microturbine generator. J. Eng. Gas Turb. Power **130**(1), 011509 (2008)
12. Haworth, D.C.: Progress in probability density function methods for turbulent reacting flows. Prog. Energ. Combust. vol. 36, pp. 168–259 (2010)
13. Jeremic, B., Scheuermann, G., Frey, J., Yang, Z., Hamann, B., Joy, K.I., Hagen, H.: Tensor visualizations in computational geomechanics. Int. J. Numer. Anal. Methods Geomech. **26**(10), 925–944 (2002)
14. Jianu, R., Demiralp, C., Laidlaw, D.H.: Exploring 3D DTI fiber tracts with linked 2D representations. IEEE Trans. Vis. Comput. Graph. **15**(6), 1449–1456 (2009)
15. Johnsen, E., Larsson, J., Bhagatwala, A.V., Cabot, W.H., Moin, P., Olson, B.J., Rawat, P.S., Shankar, S.K., Sjogreen, B., Yee, H., Zhong, X., Lele, S.K.: Assessment of high-resolution methods for numerical simulations of compressible turbulence with shock waves. J Comput. Phys. **229**(4), 1213–1237 (2010)

16. Kang, S., Iaccarino, G., Ham, F. and Moin, P.: Prediction of wall-pressure fluctuation in turbulent flows with an immersed boundary method. J Comput. Phys. **228**, 6753–6772 (2009)
17. Kindlmann, G.: Superquadric tensor glyphs. In: Proceedings of IEEE TVCG/EG Symposium on Visualization 2004, pp. 147–154 (2004)
18. Kindlmann, G., Westin, C.-F.: Diffusion tensor visualization with glyph packing. IEEE Trans. Vis. Comput. Graph. **12**(5), 1329–1336 (2006)
19. Kindlmann, G., Weinstein, D., Hart, D.: Strategies for direct volume rendering of diffusion tensor fields. IEEE Trans. Vis. Comput. Graph. **6**(2) 124–138 (2000)
20. Kuo, K.: Principles of Combustion. Wiley, Hoboken (2005)
21. Lesieur, M., Metais, O.: New Trends in Large Eddy Simulations of Turbulence. Ann. Rev. Fluid Mech. **28**, 45–82 (1996)
22. Peters, N.: Turbulent Combustion. Cambridge University Press, Cambridge (2000)
23. Poinsot, T., Veynante, D.: Theoretical and Numerical Combustion, 2nd edn. R.T. Edwards, Inc., Philadelphia (2005)
24. Pope, S.B.: Turbulent Flows, Cambridge University Press, Cambridge (2000)
25. Pope, S.B.: Advances in PDF methods for turbulent reactive flows. In: Andersson, H.I., Krogstad, P.A. (eds.) Advances in Turbulence X, pp. 529–536. CIMNE, Barcelona (2004)
26. Rhyne, T., Tory M., Munzner, T., Ward, M., Johnson, C., Laidlaw, D.: Information and scientific visualization: separate but equal or happy together at last? IEEE Visualization (Panel Proceedings), pp. 611–614. Seattle, WA (2003)
27. Robertson, G., Fernandez, R., Fisher, D., Lee, B., Stasko, J.: Effectiveness of animation in trend visualization. IEEE Trans. Vis. Comput. Graph. **14**(6), 1325–1332 (2008)
28. Richards, G.A., McMillian, M.M., Gemmen, R.S., Rogers, W.A., Cully, S.R.: Issues for low-emission, fuel-flexible power systems. Prog. Energy Combust. Sci. **27**(2), 141–169 (2001)
29. Sankaran, R., Hawkes, E.R., Chen, J.H., Lu, T., Law, C.K.: Structure of a spatially developing turbulent lean methane-air bunsen flame. Proc. Combust. Inst. **31**, 1291–1298 (2007)
30. Sherbondy, A., Akers, D., Mackenzie, R., Dougherty, R., Wandell, B.: Exploring connectivity of the brain's white matter with dynamic queries. IEEE Trans. Vis. Comput. Graph. **11**(4), 419–430 (2005)
31. Slavin, V., Pelcovits, R., Loriot, G., Callan-Jones, A., Laidlaw, D.: Techniques for the visualization of topological defect behavior in nematic liquid crystals. IEEE Trans. Vis. Comput. Graph. **12**(5), 1323–1328 (2006)
32. Vreman, A., van Oijen, J., de Goey, L., Bastiaans, R.: Direct numerical simulation of hydrogen addition in turbulent premixed Bunsen flames using flamelet-generated manifold reduction. Int. J. Hydrog. Energy. **34**, 2778–2788 (2009)
33. Tannehill, J.C., Anderson, D.A., Pletcher, R.H.: Computational Fluid Mechanics and Heat Transfer, 2nd edn. Taylor & Francis, Washington, DC (1997)
34. Tufte, E.R.: Envisioning Information. Graphics Press, Cheshire (1990)
35. Tufte, E.R.: Visual Explanations: Images and Quantities, Evidence and Narrative. Graphics Press, Cheshire (1997)
36. Tufte, E.R.: The Visual Display of Quantitative Information. Graphics Press, Cheshire (2001)
37. Wenger, A., Keefe, D.F., Zhang, S., Laidlaw, D.H.: Interactive volume rendering of thin thread structures within multivalued scientific data sets. IEEE Trans. Vis. Comput. Graph. **10**(6), 664–672 (2004)
38. Westin, C.-F., Maier, S.E., Mamata, H., Nabavi, A., Jolesz, F.A., Kikinis, R.: Processing and visualization for diffusion tensor mri. Med. Image Anal. **6**, 93–108 (2002)
39. Yilmaz, S.L., Nik, M.B., Givi, P., Strakey, P.A.: Scalar filtered density function for large eddy simulation of a premixed bunser burner. J. Propul. Power. **26**, 84–93 (2010)
40. Zhang, S., Demiralp, C., Laidlaw, D.H.: Visualizing diffusion tensor MR images using streamtubes and streamsurfaces. IEEE Trans. Vis. Comput. Graph. **9**(4), 454–462 (2003)
41. Zhang, S., Kindlmann, G., Laidlaw, D.H.: Diffusion tensor MRI visualization. In: Visualization Handbook. Academic, Amsterdam (2004)

42. Zhang, E., Yeh, H., Lin, Z., Laramee, R.S.: Asymmetric tensor analysis for flow visualization. IEEE Trans. Vis. Comput. Graph. **15**(1), 106–122 (2009)
43. Zheng, X., Pang, A.: Volume deformation for tensor visualization. In: Proceedings of the IEEE Visualization Conference '02, pp. 379–386. Boston, MA (2002)
44. Zhukov, L., Barr, A.H.: Oriented tensor reconstruction: tracing neural pathways from diffusion tensor MRI. In Proceedings of the IEEE Visualization Conference '02, pp. 387–394. Boston, MA (2002)

Shear Wave Diffusion Observed by Magnetic Resonance Elastography

Sebastian Papazoglou, Jürgen Braun, Dieter Klatt, and Ingolf Sack

Abstract Dynamic elastography is a noninvasive imaging-based modality for the measurement of viscoelastic constants of living soft tissue. The method employs propagating shear waves to induce elastic deformations inside the target organ. Using magnetic resonance elastography (MRE), components of harmonic shear wave fields can be measured inside gel phantoms or in vivo soft tissues. Soft tissues have heterogeneous elastic properties, giving rise to scattering of propagating shear waves. However, to date little attention has been paid to an analysis of shear wave scattering as a possible means to resolve local elastic heterogeneities in dynamic elastography. In this article we present an analysis of shear wave scattering based on a statistical analysis of shear wave intensity speckles. Experiments on soft gel phantoms with cylindrical glass inclusions are presented where the polarization of the shear wave field was adjusted relative to the orientation of the scatterers. A quantitative analysis of the resulting fields of shear horizontal (SH) waves and shear vertical (SV) waves demonstrates that the distribution of wave intensities in both modes obeys restricted diffusion in a similar order. This observation sets the background for quantification of shear wave scattering in MRE of body tissue where SH and SV wave scattering occur simultaneously.

1 Introduction

Magnetic resonance elastography (MRE) has been developed for measuring viscoelastic constants in living soft tissue. The method employs propagating shear waves to induce elastic deformations inside the target organ [1, 2]. Current research

S. Papazoglou · D. Klatt · I. Sack (✉)
Department of Radiology, Charité University Medicine, Charitéplatz 1, Berlin, Germany
e-mail: ingolf.sack@charite.de

J. Braun
Institute of Medical Informatics, Charité University Medicine, Hindenburgdamm 30, Berlin

D.H. Laidlaw and A. Vilanova (eds.), *New Developments in the Visualization and Processing of Tensor Fields*, Mathematics and Visualization,
DOI 10.1007/978-3-642-27343-8_8,

is focussed on human muscle [3,4], liver [5,6], breast [7], brain [8,9] and heart [10]. The acquired shear wave images allow the determination of the underlying shear elasticity of the tissue by solving an appropriate inverse problem [11]. However, the solution of the inverse problem is biased by unknown boundary conditions, noise and wave image discretization [12, 13]. At its current state, MRE relies on spatial averaging of modulus maps after wave inversion to achieve the accuracy and reproducibility required for diagnostic applications. As a consequence, information about elastic heterogeneities is lost and diagnostic conclusions are drawn based on 'global' constitutive parameters. Waves propagating through heterogeneous soft tissues are typically scattered by elastic discontinuities [14]. If shear wave scattering is very pronounced, i.e. in case of strong heterogeneity, these variations can appear as characteristic intensity speckle patterns. Up to now little attention has been paid to an analysis of shear wave scattering in terms of intensity variations [15]. Depending on the typical sample size L, shear wave lengths λ and the amount of heterogeneity, the propagation of shear waves has characteristic features and may roughly be divided into two distinct regimes [16–19]. If there is little heterogeneity and shear wave lengths are large compared with the scale of elasticity variations, scattering is weak. In this so-called *ballistic* regime, the scattering mean free path l_s is large, i.e. the average distance an incident wave can travel before being scattered exceeds sample extensions ($\lambda < L < l_s$). In case where $\lambda < l_s << L$, an incident wave is scattered many times, which results in random relationships between local phases of the wave; in other words, the typical property of a wave, its phase coherence, becomes obscured by interference of multiply scattered wave fields. In this setting the wave intensity satisfies a diffusion equation and the regime is correspondingly called the *diffusive* regime. If the amount of heterogeneity is such that $\lambda \sim l_s$, a wave cannot travel one wave length before being scattered. As a result, the diffusive spreading of shear wave intensity is increasingly slowed down and becomes confined to small areas within the region-of-interest (ROI), from which it can hardly escape. The vector field character of shear wave displacement imposes a challenge for in vivo elastography since the polarization of incident wave fields is difficult to control inside the body. The situation can be simplified in phantoms by observing plane wave scattering at plane elastic interfaces. An incident shear wave scattered at such a plane elastic discontinuity can be decomposed into two parts, which scatter independently [20]. The so-called shear horizontal (SH) wave features a polarization exclusively parallel to the surface while the shear vertical (SV) wave has a polarization component perpendicular to the plane of elastic discontinuity. In this article we present an analysis of experimental shear wave scattering based on a statistical analysis of shear wave intensity speckles observed in gel phantoms. We compare the speckle statistics of multiply scattered SH and SV waves observed in soft gel phantoms with cylindrical glass inclusions.

2 Theory

To begin with, we consider a wave propagating in the xy plane of a perfectly elastic material. A harmonic wave of this type can be described by a displacement vector field

$$\mathbf{u}(x,y,\omega) = \begin{bmatrix} u_x(x,y,\omega) \\ u_y(x,y,\omega) \\ u_z(x,y,\omega) \end{bmatrix}, \tag{1}$$

where $\omega = 2\pi f$ is the angular drive frequency. Together with the condition

$$\nabla \cdot \mathbf{u} = 0, \tag{2}$$

it defines a pure shear wave (i.e. $\mathbf{u}$ is volume preserving). Consider now a cylindrical rigid inclusion with principal axis parallel to the z direction (see Fig. 1). The shear wave given in (1) can then be decomposed into two parts that propagate independently, i.e. there is no exchange of energy between them. The part with polarization parallel to z is called shear horizontal SH wave

$$\text{SH wave}: \qquad \mathbf{u}(x,y,\omega) = \begin{bmatrix} 0 \\ 0 \\ u_z(x,y,\omega) \end{bmatrix}. \tag{3}$$

The second is made up by the two remaining components in Eq. (1) and is called the shear-vertical SV wave

$$\text{SV wave}: \qquad \mathbf{u}(x,y,\omega) = \begin{bmatrix} u_x(x,y,\omega) \\ u_y(x,y,\omega) \\ 0 \end{bmatrix}. \tag{4}$$

For a rigid inclusion all shear wave components must satisfy the boundary condition $u_{x,y,z} = 0$ at the surface of the cylinder, while the traction (first derivative of u) is indefinite. A difference between SH and SV waves is that the SV wave may change its polarization due to scattering from the inclusion, whereas the polarization of the SH wave never changes. For this reason both components of the SV wave have to be considered, when SV wave scattering is studied. Moreover, the SV wave is subject to mode conversion. Due to its component perpendicular to the cylinder surface it may also produce a pressure field. In the following the quantity of interest will be the intensity normalized to its mean value in the region of interest. For SH waves this is

$$I = \frac{|u_z(x,y,\omega)|^2}{\langle |u_z(x,y,\omega)|^2 \rangle_{\mathrm{ROI}}}, \tag{5}$$

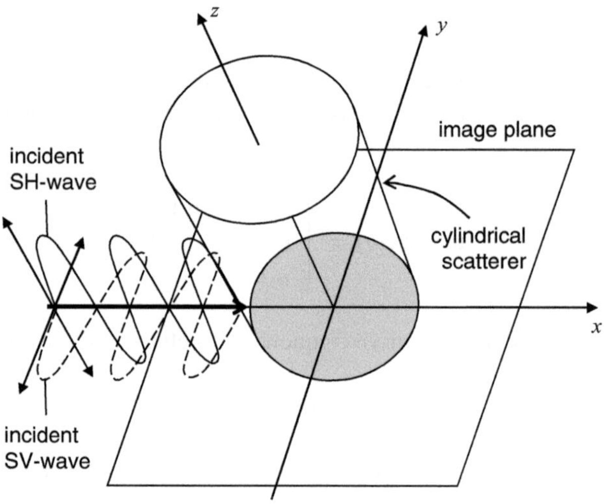

Fig. 1 Polarization of an *SH* wave and an *SV* wave relative to the surface of elastic discontinuity (boundary of the cylindrical scatterer)

and correspondingly for SV waves

$$I = \frac{|u_x(x,y,\omega)|^2 + |u_y(x,y,\omega)|^2}{\langle |u_x(x,y,\omega)|^2 + |u_y(x,y,\omega)|^2 \rangle_{\mathrm{ROI}}}. \tag{6}$$

In case when no scattering occurs the intensity distribution of a plane wave reads simply

$$P(I) = \begin{cases} 1 & \text{for } I = 1, \\ 0 & \text{else.} \end{cases} \tag{7}$$

Now consider an array of a large number of randomly distributed parallel rigid cylinders in a purely elastic matrix. The mean distance the wave can travel before it is scattered is called the scattering mean free path or elastic mean free path l_s. If l_s is much smaller than the sample size L, then a wave is scattered many times before leaving the random array behind. This results in a randomization of the local phase of the wave, which is seen as the dispersion of the coherent wave front yielding a speckled distribution of wave intensity. In this regime the intensity I obeys a diffusion equation and the intensity distribution is then described by an exponential [14]

$$P(I) \propto e^{-I}. \tag{8}$$

If scattering becomes even more pronounced, constructive interferences can eventually lead to a reduction in diffusion, confining diffusive intensity in decreasingly

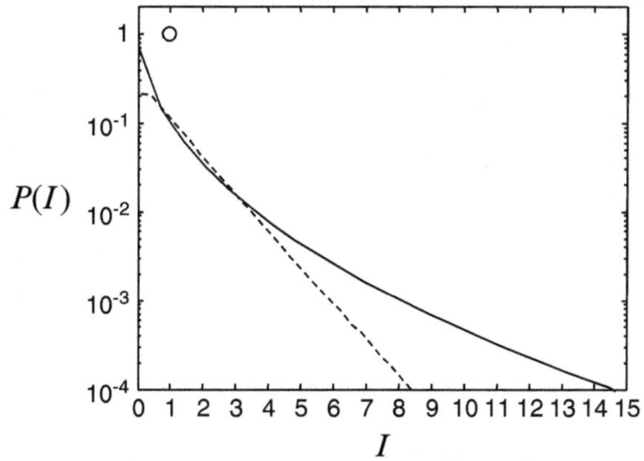

Fig. 2 Three types of intensity distributions for the ballistic regime (*open circle*), the diffusive regime (*dashed line*) and the diffusive regime with $\lambda \sim l_s$ (*solid line*) according to Eqs. (7), (8) and (9) for $g = 1$, respectively

small regions, from which it cannot escape. In this regime the distribution of intensity changes from a simple exponential to a stretched exponential given by

$$P(I) = \int_0^\infty \frac{dv}{v} \int_{-i\infty}^{i\infty} \frac{ds}{2\pi i} \exp\left[-\frac{I}{v} + sv + \Psi(s)\right], \tag{9}$$

where $\Psi(s) = g \ln^2\left(\sqrt{1+s/g} + \sqrt{s/g}\right)$ and g is the dimensionless conductance [21]. The three distributions given by Eqs. (7)–(9) are shown in Fig. 2. The dimensionless conductance g is related to a structure related quantity called the transport mean free path $l_{tr} \propto l_s$ [17]. It is known from diffusion theory as the length scale entering the diffusion constant with $D = 1/d\, v_t l_t$, where d is the spatial dimension and v_t is the energy transport velocity [19]. For two dimensional systems l_t is related to g by [22]

$$g \simeq k l_t, \tag{10}$$

where k is the wave number in the corresponding host medium. In the following the intensity distribution is analyzed in terms of g.

3 Methods

The experimental setup was adopted from standard MRE using a 1.5 T MRI scanner (Sonata, Siemens, Erlangen, Germany) and a modified echo planar imaging (EPI) sequence [23]. Motion sensitization was achieved by oscillating gradients. A phantom was made of a Wirogel (Bego Inc., Bremen, Germany) water solution with a ratio of 1:4 mass partitions. One hundred parallel glass cylinders ($L = 120$ mm, $R = 1.5$ mm) were embedded in the gel matrix at random transverse positions

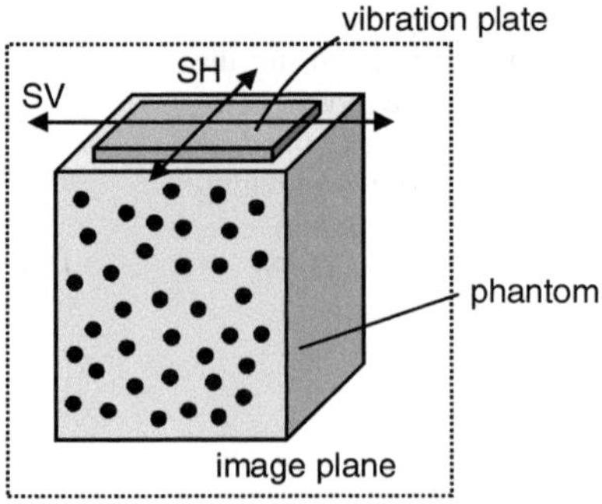

Fig. 3 Relative orientations of shear wave actuation, image plane and cylinder alignment for the *SH* and *SV* wave scenarios. See the methods section for further details

fixed by parallel plates at the ends of the cylinders, which corresponded to fixed boundary conditions at the top and bottom of the container. Random positions were determined inside a 10 cm square region using a random numbers generator (normally distributed) of Matlab (The MathWorks Inc Natick, MA) with a minimum distance between neighboring cylinders of $0.1\,R$. The resulting scatterer density was 7 %. A square vibrator plate was used for wave excitation driven by 25, 50, 75, 100, 125, 150, 175 Hz and 200 Hz time-harmonic oscillations. Two separate experiments were performed for studying the scattering of SH and SV waves. The experimental set up is shown schematically in Fig. 3. For studying the multiple scattering of SH waves the gel phantom was oriented with the cylindrical inclusions aligned parallel to the static field of the scanner. A transverse slice was chosen as image plane defining the x and y coordinates. Motion encoding was in the slice select direction, i.e. parallel to actuator motion and cylinder axes (z direction), measuring the SH wave $u_z(x, y, \omega)$. In a second experiment the phantom was oriented with the glass cylinders perpendicular to the static field. As image plane a sagittal slice was chosen again defining the xy plane. Now, motion was encoded into phase encoding direction, corresponding to x, and read-out direction corresponding to y yielding the two components of the SV wave $u_x(x, y, \omega)$ and $u_y(x, y, \omega)$. Sixteen dynamics of a wave cycle were captured at each vibration frequency. After phase unwrapping data were temporally Fourier transformed and bandpass filtered to increase the signal to noise ratio and to suppress small wavenumbers not associated with shear wave propagation. The complex wave image at vibration frequency was then used for calculating intensity speckle patterns. For the analysis, a region of interest corresponding to the location of the scatterers was chosen. The normalized intensity was determined according to Eqs. (5) and (6). The normalized intensity was then binned into 30 equally spaced bins between $[\max(I)/100, \max(I)]$ and fitted to Eq. (9). Multiple shear wave scattering was simulated employing a method introduced in [24] for soft cylinders (radius $R = 1$ mm, 40 % filling fraction, 1.8 kPa matrix and 0.18 kPa cylinders) and rigid cylinders ($R = 1.5$ mm, 4 kPa matrix and filling fraction 7 %) at an external vibration frequency of 100 Hz.

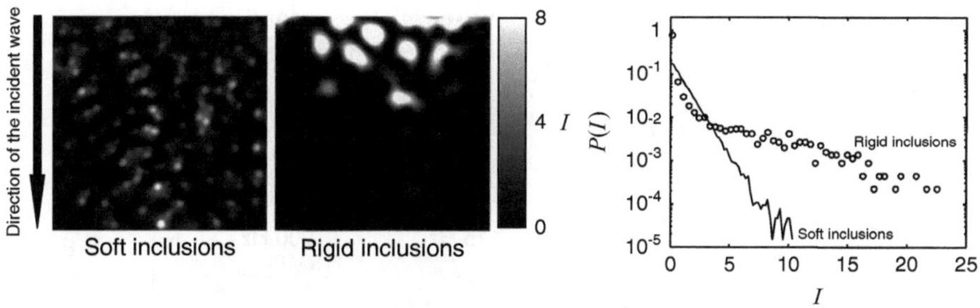

Fig. 4 Comparison of the intensity distributions of simulated SH wave scattering in a medium with soft inclusions and a medium with rigid inclusions. The graph on the *right* shows the binned intensity distributions

4 Results

Figure 4 shows results from numerical simulations of SH wave scattering in a medium with a high density of soft scatterers and a low density of rigid scatterers. Both examples show a speckled intensity distribution. It is clearly seen that in case of soft inclusions significant intensity variations occur throughout the image. In contrast, in case of rigid inclusions intensity is accumulated in regions close to the source, indicating strong suppression of shear wave propagation due to shear wave scattering. This is also reflected by the distribution of intensity shown on the right hand side of the same figure. The exponential decay towards large intensities suggests diffusive wave transport in case of soft inclusions. Instead, a stretched exponential distribution in case of rigid inclusions indicates that wave diffusion has slowed down. The measured wave images for the SH wave are shown in Fig. 5. It is seen that at low frequencies, where λ is on the order of L, wave intensity is confined near the source. In this regime, the glass cylinders resemble a homogeneous block impenetrable for shear waves. In contrast, above 50 Hz clear signatures of scattering are visible. Almost no ballistic wave front is observed at any frequency indicating that the glass cylinders scatter the shear wave very efficiently. The experiments on SV wave scattering (Fig. 6) show speckle patterns in each in-plane component similar to the out-of-plane component, represented by the SH wave scattering. The similarity between SV and SH wave scattering is even better seen in case of the normalized intensity patterns as shown in Fig. 7. Speckles seen in Fig. 7 are caused by multiple scattering since at each scattering event part of the energy of the incident wave is distributed among scattered and refracted waves which in turn interfere with the remainder of the incident wave field. Both SH and SV waves show similar speckles with respect to size and average penetration depth in the direction of shear wave propagation. Figure 8 shows the binned intensities together with the fits according to Eq. (9). The conductance was determined with $g = 2, 2, 0.3, 0.45, 0.25, 0.45, 0.85, 0.55$ for SH waves and $g = 2, 0.55, 0.3, 0.35, 0.55, 0.2, 0.25, 0.2$ for SV waves from $f = 25$–200 Hz,

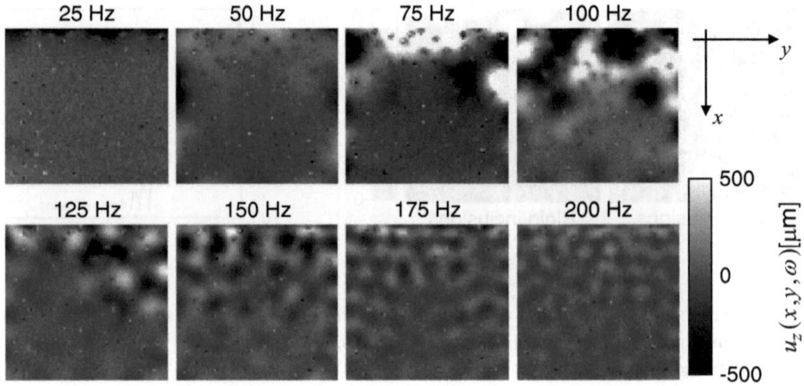

Fig. 5 Transverse slice showing measured SH wave images (for information about the coordinate axes consult Fig. 3 and the methods section)

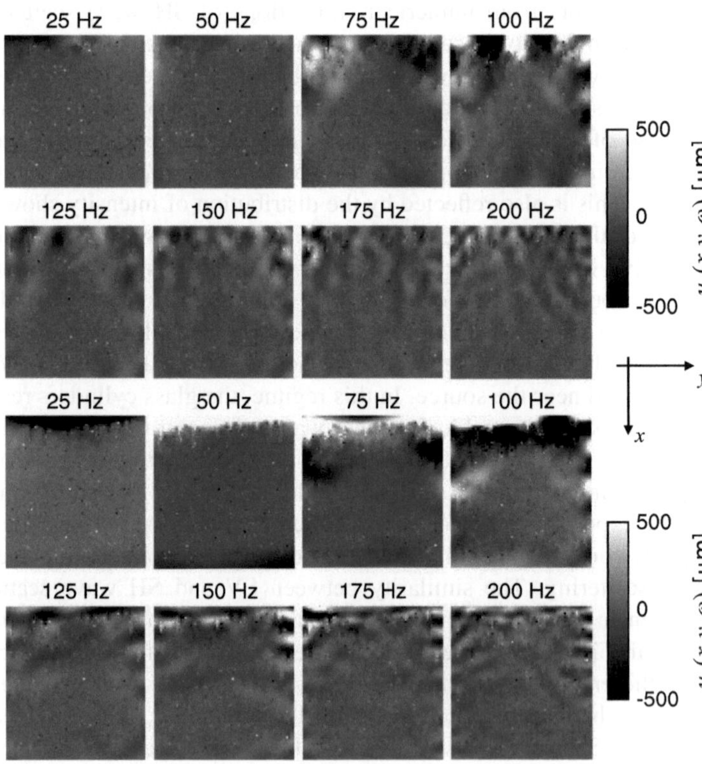

Fig. 6 Sagittal slice showing measured wave images of the SV wave components (for information about the coordinate axes consult Fig. 3 and the methods section)

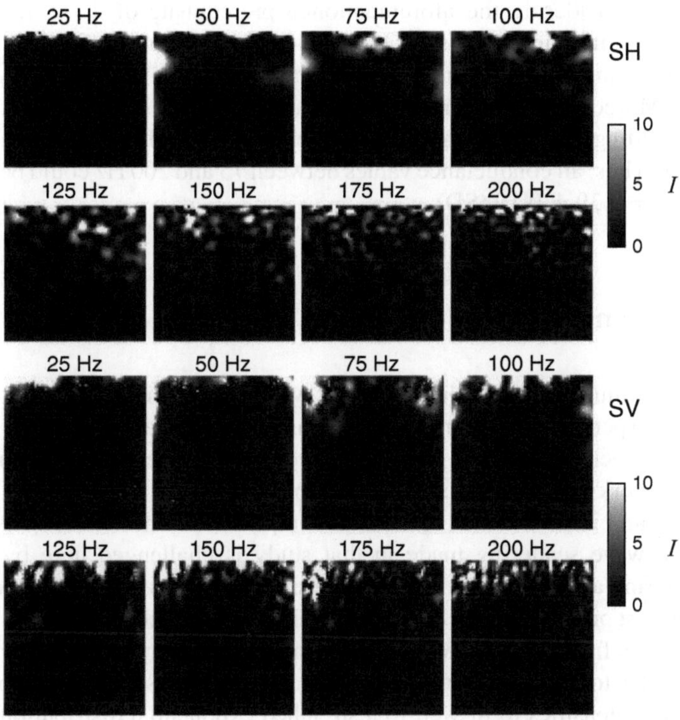

Fig. 7 Normalized intensities of the *SH* and *SV* waves according to Eqs. (5) and (6)

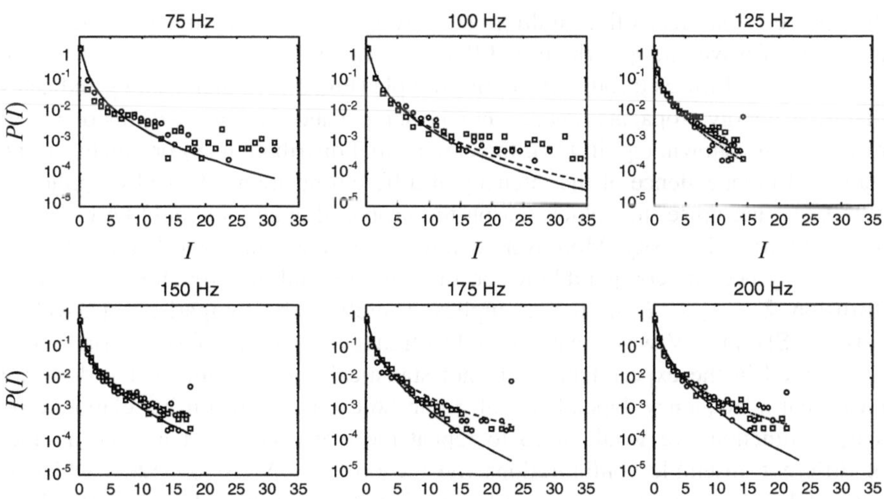

Fig. 8 Intensity distributions for the SH waves (*open squares*) and the SV waves (*open circles*). The *solid lines* and *dashed lines* correspond to fits by Eq. (9) of SH wave intensity and SV wave intensity, respectively

respectively. Considering the aforementioned prerequisite of $\lambda < l_s < L$ for observing wave diffusion, consistent g-values are only deducible for $f > 50$ Hz. In this frequency range both SH and SV wave conductance are insignificantly different (p = 0.13). Moreover, there were no significant correlations with linear regression functions yielding 3 ms slope and -1.3 ms slope for SH and SV wave conductance, respectively. Thus, all conductance values between 75 and 200 Hz could be averaged to a mean $g = 0.39 \pm 0.19$ (SD).

5 Discussion and Conclusions

In this study multiple scattering of shear waves in a disordered medium was investigated experimentally using motion-sensitive MR imaging. For the first time the scattering behavior of SH and SV waves was studied and compared. In a preliminary work on scatter-based MRE, a pure SH wave scenario was analyzed in phantoms and in vivo brain [15]. The assumption of minor differences between SH and SV wave scattering made in that study is challenged here by phantom experiments and an evaluation of wave intensity, which has not been done in MRE before. The first principal result of these experiments is that the intensity distribution clearly deviates from classical wave diffusion at the scatterer density investigated. Instead, similar to recent investigations of elastic networks using ultrasound [25], we found the intensities to fit well to a stretched exponential distribution given by Eq. (9). The relatively low value of the mean conductance $g = 0.4$ at frequencies above 50 Hz indicates the strong role of constructive interferences, which confine wave intensity to small regions not more than a few wave lengths away from the source. This means that multiple scattering from glass cylinders restricts the transport of wave energy, which is different from soft inclusions, where the wave energy freely diffuses through the entire sample (Fig. 2). An increasing g indicates that wave energy propagates deeper into the sample and shear wave diffusion is less restricted. As shown, e.g. in [14], the exponential distribution (8) originates in the statistical independence of the intensity at different positions. This also applies to pure Gaussian noise, for which reason the proposed analysis should be based on sufficient wave intensity. Moreover, at low frequencies, the wave length of shear waves may become comparable to the sample size, and the prerequisite for wave diffusion, $\lambda < l_s < L$, no longer applies. Therefore, the comparison of g-values between SH and SV waves was limited to frequencies above 50 Hz. Variations of g observed in the experiments were not statistically significant, which does not mean that a frequency dependence of g can be excluded. To further elaborate on a $g(f)$-function, we would need to repeat the experiments for a large number of different ensemble configurations (on the order of $N > 60$, [26]) which is beyond the scope of this paper. Another confounding variable which is particularly important in soft biological tissue is wave damping. Dissipation interferes with scattering by decreasing the mean free path, similar to the effect of heterogeneity. However, in gel phantoms, the typical decay length associated with dissipative loss is at least on the order of the sample size and thus of only minor importance in this study. We therefore conclude that the deviation from an exponential distribution is predominantly due to constructive interference of multiply scattered shear waves. It remains to be determined by future investigations how far the proposed conductance parameter g can be related to structure-inherent properties of living tissue. Given the well-known inversion-related artifacts in dynamic elastography, the method of intensity speckle MRE introduced here might improve the reproducibility of in vivo mechanical parameter evaluation.

In conclusion, multiple scattering of SH and SV waves was studied for the first time using motion-sensitive MRI. The results show that both shear wave modes yield similar intensity speckle patterns with characteristic deviations from classical wave diffusion. A model was employed relating the intensity distribution to a length scale on the order of tissue heterogeneities. Therefore, our results provide a perspective for applying intensity speckle MRE in vivo.

References

1. Muthupillai, R., Lomas, D.J., Rossman, P.J., Greeenleaf, J.F., Manduca, A., Ehman, R.L.: Magnetic Resonance Elastography by Direct Visualization of Propagating Acoustic Strain Waves. Science. **269**, 1854–1857 (1995)
2. Plewes, D.B., Betty, I., Urchuk, S.N., Soutar, I.: Visualizing Tissue Compliance with MR Imaging. J. Magn. Reson. Imaging, **5**, 733–738 (1995)
3. Dresner, M.A., Rose, G.H., Rossman, P.J., Muthupillai, R., Manduca, A., Ehman, R.L.: Magnetic resonance elastography of skeletal muscle. J. Magn. Reson. Imaging. **13**, 269–276 (2001)
4. Papazoglou, S., Rump, J., Braun, J., Sack, I.: Shear wave group velocity inversion in MR elastography of human skeletal muscle. Magn. Reson. Med. **56**, 489–497 (2006)
5. Huwart, L., Sempoux, C., Salameh, N., Jamart, J., Annet, L., Sinkus, R., Peeters, F., ter Beek, L.C., Horsmans, Y., van Beers, B.E.: Liver fibrosis: noninvasive assessment with MR elastography versus aspartate aminotransferase-to-platelet ratio index. Radiology. **245**, 458–466 (2007)
6. Yin, M., Talwalkar, J.A., Glaser, K.J., Manduca, A., Grimm, R.C., Rossman, P.J., Fidler, J.L., Ehman, R.L.: Assessment of hepatic fibrosis with magnetic resonance elastography. Clin. Gastroenterol. Hepatol. **5**, 1207–1213.e.2. (2007)
7. Sinkus, R., Siegmann, K., Xydeas, T., Tanter, M., Claussen, C., Fink, M.: MR elastography of breast lesions: understanding the solid/liquid duality can improve the specificity of contrast-enhanced MR mammography. Magn. Reson. Med. **58**, 1135–1145 (2007)
8. Kruse, S.A., Rose, G.H., Glaser, K.J., Manduca, A., Felmlee, J.P., Jack Jr., C.R., Ehman, R.L.: Magnetic resonance elastography of the brain. Neuroimage **39**, 231–237 (2008)
9. Sack, I., Beierbach, B., Wuerfel, J., Klatt, D., Hamhaber, U., Papazoglou, S., Martus, P., Braun, J.: The impact of aging and gender on brain viscoelasticity. Neuroimage **46**, 652–657 (2009)
10. Elgeti, T., Rump, J., Hamhaber, U., Papazoglou, S., Hamm, B., Braun, J., Sack, I.: Cardiac magnetic resonance elastography. Initial results. Invest. Radiol. **43**, 762–772 (2008)
11. Park, E., Maniatty, A.M.: Shear modulus reconstruction in dynamic elastography: time harmonic case. Phys. Med. Biol. **51**(1), 3697–3721 (2006)
12. Barbone, P.E., Gokhale, N.H.: Elastic modulus imaging: on the uniqueness and nonuniqueness of the elastography inverse problem in two dimensions. Journal: Inverse Problems not PMB (!). Phys. Med. Biol. **20**, 283–296 (2004)

13. Papazoglou, S., Hamhaber, U., Braun, J., Sack, I.: Algebraic Helmholtz inversion in planar magnetic resonance elastography. Phys. Med. Biol. **53**, 3147–3158 (2008)
14. Sheng, P.: Introduction to Wave Scattering, Localization and Mesoscopic Phenomena. In: Introduction to Wave Scattering, Localization and Mesoscopic Phenomena. Springer, Berlin (2006)
15. Papazoglou, S., Xu, C., Hamhaber, U., Siebert, E., Bohner, G., Klingebiel, R., Braun, J., Sack, I.: Scatter-based magnetic resonance elastography. Phys. Med. Biol. **54**, 2229–2241 (2009)
16. Lagendijk, Ad., van Tiggelen B.A.: Resonant multiple scattering of light. Phys. Rep. **270**, 143–215 (1996)
17. van Rossum, M.C.W., Nieuwenhuizen, Th.M.: Multiple scattering of classical waves: microscopy, mesoscopy, and diffusion. Rev. Mod. Phys. **71**, 313–371 (1999)
18. Zhang, Z.Q., Jones, I.P., Schriemer, H.P., Page, J.H., Weitz, D.A., Sheng, P.: Group Velocity in Strongly Scattering Media. Phys. Rev. E. **60**, 4843–4850 (1999)
19. Tourin, A., Fink, M., Derode, A.: Multiple scattering of sound. Waves Random Media. **10**, R31–R60 (2000)
20. Aki, K., Richards, P.G.: Quantitative Seismology. In: Quantitative Seismology. University Science Books, Sausalito (2002)
21. Nieuwenhuizen, Th.M., van Rossum, M.C.W.: Intensity Distributions of Waves Transmitted through a Multiple Scattering Medium. Phys. Rev. Lett. **74**, 2674–2677 (1995)
22. van Tiggelen B.A.: In: Fouque, J.P. (ed.) Diffusive Waves in Complex Media. Kluwer Academic, Dordrecht (1999)
23. Klatt, D., Hamhaber, U., Asbach, P., Braun, J., Sack, I.: Noninvasive assessment of the rheological behavior of human organs using multifrequency MR elastography: a study of brain and liver viscoelasticity. Phys. Med. Biol. **52**, 7281–7294 (2007)
24. Biwa, S., Yamamoto, S., Kobayashi, F., Ohno, N.: Independent scattering and wave attenuation in viscoelastic composites. Int. J. Solids Struct. **41**, 435–457 (2004)
25. Hu, H., Strybulevych, A., Page, J.H., Skipetrov, S.E., van Tiggelen, B.A.: Localization of Ultrasound in a Three-Dimensional Elastic Network. Nat. Phys. **4**, 945–948 (2008)
26. Derode, A., Tourin, A., Fink, M.: Random Multiple Scattering of Sound, I. Coherent and Ballistic Wave. Phys. Rev. E. **64**, 036605 (2001)

Part IV
Diffusion Weighted MRI Visualization

A Comparative Analysis of Dimension Reduction Techniques for Representing DTI Fibers as 2D/3D Points

Xiaoyong Yang, Ruiyi Wu, Ziáng Ding, Wei Chen, and Song Zhang

Abstract Dimension Reduction is the process of transfering high-dimensional data into lower dimensions while maintaining the original intrinsic structures. This technique of finding low-dimensional embedding from high-dimensional data is important for visualizing dense 3D DTI fibers because it is hard to visualize and analyze the fiber tracts with high geometric, spatial, and anatomical complexity. Color-mapping, selection, and abstraction are widely used in DTI fiber visualization to depict the properties of fiber models. Nonetheless, visual clutters and occlusion in 3D space make it hard to grasp even a few thousand fibers. In addition, real time interaction (exploring and navigating) on such complex 3D models consumes large amount of CPU/GPU power. Converting DTI fiber to 2D or 3D points with dimension reduction techniques provides a complimentary visualization for these fibers. This chapter analyzes and compares dimension reduction methods for DTI fiber models. An interaction interface augments the 3D visualization with a 2D representation that contains a low-dimensional embedding of the DTI fibers. To achieve real-time interaction, the framework is implemented with GPU programming.

1 Introduction

Diffusion tensor imaging (DTI) elucidates the microscopic anatomical properties of biological tissues (e.g., heart or brain) in vivo by measuring the Brownian motion of water molecules. The water diffusion can be mathematically summarized

X. Yang (✉) · R. Wu · S. Zhang
Department of Computer Science and Engineering, Mississippi State University, Mississippi, USA
e-mail: xy46@msstate.edu; rw802@msstate.edu; szhang@cse.msstate.edu

Z. Ding · W. Chen
State Key Laboratory of CAD & CG, Zhejiang University, Hangzhou, China
e-mail: dingziang@cad.zju.edu.cn; chenwei@cad.zju.edu.cn

D.H. Laidlaw and A. Vilanova (eds.), *New Developments in the Visualization and Processing of Tensor Fields*, Mathematics and Visualization,
DOI 10.1007/978-3-642-27343-8_9,

by a diffusion tensor field. By tracking the trajectories of the fastest diffusion in a diffusion tensor field with tractography [3], diffusion tensor imaging can be represented with a set of fiber tracts or three-dimensional pathways. Tractography has been widely used in the visualization and analysis of DTI datasets [19, 22, 31]. Most methods in tractography use sample points on a DTI fiber and the line segments between these sample points to represent a DTI fiber.

It is challenging to explore and analyze DTI fiber models in 3D space because the complexity of the 3D geometry causes visual clutters. Interactively exploring and selecting DTI fibers is helpful for visual inspection or statistical analysis. However, it is difficult to interact with the complex DTI fiber models. Better visual representations and more convenient user interfaces are needed to solve this problem. Interactive selection schemes [1, 6], dynamic query [28], geometric simplification [10, 14], color-mapping [12, 13], texture patterning of fiber dissimilarity [17], and uncertainty visualization [15] are some of the existing solutions that attempt to solve the problem by employing new visual forms or novel interaction methods. However, most of them operate in 3D space, where the geometry is often occluded after being projected to a 2D viewport.

Dimension reduction can help alleviate this kind of problems. If we can find a suitable transform which transfers the high-dimensional data into 2D or 3D points and maintains the original intrinsic structures, the geometry occlusion of the high-dimensional fibers can be reduced. Dimension Reduction is the process of transfering high-dimensional data into low dimensions while maintaining the original intrinsic structures. Generally speaking, the dimension reduction methods can be categorized into linear and nonlinear embedding methods. Linear embedding methods are useful when the samples on the original high-dimensional manifolds have linear relationships. With dimension reduction, we can linearly reconstruct the low-dimensional embeddings from high-dimensional data. Multidimensional Scaling (MDS) and Principal Component Analysis (PCA) are two of the linear embedding methods. However, linear embedding methods do not consider the intrinsic nonlinearity within data. In this case, the relation among data points is best represented using nonlinear embedding methods. Locally Linear Embedding (LLE) and Isometric Feature Mapping (IsoMAP) are two of the methods for nonlinear embedding.

This chapter will be organized as follows: Sect. 2 reviews the background information about dimension reduction, DTI representing, exploration and interaction methods. Section 3 describes four embedding methods in detail. Section 4 discusses dimension reduction experiments on DTI models and the results. Section 5 demonstrates the interface for DTI fiber exploration and data analysis with projected points. Section 6 concludes this chapter.

2 Background

Dimension reduction is widely used for revealing the latent low-dimensional patterns from the high-dimensional data. Numerous studies in this field have been reported, and dozens of algorithms and methods for dimensional reduction have been developed. Multidimensional Scaling(MDS) [5] takes the dissimilar matrix among high-dimensional samples as input and attempts to preserve their dissimilar information as much as possible. Principal Component Analysis(PCA) [26] assumes that the dimensions are correlated variables and preserves the dimensions representing the largest variabilities in the original data. Isometric Feature Mapping(IsoMAP) [29] uses geodesic distances as the dissimilarity information. It is a generalization of the MDS. Locally Linear Embedding(LLE) [25] is a global non-linear embedding method and it constructs local information linearly. Factor analysis(FA) [21] assumes that the observed variables depend on the linear combinations of some unknown factors. Factor analysis uncovers these interdependencies to reduce the dimensions of the data. Inspired by the idea of Johnson-Lindenstrauss lemma [2], random projection [23] uses a random matrix to project the high dimensional data into low dimension.

DTI fibers are usually integrated in the direction of the longest eigenvectors in a tensor field [3]. Streamlines [19] and streamtubes [31] are widely used to represent and display DTI fiber models. Geometrically, the shape of a DTI model composed of fiber tracts can be simplified in visual abstract forms, such as wrapped streamlines [14], topological simplification [27], or by using hierarchical principal curves [10]. A novel set of interaction techniques introduced in Sherbondy et al. [28] allows for exploring brain connectivity and interpreting pathways. In [28], a simple and flexible query language is used to enable the neuroscientists to perform the key operations: the placement and manipulation of box-shaped or ellipsoid-shaped regions. In Blaas et al. [6], a similar selection scheme is also presented. The approach is proven to be highly reproducible for fractional anisotropy (FA) calculated over the fiber bundles in clinical studies. To differentiate pathways visually, different pattern styles are defined to encode the local dissimilarity information of DTI fibers, and an online navigation tool is implemented for fiber connectivity study [17]. Coloring scheme could be used to illustrated the dissimilarities within a group of fibers in cross sections [12, 13], which makes it easy to discern clusters by colors. Other features of the fibers, such as the uncertainty arising from noise and partial volume effects [15] could also be visualized. Most of these schemes focus on limiting the amount of information in 3D models to be visualized and explored before excessive visual clutter occurs. Dimension reduction has been applied to data samples such as DTI fibers for clustering [4, 7, 8]. Only recently dimension reduction methods have been applied to DTI fibers to embed them as points in 2D for visualization and exploration [11, 18].

3 Dimension Reduction Methods

Today, the data for scientific research is usually high-dimensional and multivariate. However, not all of the variables are equally important. In addition, with the increase of the dimensions, there would be exponential growth in time complexity for information fetching and analysis. Generally speaking, there are low-dimensional structures hidden behind the very high-dimensional data for many research problems. Even when the latent relations can be found from high-dimensional data, there is still the problem of how to visualize the relations efficiently and accurately. A great amount of research has been done concerning low-dimension embedding, and dozens of embedding methods have been proposed.

Embedding techniques can be categorized into two groups: linear embedding methods (e.g. MDS, PCA) and nonlinear embedding methods (e.g. LLE, IsoMAP). Linear embedding means that each dimension from the embedding result is generated by linear combination of the original dimensions; the nonlinear embedding methods combine the original dimensions nonlinearly to generate new dimensions. The linear embedding methods work well if the relation among the high-dimensional data is linear. If the relation among original data is nonlinear, nonlinear embedding methods are more capable of preserving the geometry of nonlinear manifolds.

In this section, we briefly review MDS, LLE, PCA and IsoMAP for embedding high-dimensional DTI fiber models into 2D/3D points.

3.1 Multidimensional Scaling (MDS)

Suppose that there is a set of points $X = \{X_0, X_1, \ldots, X_n\}$ in a high-dimensional space, and the dissimilarity information (e.g. pair-wise distance) is known as $\delta_{ij}\,(1 \leq i, j \leq n)$. The dissimilarity information δ_{ij} means the distance between two DTI fiber [31]. The high-dimensional relation is described with a dissimilarity matrix D:

$$D = \begin{bmatrix} \delta_{11}, \delta_{12}, \ldots, \delta_{1n} \\ \delta_{21}, \delta_{22}, \ldots, \delta_{2n} \\ \vdots \quad \vdots \quad \cdots \quad \vdots \\ \delta_{n1}, \delta_{n2}, \ldots, \delta_{nn} \end{bmatrix}$$

The goal of MDS is to reconstruct points $Y = \{Y_0, Y_1, \ldots, Y_n\}$ in a low-dimension space to satisfy the goal function of MDS:

$$f_{MDS}^{opt} = min \Sigma_{ij} |\delta_{ij} - ||Y_i - Y_j||| \; for\ all\ i, j \in [1, n] \tag{1}$$

where $||.||$ denotes the Euclidean norm. $|.|$ denotes the absolute value. There are two major steps for MDS:

1. Compute the pair-wise dissimilarities matrix. In the experiments of this chapter, the dissimilarity information is the distance between two DTI fibers. Suppose that there are two fiber tracts, and there are p and q points on the two fibers respectively. It would take $O(pq)$ steps to calculate the dissimilarity on each DTI fiber. Moreover, the dissimilarity matrix is n by n. Therefore, the time complexity

for this step is $O(pqn^2)$. If the dissimilarity information is pre-calculated before the experiment, the time complexity is $O(n^2)$ for loading this matrix.

2. Compute the embedding results. This step needs an eigenvector analysis of the n by n matrix, resulting in a time complexity of $O(n^3)$. However, Chalmers et al. [9] proposed an algorithm which reduced the time complexity into $O(n^2)$. Therefore, the time complexity for this step is $O(n^2)$.

In summary, the time complexity for MDS is $O(n^2)$ assuming pre-calculated dissimilarity information.

3.2 Locally Linear Embedding (LLE)

LLE is a global non-linear embedding method that constructs local information linearly. Suppose there is a set of points $X = \{X_0, X_1, \ldots, X_n\}$ in a high-dimensional space, LLE preserves the local structure of the sampled points by using linear correlation coefficients to reconstruct data points from their k nearest neighbors. For reconstruction, the weight value W_{ij} for the neighbors is computed to satisfy the goal function of LLE:

$$f_{LLE}^{opt} = min \Sigma_i \Sigma_j (X_i - W_{ij} X_j)^2 \; for \; all \; i \in [1, n], \; j \in [1, k]; \tag{2}$$

There are three major steps for LLE:

1. Find k nearest neighbors for each X_i. There are n points, and for each point, the distance from X_i to every other point X_j needs to be computed. Therefore, the time complexity for this step is $O(n^2)$.
2. Solve for reconstruction weights W for X_i. A matrix of k by k consisting of all neighbors of X_i is created and manipulated for the reconstruction weights calculation. For this step, the cost would be $k \times k \times n$. If k $\ll$ n, the time complexity is $O(n)$; otherwise, the time complexity is $O(n^3)$.
3. Compute embedding coordinates using weights W. Similar to step 2, a matrix of l by l will be created and employed for the computation for each original point. Thus, if $l \ll n$, the time complexity is $O(n)$, otherwise, the time complexity is $O(n^3)$.

In summary, the time complexity of LLE is mainly based on the selection of k value. When $k \ll n$, the LLE reaches its best case run time of $O(n^2)$, and the worst case for LLE is $O(n^3)$.

3.3 Principal Component Analysis (PCA)

PCA is an eigenvector based linear embedding method designed with the assumption that there are linear correlations among different dimensions. In PCA, the first

step is to construct the covariance matrix W of different dimensions. Then, the linear projections of greatest variance from the greater eigenvectors are computed. Smaller eigenvectors are removed because greater eigenvectors of the data covariance matrix contains more information than the smaller ones do.
The major steps for PCA are:

1. Calculate the covariance matrix W. This operation needs operations of $O(n^2)$
2. Compute the embedding coordinate. This also needs $O(n^2)$ operations.

The time complexity for PCA is $O(n^2)$ [24].

3.4 IsoMap (Isometric Feature Mapping)

MDS performs low-dimensional embedding based on the distance between high-dimensional points. As a generalization of MDS to non-linear manifolds, IsoMAP uses geodesic distances. First, for any point in high dimension, its k closest neighbors are found by the Dijkstra algorithm, and the distances are calculated. Then, a weighted graph is constructed based on the neighboring and distance information. Next, the geodesic distance is calculated. The geodesic distance is defined as the sum of edge weights along the shortest path between two points (the Dijkstra's algorithm is used to find the shortest path). Finally, MDS is used for the embedding.

There are three major steps for IsoMAP:

1. Find the k nearest neighbors for all points. For this step, the distance from each point to the current is calculated, and the smallest k neighbors are found. The time complexity for this step is $O(n^2)$
2. Find the geodesic distances for all pairs of points in the graph. By taking different algorithms and data structures to calculate the geodesic distance, the time complexity is different. The Dijkstra algorithm is employed here for discussion. The edge number in the map for our problem is E = $\frac{n(n-1)}{2}$, and the time complexity for Dijkstra algorithm implemented with binary heap is $O(|E| + |V|log|V|)$, where E is the edge number and V is the vertex number. For step2 of IsoMAP, the time complexity is $O(\frac{n(n-1)}{2} + nlogn) = O(n^2)$
3. Construct the embedding. The time complexity for this step is $O(n^2)$.

In summary, the time complexity for IsoMAP is $O(n^2)$.

4 Experiment

We have conducted experiments for embedding DTI fibers as 2D and 3D points with the aforementioned methods. We will discuss the data, methods, and results of the experiments followed by a discussion in this section.

4.1 Data

Two normal human brain DTI models are used to generate four brain datasets (B1, B2, B3, and B4) and one normal porcine heart model is used to generate two heart datasets (H1 and H2) for the experiment. B2, B4, and H2 are generated from B1, B3, and H1 by culling fiber tracts whose average curvature is smaller than 0.3. The purpose is to explore the relation between fiber number and information loss during embedding. The voxel size is 1.7 mm × 1.7 mm × 1.7 mm for the brain datasets and 1.17 mm × 1.17 mm × 2.4 mm for the porcine hearts datasets. The DTI models were generated with a seeding density of 1 seed per 4 by 4 voxels in the data volume. The fiber numbers are 1248, 192, 1622, 458 for B1, B2, B3, B4 and 1066, 347 for H1 and H2 respectively.

4.2 Method

Two individual experiments are carried out. The first experiment is to embed the high-dimensional models into 2D points. The second experiment is to embed the high-dimensional models into 3D points. In both experiments we compare the embedding errors from high dimension into low dimension using different embedding methods on the DTI datasets. To measure the embedding error, Kruskal in [20] proposed a badness-of-fit standard called stress. The stress is defined as:

$$Stress = \sqrt{\frac{\sum_i \sum_j (D_{ij} - \mathring{D}_{ij})^2}{\sum \sum D_{ij}^2}}, \tag{3}$$

where D_{ij} is the dissimilar value before embedding, and $\mathring{D}_{ij}$ is the dissimilar value after embedding. Assume δ_{ij} is distance between the *ith* and *jth* points in high dimension. In our experiment, it is the distance between two DTI fibers [31]. n is the total number of points. x_i, x_j, y_i, y_j, z_i, and z_j indicate the x, y ,z values of the points after embedding. Before dimension reduction, the sum distance is $sum_1 = \sum_{i,j} \delta_{i,j}$ while after the dimension reduction, the sum distance between all pair of point are $sum_2 = \sum_{j=1}^{j=n} \sum_{i=1}^{i=n} \sqrt{(x_i - x_j)^2 + (y_i - y_j)^2}$ if the embedding results are 2D points. Otherwise, $sum_2 = \sum_{j=1}^{j=n} \sum_{i=1}^{i=n} \sqrt{(x_i - x_j)^2 + (y_i - y_j)^2 + (z_i - z_j)^2}$ if the embedding results are 3D points; $R = sum_1/sum_2$. The usage of R is to make sure that the change of scale in distance caused by the embedding method is accounted for. We define the embedding error for our experiments as:

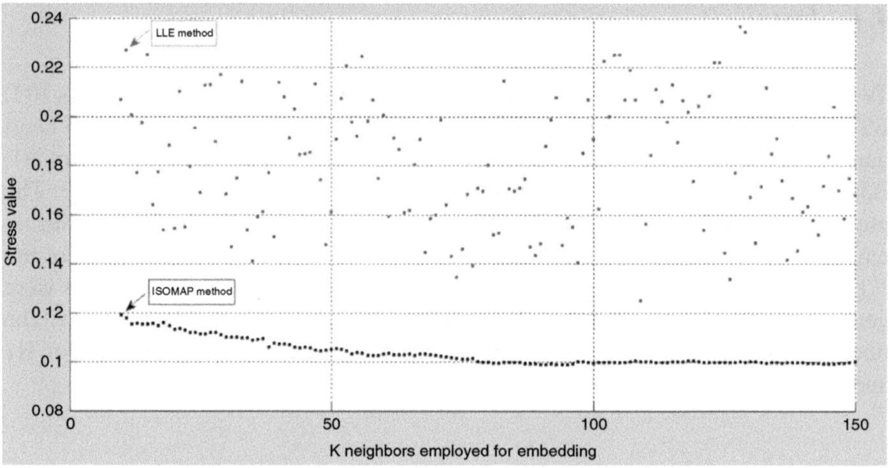

Fig. 1 Stress values for B1 from LLE and IsoMAP embeddings with increasing k values. *Red points* are results from LLE, *blue points* are results from IsoMAP

$$Stress_{2D} = \sqrt{\frac{\sum_{j=1}^{j=n}\sum_{i=1}^{i=n}(\frac{\delta_{i,j}}{R} - \sqrt{(x_i - x_j)^2 + (y_i - y_j)^2})^2}{\sum_{j=1}^{j=n}\sum_{i=1}^{i=n}\frac{\delta_{i,j}^2}{R^2}}} \quad (4)$$

$$Stress_{3D} = \sqrt{\frac{\sum_{j=1}^{j=n}\sum_{i=1}^{i=n}(\frac{\delta_{i,j}}{R} - \sqrt{(x_i - x_j)^2 + (y_i - y_j)^2(z_i - z_j)^2})^2}{\sum_{j=1}^{j=n}\sum_{i=1}^{i=n}\frac{\delta_{i,j}^2}{R^2}}} \quad (5)$$

Both LLE and IsoMAP need a neighborhood size as parameter k. Different k values will lead to dramatically different embedding results and stress values. There appears to be a limited amount of work that has been done before about the most reasonable k value for a manifold. When k value is too small, large continuous shapes in high dimension will become small discrete shapes; when k value is too large, the local topology is not correctly reflected. Furthermore, for different DTI models and different embedding methods (LLE/IsoMAP), the best suitable k values are different. Figure 1 is the test result about the different stress values by LLE and IsoMAP with increasing k on B1. From Fig. 1, we see that the stress values for LLE fluctuate when the k values are increased. For IsoMAP, the stress values become smaller when the k values are increased because of shorter geodesic distances obtained from the Dijkstra algorithm.

4.3 Results

Different k values (30, 70, 110, 150, and 190) are used to generate a group of testing results for embedding DTI fibers to 2D points using LLE and IsoMap. The results are shown in Fig. 2. We test different k values around 110, and select $k = 106$, where LLE method achieves the minimum stress value in Fig. 1, for the stress value experiments below. In [30], Xie et al. use the LLE method for ear recognition. In their paper, similar strategy about the k value selection is used for LLE.

Tables 1 and 2 show the stress values for MDS, LLE, PCA and IsoMAP embedding in 2D and 3D. B1 and B2 are from brain model 1, while B3 and B4 are from brain model 2. H1 and H2 are from the porcine heart model. The difference between the related values is that the former one is from a dataset without culling fiber tracts with similar curvature, while the latter one culls fiber tracts whose average curvature is less than 0.3.

Tables 3 and 4 show the run time for completing the experiments on 2D and 3D embeddings.

4.4 Discussion on Dimension Reduction

From the experiments above, some general trends are discovered for DTI models embedding:

1. Embedding the high-dimensional manifolds into 3D space can lead to smaller stress values than into 2D. Using one more dimension, more information will be preserved. The navigation and rotation in the 3D space could also avoid the problem of occlusion.
2. The k value selection (for LLE and IsoMAP) is still an open problem.The selection of k value should depend on the structure of a given manifold. However, the topology of DTI fibers is not clear. For different models (brains/hearts), different sample rates (dense/sparse), and different embedding methods (LLE/IsoMAP), the best k value is different.
3. The 3D stress values for MDS and IsoMAP are similar.
4. The stress values for Brain_1 and Brain_2 are similar, while the stress values for the heart DTI models are relatively larger. Anatomically, the structures of human brains and porcine hearts are quite different. Connected by the corpus callosum, the left and right cerebral hemispheres in the DTI models are symmetrical. However, the hearts have layers of muscle structure, and most of the tracts in DTI models have a helical style. The structural difference between them might be the reason for lower stress values of brain models than the heart models.
5. We believe that the DTI fibers have a nonlinear structure, which means that they can be best represented by some nonlinear embedding methods. However, from the comparison of the stress values, no significant differences are found between the linear and nonlinear methods. In [16], Goldberg et al. show that for some

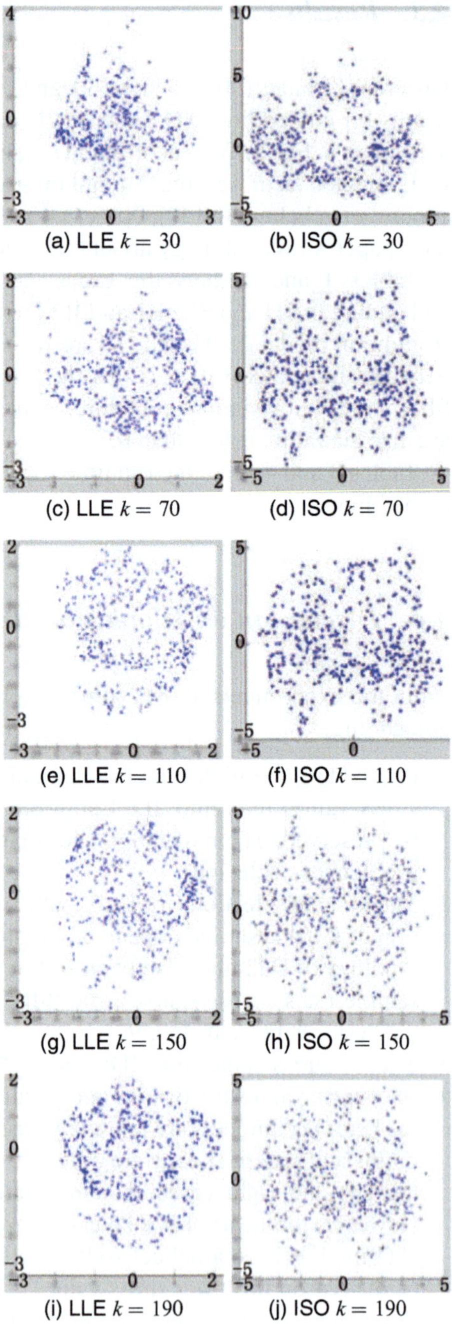

(a) LLE $k = 30$ (b) ISO $k = 30$
(c) LLE $k = 70$ (d) ISO $k = 70$
(e) LLE $k = 110$ (f) ISO $k = 110$
(g) LLE $k = 150$ (h) ISO $k = 150$
(i) LLE $k = 190$ (j) ISO $k = 190$

Fig. 2 *First column*: LLE 2D embedding results. *Second column*: IsoMAP 2D embedding results

Table 1 Stress values for 2D embeddings

Category	Methods	B1	B2	B3	B4	H1	H2
Linear	MDS	0.06	0.07	0.10	0.04	0.13	0.09
Linear	PCA	0.11	0.12	0.11	0.12	0.22	0.24
Nonlinear	LLE	0.14	0.10	0.14	0.13	0.32	0.30
Nonlinear	IsoMAP	0.11	0.10	0.10	0.05	0.36	0.22

Table 2 Stress values for 3D embeddings

Category	Methods	B1	B2	B3	B4	H1	H2
Linear	MDS	0.02	0.03	0.03	0.02	0.04	0.03
Linear	PCA	0.03	0.05	0.05	0.04	0.08	0.13
Nonlinear	LLE	0.06	0.03	0.06	0.05	0.08	0.07
Nonlinear	IsoMAP	0.03	0.03	0.02	0.02	0.09	0.06

Table 3 Run time for 2D embeddings (in seconds)

Category	Methods	B1	B2	B3	B4	H1	H2
Linear	MDS	81.00	2.92	143.05	11.36	60.04	7.49
Linear	PCA	54.67	1.11	105.38	5.83	37.97	3.12
Nonlinear	LLE	227.24	3.11	729.57	31.54	145.25	9.76
Nonlinear	IsoMAP	60.23	1.66	111.63	6.92	42.10	3.65

Table 4 Run time for 3D embeddings (in seconds)

Category	Methods	B1	B2	B3	B4	H1	H2
Linear	MDS	86.70	3.81	155.29	13.08	61.64	8.15
Linear	PCA	57.99	1.46	106.66	6.73	40.97	3.18
Nonlinear	LLE	228.13	3.56	768.68	34.74	158.21	15.68
Nonlinear	IsoMAP	61.86	1.69	114.12	7.43	42.28	3.68

very simple manifolds, if some conditions are violated, the embedding methods (e.g., LLE) cannot recover the underlying manifolds. This is the same for the DTI models. The DTI models are high-dimensional and their structures vary model by model. It cannot be guaranteed that a specific embedding method can always work on the different DTI models.

5 User Interface

We designed a user interface for the ease of exploring DTI fibers. The exact geometry in a standard 3D view (DTI fiber models) and a simplified view of fibers (2D/3D embedded points) are shown simultaneously, and the user interacts with the simplified view to reduce the workload. Also, the interface program can be used in other aspects of DTI fiber study such as interactive fiber tract selection. Our method simplifies the high-dimensional DTI fibers and enhances interaction, visualization, and exploration with these fibers.

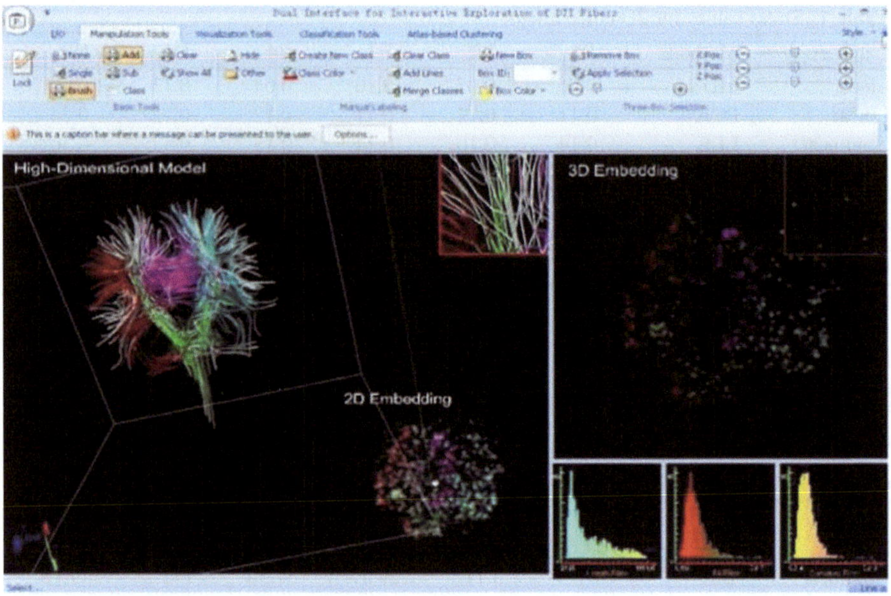

Fig. 3 The basic layout of the user interface

The layout shown in Fig. 3 enables efficient manipulation, navigation, and quantitative analysis of DTI fiber tracts. Three major components are shown. On the left side, the fiber tracts are shown and interactions such as rotation, lens viewing, coloring, slicing, and selection are supported. At the right bottom of the left side is the 2D embedding results; on the right side, a 3D embedding of the DTI fibers is shown. There are several other views to enable interactive filtering the fiber tracts and make numerical analysis.Free fiber selection and multiple-box ROIS are supported. An advantage of our interface is its capability of describing the fiber model in multiple aspects. To enable interactively exploring the properties of the DTI models, additional histogram views are added. Properties of the DTI fibers, such as the fiber length, the average linear anisotropy (LA), the average fractional anisotropy (FA), and the average curvature along each fiber tract are shown in separate views. The users can select groups of fibers based on their observations and the histograms. In Figs. 3 and 4, the selected fiber tracts within a single group share the same color.

Figure 4 includes examples of using the embedding methods to help with brain DTI fiber model study. Visualization of the heart and brain models could be helpful for research. However, these models consist of many spatially close fibers. As shown in Fig. 4, the high-dimensional manifolds are embedded into 2D/3D space as points. This low-dimension embedding maintains the original DTI models' structural information with an uncluttered 2D/3D representation of the data, which reduces the user time and mental workload in recognizing 3D DTI fibers. Although

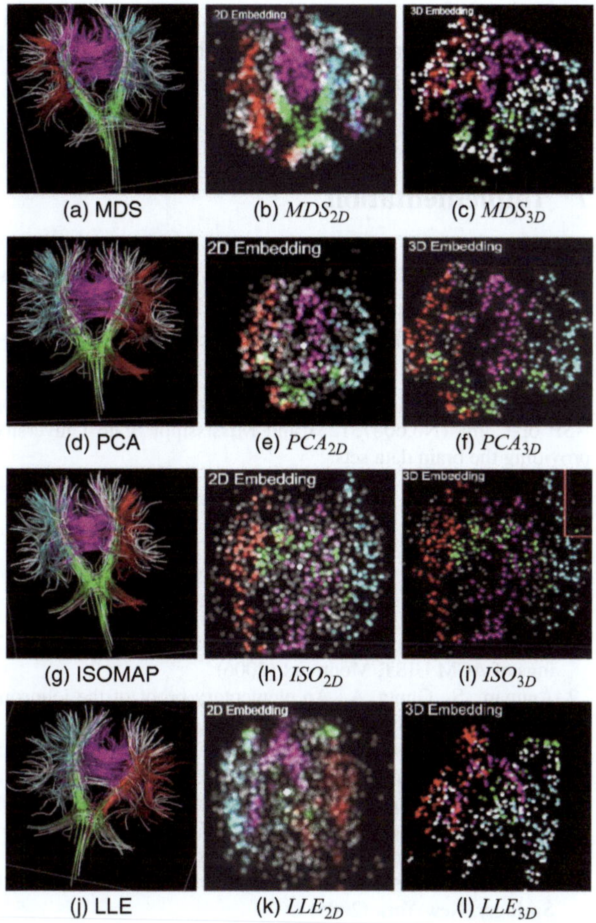

(a) MDS (b) MDS_{2D} (c) MDS_{3D}
(d) PCA (e) PCA_{2D} (f) PCA_{3D}
(g) ISOMAP (h) ISO_{2D} (i) ISO_{3D}
(j) LLE (k) LLE_{2D} (l) LLE_{3D}

Fig. 4 *First column*: human brain DTI models with 458 fiber tracts; *second column*: the 2D points representation of the first column. *Third column*: the 3D points representation of the first column. For each row from top to down, the figures are for MDS, PCA, IsoMAP and LLE respectively

the 2D points reduce the user time and mental workload in recognizing 3D DTI fibers, it will lose much information from the DTI fibers. A 3D points embedding is shown in the last column. Although 3D points also creates clustering and occlusion, we can use rotation to navigate in the 3D space for a better understanding. From previous tables, 3D embedding preserves more proximity than the 2D embedding does.

6 Conclusions

In this chapter, we review the functionalities and the time complexities of MDS, LLE, PCA, IsoMAP and quantitatively compare different embedding methods' stress values and run time on DTI datasets. Moreover, a novel interface is designed to achieve more effective representations and visual explorations of DTI fiber models. Users can view the DTI fiber tracts and their 2D/3D embedding results simultaneously for an efficient understanding of the DTI fibers.

7 Implementation

We implemented the calculation of MDS, LLE, PCA and IsoMAP with Matlab. The PC is equipped with an Inter Core 2 Duo 2.4 GHz CPU, 4 G host memory. The user interface was implemented with Microsoft Visual C++ 2008 and OpenCV.

Acknowledgements This work is partially supported by 973 program of China (2009CB320800), NSF of China (No.60873123) and Mississippi State University. We thank Dr. David Tate for providing the brain data sets.

References

1. Akers, D.: A cooperatively designed marking interface for 3D pathway selection. In: Proceedings of ACM UIST, Montreux (2006)
2. Anupam, S., Gupta, A.: An elementary proof of the johnson-lindenstrauss lemma. Technical report 99-006, University of California, Berkeley (1999)
3. Basser, P.J., Pajevic, S., Pierpaoli, C., Duda, J., Aldroubi, A.: In vivo fiber tractography using dt-mri data. Magn. Reson. Med. **44**, 625–632 (2000)
4. Belkin, M., Niyogi, P.: Laplacian eigenmaps and spectral techniques for embedding and clustering. In: Diettrich, T.G., Becker, S., Ghahramani, Z. (eds.) Advances in Neural Information Processing Systems 14. MIT Press, Cambridge (2002)
5. Borg, I., Groenen, P.J.F.: Modern Multidimensional Scaling: Theory and Applications. Springer, New York (2005)
6. Blaas, J., Botha, C., Peters, B., Vos, F., Post, F.: Fast and reproducible fiber bundle selection in dti visualization. In: Proceedings of IEEE Visualization, Minneapolis, pp. 59–64 (2005)
7. Brun, A., Knutsson, H., Park, H.J., Shenton, M.E., Westin, C.-F.: Clustering fiber tracts using normalized cuts. In: Seventh International Conference on Medical Image Computing and Computer-Assisted Intervention (MICCAI'04), pp. 368–375, Rennes – Saint Malo. Springer, Berlin/Heidelberg (2004)
8. Brun, A., Westin, C.-F., Herberthson, M., Knutsson, H.: Fast manifold learning based on riemannian normal coordinates. In: Proceedings of the 14th Scandinavian Conference on Image Analysis (SCIA'05), Joensuu (2005)
9. Chalmers, M.: A linear iteration time layout algorithm for visualising high-dimensional data. In: IEEE Symposium on Information Visualization, San Francisco, pp. 127–132 (1996)
10. Chen, W., Zhang, S., Correia, S., Ebert, D.S.: Abstractive representation and exploration of hierarchically clustered diffusion tensor fiber tracts. Comput. Graph. Forum **27**(3), 1071–1078 (2008)
11. Chen, W., Ding, Z., Zhang, S., Brandt, A.M., Stephen Correia, S., Qu, H., Crow, J.A., Tate, D.F., Yan, Z., Peng, Q.: Exploring 3D DTI fiber-tracts with linked 2D representations. IEEE Trans. Vis. Comput. Graph. (Proc. Visualization '09) **15**(6), 1433–1440 (2009)
12. Demiralp, C., Laidlaw, D.H.: Similarity coloring of dti fiber tracts. In: Proceedings of DMFC Workshop at MICCAI, London (2009)

13. Demiralp, C., Zhang, S., Tate, D., Correia, S., Laidlaw, D.H.: Connectivity-aware sectional visualization of 3D dti volumes using perceptual flat-torus coloring and edge rendering. In: Proceedings of Eurographics, Vienna (2006)
14. Enders, F., Sauber, N., Merhof, D., Hastreiter, P., Nimsky, C., Stamminger, M.: Visualization of white matter tracts with wrapped streamlines. In: Proceedings of IEEE Visualization, Minneapolis, pp. 51–58 (2005)
15. Friman, O., Westin, C.-F.: Uncertainty in white matter fiber tractography. In: Proceedings of MICCAI, Palm Springs, pp. 107–114 (2005)
16. Goldberg, Y., Zakai, A., Kushnir, D., Ritov, Y.: Manifold learning: the price of normalization. Mach. Learn. Res. **9**, 1909–1939 (2008)
17. Jianu, D., Zhou, W., Demiralp, C., Laidlaw, D.H.: Visualizing spatial relations between 3D-dti integral curves using texture patterns. In: Proceedings of IEEE Visualization Poster Compendium, Sacramento (2007)
18. Jianu, R., Demiralp, C., Laidlaw, D.H.: Exploring 3D DTI fiber-tracts with linked 2D representations. IEEE Trans. Vis. Comput. Graph. (Proc. Visualization '09) **15**(6), 1449–1456 (2009). http://vis.cs.brown.edu/docs/pdf/g/Jianu-2009-E3D.pdf.html(pdf) http://vis.cs.brown.edu/results/bibtex/Jianu-2009-E3D.bib(bibtex: Jianu-2009-E3D)
19. Kindlmann, G.: Visualization and analysis of diffusion tensor fields. Ph.D. Thesis, University of Utah (2004)
20. Kruskal, J.B.: Nonmetric multidimensional scaling: a numerical method. Psychometrika **29**, 115–129 (1964)
21. Mardia, K.V., Kent, J.T., Bibby, J.M.: Multivariate Analysis. Probability and Mathematical Statistics. Academic, London (1995)
22. Moberts, B., Vilanova, A., vanWijk, J.J.: Evaluation of fiber clustering methods for diffusion tensor imaging. In: Proceedings of IEEE Visualization, Minneapolis (2005)
23. Vempala, S.: The Random Projection Method, volume 65 of DIMACS Series in Discrete Mathematics and Theoretical Computer Science. American Mathematical Society, Providence (2004)
24. Robila, S.: Distributed Processing of Hyperspectral Images, Proceedings ASPRS Annual Conference, Denver, CO (2004)
25. Roweis, S.T., Saul, L.K.: Nonlinear dimensionality reduction by locally linear embedding. Science **290**(5500), 2323–2326 (2000)
26. Pearson, K.: On Lines and Planes of Closest Fit to Systems of Points in Space. Philosophical Magazine **2**(6), 559–572 (1901)
27. Schultz, T., Theisel, H., Seidel, H.-P.: Topological visualization of brain diffusion mri data. IEEE Trans. Vis. Comput. Graph. **13**(6), 1496–1503 (2007)
28. Sherbondy, A., Akers, D., Mackenzie, R., Dougherty, R., Wandell, B.: Exploring connectivity of the brainś white matter with dynamic queries. IEEE Trans. Vis. Comput. Graph. **11**(4), 419–430 (2005)
29. Tenenbaum, J.B.: Mapping a Manifold of Perceptual Observations. MIT Press, Cambridge (1998)
30. Xie, Z.X., Mu, Z.C., Xie, J.J.: Multi-pose ear recognition based on locally linear embedding. CAAI Trans. Intell. Syst. **3**(4), 454–462 (2008)
31. Zhang, S., Demiralp, C., Laidlaw, D.H.: Visualizing diffusion tensor mr images using streamtubes and streamsurfaces. IEEE Trans. Vis. Comput. Graph. **9**(4), 454–462 (2003)

Exploring Brain Connectivity with Two-Dimensional Maps

Çağatay Demiralp, Radu Jianu, and David H. Laidlaw

Abstract We present and compare two low-dimensional visual representations, *2D point* and *2D path*, for studying tractography datasets. The goal is to facilitate the exploration of dense tractograms by reducing visual complexity both in static representations and during interaction. The proposed planar maps have several desirable properties, including visual clarity, easy tract-of-interest selection, and multiscale hierarchy. The 2D path representations convey the anatomical familiarity of 3D brain models and cross-sectional views. We demonstrate the utility of both types of representation in two interactive systems where the views and interactions of the standard 3D streamtube representation are linked to those of the planar representations. We also demonstrate a web interface that integrates precomputed neural-path representations into a geographical digital-maps framework with associated labels, metrics, statistics, and linkouts. We compare the two representations both anecdotally and quantitatively via expert input. Results indicate that the planar path representation is more intuitive and easier to use and learn. Similarly, users are faster and more accurate in selecting bundles using the path representation than the 2D point representation. Finally, expert feedback on the web interface suggests that it can be useful for collaboration as well as quick exploration of data.

1 Introduction

Diffusion-weighted MRI (DWI) enables neural pathways in the in vivo brain to be estimated as a collection of space curves, called a tractogram. The study of tractograms (i.e., tractography) has important applications in both clinical and basic neuroscience research on the brain. Tractograms have visual complexity

Ç. Demiralp (✉) · R. Jianu · D.H. Laidlaw
Brown University, Providence, RI, USA
e-mail: cad@cs.brown.edu; jr@cs.brown.edu; dhl@cs.brown.edu

D.H. Laidlaw and A. Vilanova (eds.), *New Developments in the Visualization and Processing of Tensor Fields*, Mathematics and Visualization,
DOI 10.1007/978-3-642-27343-8_10, © Springer-Verlag Berlin Heidelberg 2012

Fig. 1 *2D point* representation linked with a streamtube representation in an interactive tractography visualization tool

proportional to the intricacy of the axonal brain connectivity and, with increasing DWI resolutions, this complexity is becoming greater and greater. It is thus often difficult for practitioners to see tract projections clearly or identify anatomical and functional structures easily in these dense curve collections. This is important because, for example, a clinical study of a neurodegenerative disease sometimes involves selecting more than 30 tracts of interest (TOIs) manually across different datasets. Therefore, it is necessary for tractography visualization tools to provide means to reduce and help cope with visual complexity at both the interaction and data representation levels.

We believe two concepts, *abstraction* and *filtration*, can be applied to representation of datasets to help users overcome the difficulties of visual complexity. While abstraction involves simplification and generalization, filtration here entails clustering and hierarchization. With these ideas in mind, we proposed low-dimensional point representations for better interaction with fiber tracts in [19], along with [7]. Driven by known embedding methods, embedding in two-dimensional space provides an interesting window into the manifold space of neural connectivity and helps in fine selection of tracts. Figure 1 shows a snapshot from a tractography visualization tool that uses the point representation. A drawback of point representations is, however, that coordinate axes in the low-dimensional space lack an anatomical interpretation. It is clear from evaluations in [7, 19] that having a frame of reference, anatomical or otherwise, is important for users. Motivated by this problem, we introduced in [20] two-dimensional neural paths that have the desirable

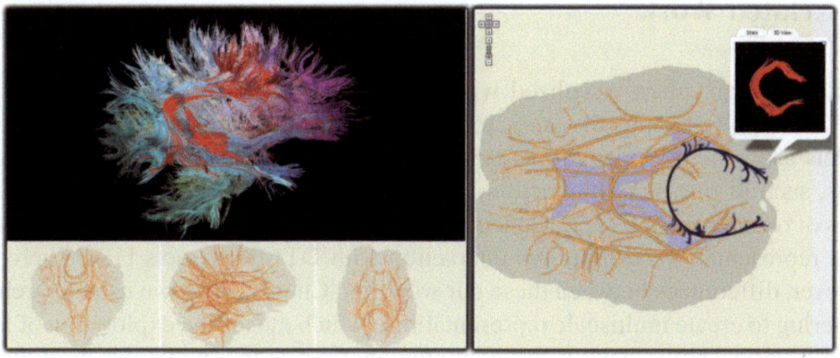

Fig. 2 Schematic *2D path* projections of tractograms as part of a standalone interactive system (*left*) and as a web-accessible digital map (*right*). The digital map interface easily incorporates any tract-associated information, including labels, links, metrics, and statistics. Shown in the pop-up window on the *right* is the "brain view" of the selected tract

properties of low-dimensional representations while preserving meaningful and familiar coordinates. Figure 2 shows screenshoots of a tool and its web interface that uses the neural path representation.

Here, we provide a unified discussion of our earlier work on planar point and path representations of tractograms [19, 20]. Since diffusion imaging is the source of all the data used in our work, we give an elementary introduction to DWI in the following section. We then discuss related work on tractography and web-based visualizations.

2 DWI

Diffusion-weighted magnetic resonance imaging (DWI) measures the diffusion rate of water molecules in biological tissues in vivo [25]. Since tissue characteristics, geometric or otherwise, at a given point affect the diffusion rate, measured diffusion-rate information is an indicator of the tissue characteristics at the point. In particular, water in fibrous tissues such as brain white matter (a collection of myelinated axons) diffuses faster along fibers than orthogonal to them. Therefore, it is possible to estimate fiber trajectories computationally using diffusion models such as the tensor model that quantify anisotropic diffusion. Diffusion imaging based on fitting second-order tensors to DWI sequences is known as diffusion-tensor magnetic resonance imaging (DTI) [4]. Fiber trajectories are computed from DTI data by integrating bidirectionally along the principal eigenvector of the underlying tensor field. This process, called fiber tracking, yields a dense collection of integral curves (i.e., a tractogram). All the tractograms used in our work were obtained using a deterministic fiber-tracking algorithm in DTI volumes.

3 Related Work

Tractograms are often visualized with streamlines or variations of streamlines in 3D [23, 29]. Reflecting the intricacy of the connectivity in the brain, these 3D models are generally visually dense. Consequently, typical interaction tasks over tracts, such as fine bundle selection, are often difficult to perform and have been a focus of recent research [1,2]. Concurrently with Chen et al. [7], we proposed planar point representations to improve interaction with DTI fiber tracts [19]. There are, however, differences between these our work and Chen's. First, we use hierarchical clustering to create multiscale representations, which makes the exploration of large datasets easier, both visually and computationally. Second, while our work uses a simple force-based embedding method, Chen et al. uses the SMACOF algorithm, an iterative method that minimizes the metric stress of multidimensional scaling (MDS) using majorization [9]. And third, we use the embedding procedure also to create a "nice" coloring in which colors of data points perceptually reflect the relationships among them.

To address concerns of experts about insufficient anatomical context in the 2D point representation, we recently introduced 2D path representations [20], which are also projections of fiber tracts into a plane, but as planar curves rather than points. One of the advantages of 2D projections is that they can be naturally integrated into a web-based digital geographic map framework. Basic data visualization has been available on the web for many years but was usually limited to traditional techniques such as bar graphs and charts. More recently, however, visualization research started targeting this environment and advanced applications have emerged. ManyEyes [26] paved the way for everyday data visualization, with subsequent studies such as [27] and [8] proving the need for accessible web visualization. While web-development toolkits such as [5] greatly aid web visualization development, large-scale web-visualization is limited by inherent browser capabilities, as demonstrated in [21]. Alternatively, stand-alone systems have been made available as applets or can be run as client applications directly from websites. However, users still must control the parameters involved in producing visualizations, specify their data queries and learn the system features. This often constitutes an undesirable overhead. Yet another approach, most similar to our work from an implementation standpoint, is to use Ajax (asynchronous JavaScript and XML) technology to perform the rendering on the server side and serve images asynchronously to the client browser. The essential difference between the present work and traditional offline visualization systems is that we separate interaction and display from rendering and computation. Our brain maps differ by eliminating user effort in creating visualizations, instead assigning this task to experts, and by using the Google Maps API, which is an Ajax framework for interactive display of pre-rendered images. Closest to our work in this latter aspect are X:MAP [28] and Genome Projector [3], which present genome browser tools implemented using the Google Maps API. We extend this idea to a new domain and demonstrate its usefulness for tractography datasets.

4 Methods

Both point and path representations are projections of fiber tracts onto the plane: Each tract is represented with a 2D point in the former and a 2D curve in the latter. Formally, given a polyline form of a fiber tract $C \in \mathbb{R}^{n\times 3}$ with n vertices in 3-space, its point representation is obtained with a map $\pi_{point} : \mathbb{R}^{n\times 3} \rightarrow \mathbb{R}^2$ and its path representation is obtained with another map $\pi_{path} : \mathbb{R}^{n\times 3} \rightarrow \mathbb{R}^{m\times 2}$.

Generation of these two representations shares three common steps. First, we obtain a whole-brain tractogram by fiber tracking in a diffusion-tensor volume fitted to a given DWI brain sequence. Second, we compute similarities between all pairs of tracts within the tractogram, obtaining a similarity (or affinity) matrix. Third, using the similarity matrix from the previous step, we run a hierarchical clustering algorithm on the tractogram, obtaining a clustering tree (dendrogram).

We create the 2D point representation of the tractogram by embedding the tracts in the plane with respect to the similarity matrix, using a simple iterative force-directed method. We use the hierarchical clustering tree to create multiscale point representations. For the path representation, we first pick a cut on the clustering tree and obtain a clustering. Then, by treating cluster centroids as pivots, we create projections of tractograms onto the major orthogonal planes as curves. We render these 2D curves stylistically using heuristics determined by the topology and geometry of the corresponding tracts and tract clusters.

We give details of these steps in subsequent sections.

4.1 *Image Acquisition and Fiber Tract Generation*

DWI brain datasets used in this paper were acquired from healthy volunteers on a 1.5T Siemens Symphony scanner with the following acquisition parameters in 12 bipolar diffusion-encoding gradient directions: thickness = 1.7 mm, FOV = 21.7 cm × 21.7 cm, TR = 7,200 ms, TE = 156 ms, b = 1,000, and NEX = 3. For each DWI sequence, the corresponding DTI volume was obtained by fitting six independent parameters of a single second-order tensor at each voxel to the 12 measurements from the DWI sequence [4]. We generate fiber-tract models of the whole brain by integrating the major eigenvector field of the diffusion tensor field bidirectionally starting at seed points. We use the second-order Runge-Kutta integration method. Since the tensor field is sampled on a volumetric grid, we evaluate its value at non-grid positions using tricubic interpolation during the integration. We use a constant integration step size of 0.5 mm and stop the integration when we reach an area of gray matter, low linear anisotropy, or low signal-to-noise (SNR) ratio.

4.2 Measuring Similarities Between Fiber Tracts

We quantify the similarity between two tracts using the distance measure discussed in [10]. This measure tries to capture how much any given two tracts follow a similar path, while giving more weight to the points closer to tract ends. Given two integral curves $C_i = \{C_i^1, \ldots, C_i^m\}$ and $C_j = \{C_j^1, \ldots, C_j^n\}$ that are represented as polylines with m and n vertices respectively, we first find mean weighted distances d_{ij} and d_{ji}, and then determine the maximum of these two distances as the distance D_{ij} between the two curves:

$$d_{ij} = \frac{1}{m} \sum_{k=1}^{m} {\alpha_i}^k \text{dist}(C_i^k, C_j) \tag{1}$$

$$d_{ji} = \frac{1}{n} \sum_{k=1}^{n} {\alpha_j}^k \text{dist}(C_j^k, C_i) \tag{2}$$

$$D_{ij} = D_{ji} = \max(d_{ij}, d_{ji}) \tag{3}$$

The function $\text{dist}(p, C)$ returns the shortest Euclidean distance between the point p and curve C. Also, $\alpha_k = \frac{1}{Z} e^{|k-(m+1)/2)|^2/\sigma^2}$, where the normalizing factor $Z = \sum_{k=1}^{m} e^{|k-(m+1)/2|^2/\sigma^2}$. We set the parameter σ automatically, proportional to L_C, the length of the fiber tract, such that $\sigma = \lambda L_C$, where $\lambda \in (0, 1]$. We set $\lambda = 0.5$ for the datasets used for this paper.

We compute the distance between each pair of integral curves as explained and assemble the measures to create a distance matrix. Note that our measure is symmetric and positive definite but does not necessarily satisfy the triangle inequality and, therefore, is not a metric. While our approach is independent of a particular similarity measure, good results in practice require a good similarity measure—one that reflects users' understanding of the similarity between data points (i.e., tracts) and works well for the task at hand.

4.3 Clustering

For a given tractography dataset we compute a clustering tree using an average-linkage hierarchical clustering algorithm on the tract distance matrix (e.g., [11]). We choose the average-linkage criterion because it is less sensitive than the minimum-linkage to broken tracts that might occur because of tracking errors. The output of the clustering algorithm is a hierarchical tree called a dendrogram. The height of the tree can be thought as the radius of the bounding ball of the dataset—in the units of the similarity measure used. Any horizontal cut on this tree provides a clustering of the dataset. Therefore, for example, the root node represents a clustering with a single cluster containing all the data points. Conversely, the leaf nodes correspond to a clustering where every data point is a cluster.

We obtain a clustering of tracts by manually setting a cut threshold on the dendrogram. This threshold can be also interactively changed by user to control the coarseness of the clustering. A constant cut at 60 % of the clustering tree's height gave consistent results across the six datasets we experimented with.

4.4 Planar Projections of Fiber Tracts

4.4.1 Fiber Tracts as Embedded Points

We use a simple iterative force-directed method for embedding tracts in the plane [12]. Embedding is a one-to-one smooth mapping from fiber tracts to points in the plane that preserves the "structure" of the fiber tracts. In this context, the distance matrix computed can be considered a manifestation of the structure in the fiber tract space.

For a given dataset with M tracts, we start with M corresponding points, all initially placed at the origin. We then iteratively adjust the positions of these points by moving the pairs of points closer to or further from each other to match with the corresponding D_{ij} entries in the similarity matrix. To achieve interactive performance, we use a stochastic sampling technique described in [6] for updating the "forces" between data points. Briefly, instead of computing forces on a point x_i from every other point in the dataset, we limit the points acting on x_i to $x_j \in \mathscr{F} = \{N_i \cup S_i\}$, where N_i and S_i are disjoint sets with a constant size. We iterate over data points and resample S_i each time by uniformly randomly selecting points from the whole dataset. For a randomly selected point x_k, if D_{ik} is smaller than $\max_{x_l \in N_i} D_{il}$, the maximum distance from x_i to any member of N_i, then x_k is assigned to N_i, otherwise it is assigned to S_i. N_i loosely represents the neighborhood of the point iterated.

4.4.2 Using Embedding to Color Fiber Tracts

Given a similarity measure, a good coloring of fiber tracts should reflect the similarities between the tracts such that similar tracts are assigned to similar colors and different tracts are assigned to different colors. Embedding fiber tracts in perceptually uniform color spaces, which are subsets of $\mathbb{R}^3$, provides a practical way to approximately achieve this goal. A perceptually uniform color space is an empirically constructed color space in which the Euclidean distances between color triplets are approximately proportional to the perceptual differences between them. L*a*b* and Luv are two common examples of such color spaces [13].

Embedding fiber tracts in the L*a*b* space is the general coloring scheme used in the interactive tools presented here. For this, we compute an approximation of the

L*a*b* color gamut, as visible on the right panels of Fig. 6, and use it as a container for force-directed embedding. To avoid having to adjust a repulsive container force, which would likely need a steep, hard-to-control gradient, we perform a physically accurate simulation with container contact detection. The embedding begins in the center of the gamut and is gradually expanded until most of the space is filled. During implementation we observed that the largest distances are often embedded along the luminance axis, the vertical (in paper coordinate frame) axis of the color gamut. This is problematic because luminance offers little resolution and can be interpreted as a lighting effect. We therefore apply a "flattening" force at the beginning of a simulation cycle to force large distances to lie in the horizontal plane (a*b*-plane). These force components, acting on the luminance axis towards the center of the gamut, wear off as the embedding moves towards a steady state. The force computation used is the same as for the 2D embedding, with straightforward 3D modifications. In terms of interaction, the color embedder supports only collapsing and color grabbing.

4.4.3 Fiber Tracts as Planar Curves

For a given tractography dataset, we create schematic views of fiber tracts projected on the sagittal, coronal, and transverse planes.

We start by obtaining a clustering of the dataset by picking a cut on the already computed hierarchical clustering tree. We then create simple orthogonal projections of tracts on each plane. Suppose the sagittal plane is aligned with the xy-plane and let $v = (x, y, z) \in \mathbb{R}^{n \times 3}$ be a vertex of a tract. Then the projection onto the sagittal plane is given by the simple equation $\pi(v) = (x, y)$. We cull out tracts that do not contribute significantly to the projection. If the ratio of projected tract length to true tract length is under a threshold value, we remove the tract from the corresponding cluster. We set the culling threshold to 0.65 for the projections used in our experiments. Finally, we compute a centroid for each cluster by choosing the tract with the smallest maximum distance to any other tract in the cluster. We found that for illustration purposes it is desirable to avoid broken tracts. We therefore weigh the centroid selection to favor longer tracts by dividing the maximum distance from each tract to any other tract by the tract's length.

We opted for a non-photorealistic rendering of brain projections to avoid their interpretation as 3D views and to harness users' intuitions about 2D maps. The rendering assumes a given clustering with assigned centroid tracts, which can be computed as described in the previous section. The centroid tracts will define a schematic neural skeleton on top of which the non-centroid tracts are scaffolded. Projections of centroid curves are smoothed prior to rendering to achieve a schematic representation and to reduce clutter. This is done by sampling a number of evenly distributed control points (five in our implementation) along the tract projection and using them as control points for a piecewise cubic spline with 30 segments. The thickness of a centroid curve is proportional to the square root of the number of tracts in the bundle. Once centroid tracts are represented as 2D

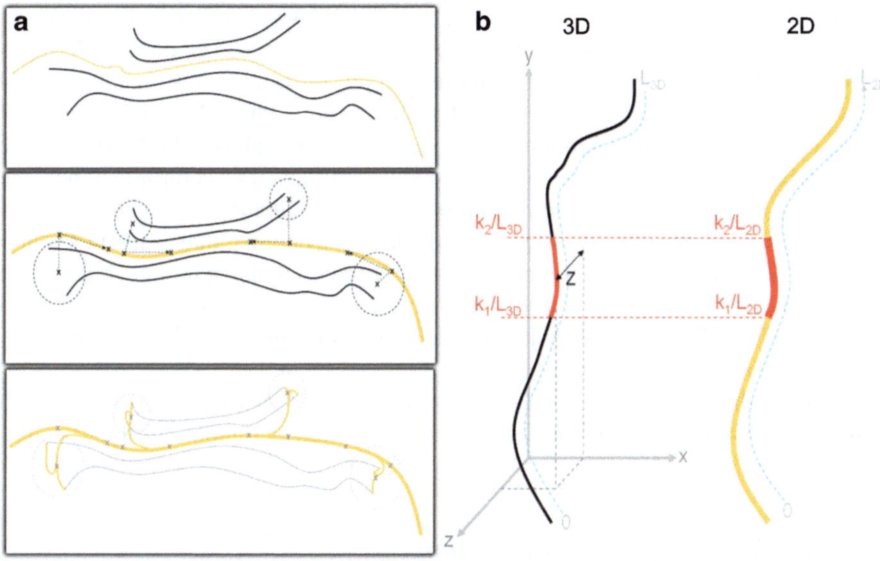

Fig. 3 (**a**) Schematic tract-cluster representation. (*Top*) 2D projections of a tract bundle, with an associated centroid curve (*yellow*), are determined from a hierarchical clustering of initial 3D tracts. (*Middle*) The centroid curve is smoothed by a spline and the endpoints of non-centroid curves are clustered using their initial 3D coordinates (four clusters); for each cluster, three control points linking the center of the cluster to the centroid spline are computed. (*Bottom*) Splines are run from each curve endpoint through the control points of its corresponding cluster. (**b**) Depth ordering of 2D paths. For each segment of a 2D spline, we locate a corresponding segment on the 3D curve from which the spline was derived by traveling the same fractional distance along both curves. The depth of the 2D segment is the same as the depth of the middle of its corresponding 3D segment

splines, endpoints of non-centroid curves are linked to their cluster's centroid spline following the procedure illustrated in Fig. 3a. First, the endpoints of non-centroid curves in a bundle are clustered based on the endpoints' initial 3D coordinates. Two endpoints are placed in the same cluster if the distance between them is less than 2 mm. Then, for each such endpoint cluster we compute three control points that link the geometrical center of the endpoint cluster to the centroid spline: the first point is the center itself, the second is a point on the centroid spline closest to the center point, and the third is determined by traveling from the second point down the centroid spline, towards each curve's other endpoint, for a predefined distance (e.g., half of the distance between the first two points). Ultimately, splines are run from each tract endpoint through its cluster's three control points, thus linking each endpoint to the centroid path. The thickness of these endpoint linkage splines gradually increases from unit thickness (i.e., single-tract thickness) at the tract endpoint to a thickness proportional to the square root of the endpoint cluster size, where it merges with the centroid spline.

We depth-order spline segments so that 2D centroid splines crossings can indicate the depth ordering of their corresponding 3D shapes. The depth ordering is done differently for centroid splines and non-centroid splines, since while centroid curves are close representations of actual 3D tracts, non-centroid curves are abstract representations obtained through the process described above. Furthermore, the depth ordering is approximate and may produce artifacts. For centroid splines, the depth of a spline segment is computed by finding a matching segment on the 3D tract from which the spline was derived and taking the depth of that segment's center (see Fig. 3b). The matching segment on the 3D tract has its endpoints at the same fractional distance from the start of the 3D tract as the 2D segment's distance from the start of the 2D spline. The depth of any non-centroid spline is determined by averaging the depth of the corresponding 3D tract.

In the following two sections, we give details on how we use 2D neural path representations as part of an interactive application and as standalone digital maps.

4.5 Linked Multi-view Interaction

We expect a typical use of low-dimensional representations to be as part of interactive applications where views and interactions of conventional representations are linked with those of low-dimensional representations. We have developed two interactive visualization systems using the 2D point and path representations, respectively, to demonstrate this mode of use (see Figs. 4 and 5).

Both applications have a view of tractography data visualized using 3D streamtubes. Coloring is generated through the embedding of tract similarity into the L*a*b* color space. In addition to the standard 3D viewing interactions, we have two basic 3D selection/deselection interactions on streamtube models: sphere selection and brushing. Sphere selection, like box selection, enables the users to select the intersecting tracts by moving a sphere of desired radius. Brushing lets users draw 2D curves on the viewing plane and select the intersecting tracts.

Both sphere-selection and brushing can be used to further prune the current selection but they cannot be used to grow it. For that purpose, we provide a selection-growing interaction that gradually adds tracts closest to the current selection. Proximity is again determined by the distance measure discussed above.

On the 2D point representation, we provide point selection and point collapsing. Selection is performed by clicking and dragging; multiple selection can be performed to select points from non-adjacent regions. Collapsing groups a set of points into a single clustered representation. This can be used either for easier tract bundle selection or as a mechanism for manually refining embeddings: points belonging to the same tract bundle can be grouped together if the embedding algorithm places them apart. The centroid of the grouping is used in subsequent embedding iterations. The hierarchical clustering tree provides a filtration of the dataset via cuts. Figure 6 shows how a cut on the dendrogram, which results in a clustering, provides a coarser representation.

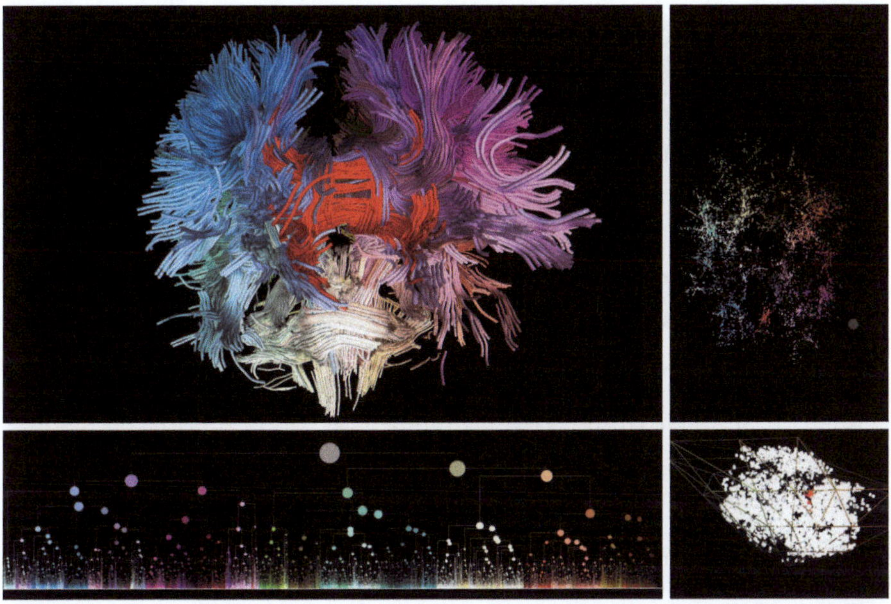

Fig. 4 Coordinated DTI tractogram model exploration in lower-dimensional visualizations: 2D embedding (*upper right*), hierarchical clustering (*lower left*), and L*a*b* color embedder (*lower right*). Selection of a fiber bundle (*red*) in the hierarchical clustering is mirrored in the other views

In the 2D path tool, we link projective views on the sagittal, coronal, and transverse planes to a standard 3D streamtube model. The clustering cut threshold that defines the specificity of the projected bundles can be altered interactively during visualization. Tract clusters in the planar projections can be selected by drawing line segments that select intersecting bundles. A selection in any of the planar views is mirrored in the 3D model view as well as in all other 2D projections.

4.6 Digital Map Interface

Brain mapping is one of the quintessential problems in neurosciences. We believe that a geographical map metaphor is well suited to the visualization and analysis of results obtained in that area. Therefore, producing a representation of the brain that is viewed, interacted, queried, and enriched like an online geographical map was one of the motivations behind our creation of the 2D path representation.

For this, we use the Google Maps API, an Ajax framework used to render large maps, to interactively display our tractogram maps on the web. The Google Maps API receives input image data in the form of a set of small images, called tiles, that when assembled together form the different zoom levels of the map. Each zoom level z consists of a rectangular grid of tiles of size $2^z \times 2^z$. The API decodes the zoom

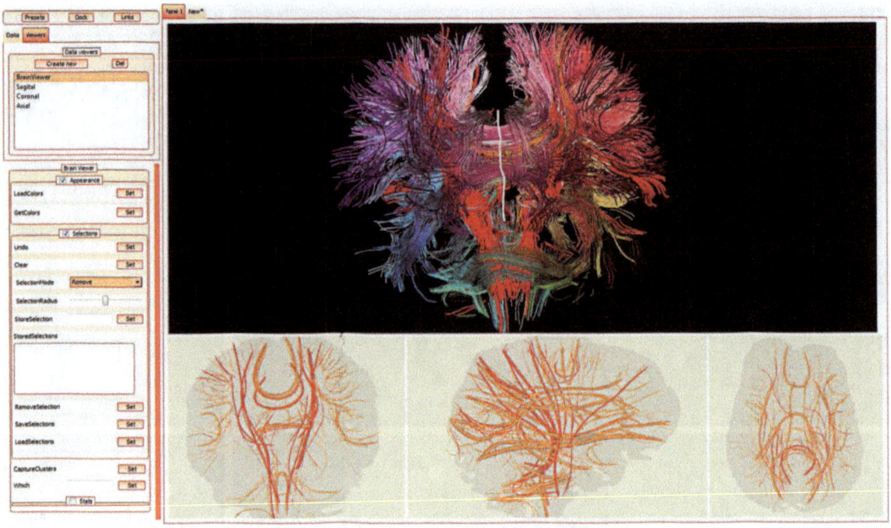

Fig. 5 An interactive analysis system using linked views and planar tract-bundle projections. Three planar representations, along the coronal, transverse and sagittal planes (*bottom panels*), are linked to a 3D streamtube model (*upper left*) and a 2D point embedding of tract similarities (*upper right*). Selections in the projection views can be made by clicking or cutting across cluster curves and are mirrored in the 3D view. Points corresponding to the selected tracts are interactively embedded into the plane and used to refine selections at the tract level

level and coordinates of the currently viewed map region to retrieve and display the visible tiles. The developer can load a custom set of tiles in the API by implementing a callback function that translates numerical tile coordinates and zoom level into unique paths to the custom tiles.

The API provides basic functionality such as zooming and panning and allows programmatic extension or customization with markers and polyline overlays, information pop-ups and event management. The API can easily be integrated into any webpage supporting Javascripts.

Our visualization system can render our 2D projections into a set of image tiles instead of the screen. For each cluster, including both tract-bundled and endpoint clusters, we export information required for interaction and browsing. Selection information consisting of evenly spaced points along splines and thickness radii for splines contained in a cluster is exported. In line with the tile paradigm, instead of exporting this information to a single large file, we divide it geometrically across corresponding tiles and write it as multiple tile-content text files. Upon user selection, the content file of a clicked tile is fetched from the server and its data analyzed for an intersection. This approach avoids loading and searching through large files. A valid cluster selection is marked on the map with polyline overlays running over tract splines contained in the selected cluster (see Fig. 7). For this purpose, spline coordinates for each cluster are exported to files indexed by a unique cluster identifier.

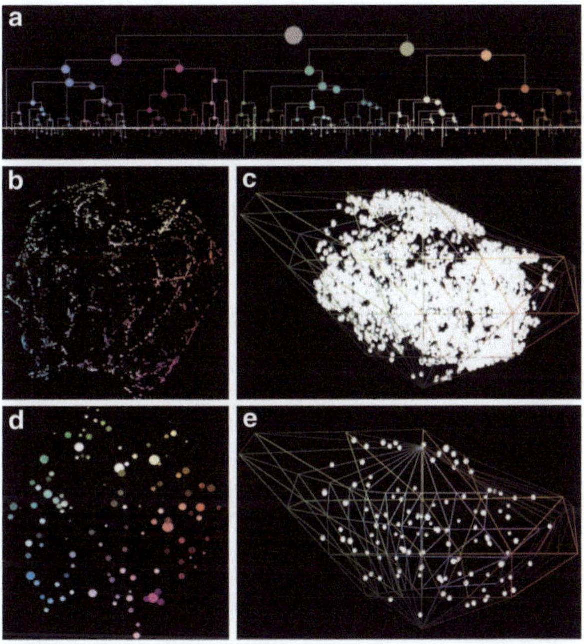

Fig. 6 A clustering cut in the dendrogram view (*top row*) is applied to the linked 2D embedding and 3D colorer (*middle row*). Points belonging to the same cluster are collapsed to their centroids (*bottom row*)

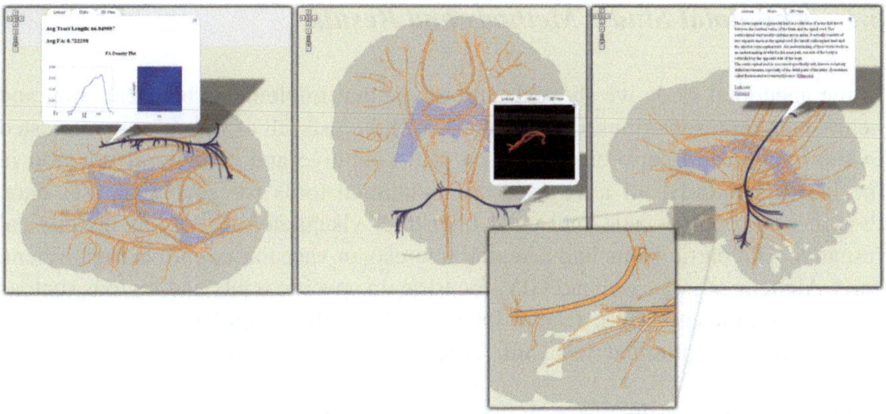

Fig. 7 DTI tractography data projected onto the sagittal, coronal and transverse planes. Major tract bundles are represented schematically by their centroid tract; individual tracts in bundles are linked from the centroid bundle to their projected endpoints. Zooming in allows access to smaller clusters of tracts. Bundles can be selected and pre-computed statistical data along with 3D views of the tract bundle ("brain view") can be displayed

Finally, for each tract cluster we export a variety of metadata accessible during map browsing in information boxes, as shown in Fig. 7. A short description and links to the most relevant publications or research can be added manually for major tracts. A few 3D poses of each tract bundle are prerendered and exported as animated GIF images, indexed by the cluster identifier. Statistical data, in both textual and graphical form, are computed for each cluster and written as HTML content to cluster-indexed files. This information is loaded and displayed in tabbed information boxes at the user's request.

4.7 Implementation

We implemented both interactive systems in C++ using G3D and Qt libraries [14, 24]. We created the web interface for the neural path representation using the Google Maps API [15]; it can be accessed via the url link [17].

5 User Evaluation

We compared the two representations both anecdotally and quantitatively.

5.1 Anecdotal Study: Methods and Results

In the anecdotal study we showed a prototype that implements both 2D point and neural path representations to three neuropsychologists, all of whom were interested in the relationship between fiber tracts and cognitive and behavioral function in the brain. Similarly, all have used computational tools for analyzing DTI data, though only one of them had used fiber-tract visualization tools in his clinical research. The participants had research interests in vascular cognitive impairment, early Alzheimer's disease, and HIV, focusing on specific tracts and regions such as the corpus callosum (CC), frontal lobe, basal ganglia, cingulate bundle, superior and inferior longitudinal fasciculi, anterior internal capsule, and the uncinate fasciculus.

Our anecdotal evaluation protocol was straightforward: we demonstrated the prototype while asking questions and collecting participants' feedback. Two of the experts also tried both interfaces themselves by selecting a set of major TOIs, the CC, cingulate bundle, uncinate anterior internal capsule, and the corticospinal tract. There was agreement that the 2D neural path representation was more intuitive and easier to use and learn than the 2D point representation.

Our experts also found the web interface with the digital map interaction useful. Although they believed that the standalone application with linked representations would remain necessary for quantitative analyses requiring interactive fine selection, they thought the web accessibility opened up interesting possibilities. They were particularly excited about browsing through datasets while commuting or at home, of quickly inspecting unfamiliar datasets, and of sharing such visualizations with collaborators.

5.2 *Quantitative Study*

In the quantitative evaluation, we compared the point and path representations by measuring user performance on a bundle selection task.

5.2.1 Applications

We used our interactive path representation system (Sect. 4.5) and the 2D point representation tool in [7] in running the comparative study. The reason that we used this tool instead of ours is that it had already been compared to other known tractography tools such as CINCH, MedINRIA, and BrainApp, and was reported to be preferable to them. Using it would give us some idea about, the relative merits, if any, of the 2D path representation tool over not only the 2D point representation but also these other tractography applications already compared. Chen et al.'s application offers a brush tool that works similarly to ours in 3D and as a lasso tool on the 2D point representation. Users can select tracts or points and then remove them or, conversely, remove everything else from a current selection.

5.2.2 Participant Pool

Our four subjects were all familiar with neuroanatomy and tractography. They also had experience with one or more tractography visualization tools. Our first subject was a neuroscience graduate student working on tracing white-matter tracts from frontal subregions to basal ganglia and the medial temporal lobe. Our second user was a neuropsychology postdoc with 5 years' experience with DWI in clinical research. This user, who participated in the anecdotal study as well, studied white-matter variation in the neurodegenerative diseases specified above. Our third subject was a biomedical engineering graduate student who had significant tract-selection experience working as a rater for a neuroscientist. Our last subject was a computer-science graduate student doing research on computational DWI algorithms. Two of the users were male and two female.

5.2.3 Task

We measured user performance on bundle selection, a typical real-world task tractography tools. Users were asked to select three major bundles, the cingulate bundle (cb), corticospinal tract (cst), and right superior longitudinal fasciculus (slf), in two different brain datasets. We chose these bundles because they represent the easy-to-hard selection-difficulty range well and were used for evaluation in [7].

For each system, we explained to users the underlying visualization concepts and demonstrated the basic interactions, mainly involving brushing on 2D and streamtube representations. After this introduction, users were asked to select the bundles (cb, cst, and slf) on two different training datasets. Following training, the users performed the task on two different test datasets while we collected their task-completion times. After each selection they provided subjective confidence estimates in the range 1–5 (1: not confident, 5: very confident) for their selection (fractional estimates were allowed). After completing the task on both systems, users were asked to fill out a post-questionnaire qualitative feedback on their experience. Half the users performed the task first on the 2D point-representation tool and the other half on the 2D path tool.

5.2.4 Factors and Measures

The sole factor considered in our quantitative experiment was the type of low-dimensional representation: 2D point and 2D path. All subjects used both types of representation. We recorded the users' bundle-selection times and subjective confidence values as measures of performance.

5.2.5 Results

In order to understand if the differences between user performances on the two tools were significant, we ran the paired t-test on our measurements. Results show that users were significantly faster on the 2D path tool than the 2D point tool ($p = 0.02$). Users were also significantly more confident using the 2D path representation than the 2D point representation ($p = 0.01$). Table 1 summarizes users' overall and per-bundle mean performances on each tool. Figure 8 shows the difference between the means of performance measures per user (2D-path-performance values are subtracted from 2D-point-performance values) and the mean over users. Error bars indicate the standard error in per-user differences.

We observed some interaction patterns worth reporting. Two distinct selection strategies were used with the 2D path tool. Two users consistently brushed over large areas of the projection to ensure that the targeted bundle was selected and then relied on the 3D view to clean up the selection. The other two users aimed for fine selections in the 2D projections and then inspected the 3D view to determine whether any fibers had been left out. They added the missing tracts using short,

Table 1 User performance on bundle selection task

	Time (s)				Confidence			
	cb	cst	slf	Mean	cb	cst	slf	Mean
2D point	227	361	234	274	4.1	3.3	3.1	3.5
2D path	136	165	215	172	4.1	3.8	3.7	3.9

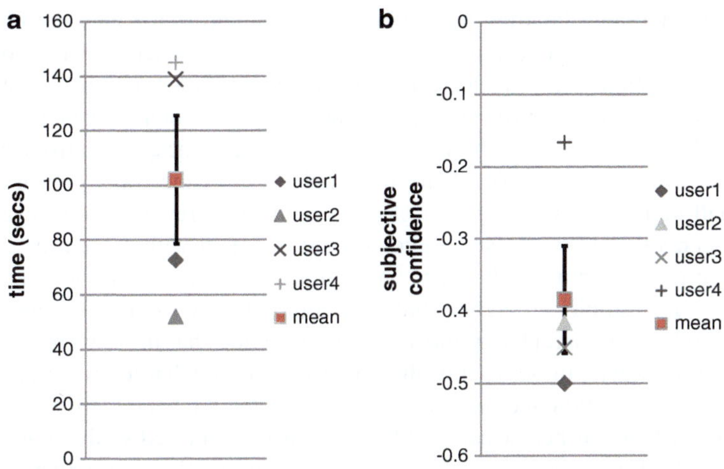

Fig. 8 Per-user differences between (**a**) time and (**b**) confidence measurements with the two tools. Differences are obtained by subtracting 2D-point-tool performance values from 2D-path-tool performance values. *Red squares* show the mean performance difference between the tools. Error bars around the *red squares* indicate the standard error of per-user differences

targeted brush strokes and then removed tubes that had been added erroneously during this operation. These latter users seemed to have a better understanding of the mapping between the 3D view and the 2D projections, perhaps explaining the difference in strategies.

All subjects used the 2D point representation relatively rarely. The most common operation was to remove points they were completely confident were not part of the selection (e.g., half of the brain, or peripheral U-shaped bundles). However, in the absence of a clear contextual mapping between the 2D point and streamtube views, subjects were hesitant to perform bold operations in 2D, at least in the short run.

6 Discussion

It is important to note that our representations rely on the anatomical fidelity of the intermediate results at each step. For example, broken trajectories due to fiber tracking errors can reduce the effectiveness of the representation. Similarly, our method expects the clustering algorithm and similarity measure to provide anatomically plausible results. However, it is difficult for a single distance measure to capture the anatomical similarity completely. Furthermore, on the same data, a good similarity measure for one purpose can be entirely irrelevant for another. While the choice of similarity measure makes clustering a subjective task, clustering algorithms themselves also have intrinsic limitations [22].

One potential limitation of the planar path representation is that bundles surrounded by the other bundles similar in orientation and shape may not be clearly visible. While we have not found this to be an issue in practice, moving projection planes along major axes while restricting the projecting tracts to a volumetric window moving with the projection plane can help solve potential problems.

Also, although we believe the subjects participating in the quantitative evaluation constituted a realistic sample of the potential users of the tools presented here, their small number limits the power of the conclusions drawn from the evaluation. And, in general, results of user studies should be taken with a grain of salt. It is difficult to run experiments that vary one factor while keeping all the other factors constant. For example, an earlier study [7] compared the 2D point representation tool with other tractography applications and reported that users were faster with the former. However, in our user evaluation, we observed that users rarely used the 2D point representation and the brush tool dominated their interaction, raising the question whether the performance difference in the reported evaluation was due mainly to the brush tool or to the 2D point representation. An experiment that replaced the brush tool with a more standard box-selection tool, say, might resolve this question. In either case, we believe that abstract representations, including the 2D point representation, are useful in the long run, as users gain more experience with the mapping between brain tractograms and low-dimensional representation primitives. In general, however, it is not realistic to expect practitioners to learn the correspondence between the new representation and the actual fiber-tract collection quickly, unless the tools are easily interpretable using a conventional anatomical framework. Furthermore, in order for any tool using a new representation of tractograms to have clinical relevance, it should provide anatomical context and intuitive functionality for region-of-interest analysis on both conventional and new representations.

While we have focused on planar spatial representations here, it is possible to create abstract representations of tractograms. For example, the hierarchical clustering tree itself can be considered as a representation of the tractogram. Or consider the circular map of connectivities (or dependencies) in Fig. 9. We obtain this dependency graph representation by first clustering tract endpoints using hierarchical clustering and setting an implicit dependency between the endpoints of each tract. We then visualize the resulting hierarchical tree with pairwise connectivities using hierarchical edge bundling [16], which feedback from a neuropsychologist suggest might be useful for understanding connectivity densities and profiles.

Although DWI is the only imaging protocol to estimate the brain neural architecture in vivo, there are in vitro imaging techniques, such as the three-dimensional electron microscopy used particularly in the emerging field of connectomics, with

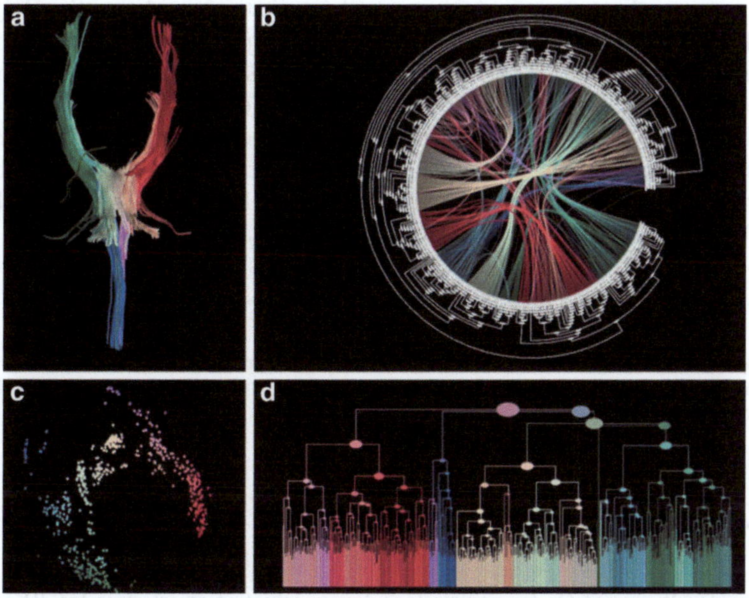

Fig. 9 Low-dimensional representations can have varying degrees of abstraction. The internal capsule is represented with (**a**) streamtubes, (**b**) a circular connectivity map, (**c**) a set of embedded points in the plane, and (**d**) a hierarchical clustering tree

which neural structures can be imaged on much smaller scales (e.g., individual axon bodies) [18]. We believe that the general ideas as well as the specific techniques presented in this paper can extend to the visualization and analysis of visually complex axonal structures originating from these high-throughput imaging techniques.

7 Conclusions

Combining traditional 3D model viewing and intuitive low-dimensional representations with anatomical context can ease navigation through complex fiber tract models, improving exploration of the connectivity in the brain. We presented two planar maps, point and path representations of tractograms, that facilitate exploration and analysis of brain connectivity. Essentially, both representations are created by applying abstraction and filtration on tractograms. We achieve abstraction by simplifying and generalizing fiber tracts with points and schematic curves in the plane. We create filtrations of tractograms by computing hierarchical clustering trees. These help create better abstractions and also provide multiscale views of data, which is important in reducing visual complexity and noise.

We compare the two representations both qualitatively and quantitatively with help of experts. Results suggest that the 2D path representation is more intuitive and easier to learn and use than the 2D point representation. We also introduced a novel way of making tractography data accessible by publishing neural maps online through a digital map framework. Our representation is conducive to such a geographic map interface by construction. This interface leads to new possibilities for enriching tractography datasets using the mass knowledge base available on the web. User feedback indicates that our web interface can be particularly useful for browsing unfamiliar datasets quickly, for analysis tasks that do not require fine selection and for sharing data in collaborative settings.

Acknowledgements We thank Song Zhang for generously providing us the 2D point representation tool used for comparison in our quantitative evaluation. This work was supported by NIH grant 1R01EB00415501A1.

References

1. Akers, D.: Wizard of Oz for participatory design: inventing an interface for 3D selection of neural pathway estimates. In: Proceedings of CHI 2006 Extended Abstracts, Montréal, pp. 454–459. ACM Press, New York (2006)
2. Akers, D., Sherbondy, A., Mackenzie, R., Dougherty, R., Wandell, B.: Exploration of the brain's white matter pathways with dynamic queries. In: Proceedings of Visualization, Austin, TX, pp. 377–384 (2004)
3. Arakawa, K., Tamaki, S., Kono, N., Kido, N., Ikegami, K., Ogawa, R., Tomita, M.: Genome Projector: zoomable genome map with multiple views. BMC Bioinformatics **10**(1), 31 (2009)
4. Basser, P.J., Mattiello, J., LeBihan, D.: Estimation of the effective self-diffusion tensor from the NMR spin echo. J. Magn. Reson. B **103**(3), 247–254 (1994)
5. Bostock, M., Heer, J.: Protovis: A graphical toolkit for visualization. IEEE Trans. Vis. Comput. Graph. **15**(6), 1121–1128 (2009)
6. Chalmers, M.: A linear iteration time layout algorithm for visualising high-dimensional data. In: Proceedings of the 7th Conference on Visualization'96. IEEE Computer Society Press, Los Alamitos (1996)
7. Chen, W., Ding, Z., Zhang, S., MacKay-Brandt, A., Correia, S., Qu, H., Crow, J.A., Tate, D.F., Yan, Z., Peng, Q.: A novel interface for interactive exploration of DTI fibers. IEEE Trans. Vis. Comput. Graph. (Proc. of Visualization) **15**(16), 1433–1440 (2009)
8. Danis, C., Viegas, F., Wattenberg, M.: Your place or mine? Visualization as a community component. In: Proceedings of CHI, Florence. ACM, New York (2008)
9. de Leeuw, J.: Applications of convex analysis to multidimensional scaling. In: Barra, J., Brodeau, F., Romier, G., Cutsem, B.V. (eds.) Recent Developments in Statistics, pp. 133–146. North Holland Publishing Company, Amsterdam/New York (1977)
10. Demiralp, C., Laidlaw, D.H.: Similarity coloring of DTI fiber tracts. In: Proceedings of DMFC Workshop at MICCAI, London, UK (2009)
11. Duda, R.O., Hart, P.E., Stork, D.G.: Pattern Classification, 2nd edn. Wiley-Interscience, New York (2000)
12. Eades, P.: A heuristic for graph drawing. Congr. Numer. **42**(149160), 194–202 (1984)
13. Fairchild, M.D.: Color Appearance Models. Wiley-IS&T, Chichester, UK (2005)
14. G3D. http://g3d-cpp.sourceforge.net/ (2012)
15. GoogleMapsAPI. http://code.google.com/apis/maps/ (2012)

16. Holten, D.: Hierarchical edge bundles: visualization of adjacency relations in hierarchical data. IEEE Trans. Vis. Comput. Graph. **12**, 741–748 (2006). doi:http://doi.ieeecomputersociety.org/10.1109/TVCG.2006.147
17. http://graphics.cs.brown.edu/research/sciviz/newbraininteraction/ (2012)
18. Jeong, W.K., Beyer, J., Hadwiger, M., Vazquez, A., Pfister, H., Whitaker, R.T.: Scalable and interactive segmentation and visualization of neural processes in EM datasets. IEEE Trans. Vis. Comput. Graph. **15**, 1505–1514 (2009)
19. Jianu, R., Demiralp, C., Laidlaw, D.: Exploring 3D DTI fiber tracts with linked 2D representations. IEEE Trans. Vis. Comput. Graph. (Proc. of Visualization) **15**(6), 1449–1456 (2009)
20. Jianu, R., Demiralp, C., Laidlaw, D.H.: Exploring brain connectivity with two-dimensional neural maps. In: IEEE Visualization 2010 Poster Compendium, Salt Lake City, UT (2010)
21. Johnson, D., Jankun-Kelly, T.: A scalability study of web-native information visualization. In: Proceedings of Graphics Interface, Windsor, pp. 163–168 (2008)
22. Kleinberg, J.M.: An impossibility theorem for clustering. In: Advances in Neural Information Processing Systems (NIPS), Vancouver, BC, pp. 446–453. MIT Press, Cambridge (2002)
23. Mori, S., Van Zijl, P.: Fiber tracking: principles and strategies – a technical review. NMR Biomed. **15**(7–8), 468–480 (2002)
24. Qt. http://www.qtsoftware.com/ (2012)
25. Tanner, J.E.: Transient diffusion in system partitioned by permeable barriers. Application to NMR measurements with a pulsed field gradient. J. Chem. Phys. **69**(4), 1748–1754 (1978)
26. Viégas, F., Wattenberg, M., Van Ham, F., Kriss, J., McKeon, M.: ManyEyes: a site for visualization at internet scale. IEEE Trans. Vis. Comput. Graph. **13**(6), 1121 (2007)
27. Viégas, F., Wattenberg, M., McKeon, M., Van Ham, F., Kriss, J.: Harry Potter and the meat-filled freezer: a case study of spontaneous usage of visualization tools. In: Proceedings of HICSS, Waikoloa. IEEE Computer Society Press, Los Alamitos (2008)
28. Yates, T., Okoniewski, M., Miller, C.: X: map: annotation and visualization of genome structure for Affymetrix exon array analysis. Nucleic Acids Res. **36** (Database issue), D780 (2008)
29. Zhang, S., Demiralp, C., Laidlaw, D.: Visualizing diffusion tensor MR images using streamtubes and streamsurfaces. IEEE Trans. Vis. Comput. Graph. **9**(4), 454–462 (2003)

Uncertainty Propagation in DT-MRI Anisotropy Isosurface Extraction

Kai Pöthkow and Hans-Christian Hege

Abstract Scalar anisotropy indices are important means for the analysis and visualization of diffusion tensor fields. While the propagation of uncertainty and errors has been studied for a variety measures, this chapter additionally considers the extraction of isosurfaces from anisotropy fields. We use the numerical condition to estimate the uncertainty propagation from the diffusion tensor eigenvalues via fractional (*FA*) and relative anisotropy (*RA*) to the position and shape of isosurfaces. Using level crossing probabilities we quantify and visualize the spatial distribution of uncertain isosurfaces. The superiority of *FA* to *RA* in terms of uncertainty propagation that was shown for anisotropy *images* in the literature does *not* hold for *isosurfaces* extracted from these images. Instead, our results indicate that for the purpose of isosurface extraction both measures perform approximately equally well.

1 Introduction

Diffusion Tensor MRI provides estimates for the major orientations of water diffusion within tissue. From these, conclusions about the microstructure of the tissue can be drawn. For example, the dominant direction of anisotropic diffusion in white matter of the brain corresponds to the orientation of neural axons. Several data acquisition and processing techniques for DTI have been established and used to assess the development or pathology of white matter for a variety of diseases, see, e.g., [8, 15] and, for a current overview, [2].

Diffusion tensor fields are computed from diffusion weighed MR images and usually defined on some regular grid. In Euclidean space each tensor $D(\mathbf{x}_j)$ associated with a point $\mathbf{x}_j \in \mathbb{R}^3$ can be represented by a symmetric matrix. Note

K. Pöthkow (✉) · H.-C. Hege
Zuse Institute Berlin, Berlin, Germany
e-mail: poethkow@zib.de; hege@zib.de

D.H. Laidlaw and A. Vilanova (eds.), *New Developments in the Visualization and Processing of Tensor Fields*, Mathematics and Visualization,
DOI 10.1007/978-3-642-27343-8_11,

that we will drop the argument $\mathbf{x}_j$ where applicable to simplify notation. In order to extend the discretely sampled tensor field to a field in a continuous domain, various reconstruction schemes have been proposed, see, e.g., reference [11]. Each tensor is uniquely described by its eigenvalues $\lambda_1, \lambda_2, \lambda_3$ and its eigenvectors $\mathbf{e}_1, \mathbf{e}_2, \mathbf{e}_3$ that satisfy $D\mathbf{e}_i = \lambda_i \mathbf{e}_i$ for $i \in \{1, 2, 3\}$. Diffusion tensors are positive definite, i.e. all eigenvalues are positive.

The large amount of information makes analysis and visualization of tensor fields difficult. A simplified representation of a tensor field can be achieved by mapping the tensor values to scalar quantities. There is a variety of such quantities that are all based on the tensors' eigenvalues, e.g., total and mean diffusivity, relative (RA) and fractional anisotropy (FA) among others [7]. All these measures are invariant under rotation and scaling of the coordinate system, as well as sorting of the eigenvalues. This chapter focuses on anisotropy measures and their isosurfaces. The FA and RA are given by

$$FA = \sqrt{\frac{1}{2}} \sqrt{\frac{(\lambda_1 - \lambda_2)^2 + (\lambda_1 - \lambda_3)^2 + (\lambda_2 - \lambda_3)^2}{\lambda_1^2 + \lambda_2^2 + \lambda_3^2}} \tag{1}$$

and

$$RA = \sqrt{\frac{1}{2}} \frac{\sqrt{(\lambda_1 - \lambda_2)^2 + (\lambda_1 - \lambda_3)^2 + (\lambda_2 - \lambda_3)^2}}{\lambda_1 + \lambda_2 + \lambda_3}. \tag{2}$$

We use the term *anisotropy index* (AI) to denote both *FA* and *RA*. The level sets of the AI with respect to a threshold θ are the sets of all locations $\mathbf{x}$ where $AI = \theta$, also written as $AI^{-1}(\theta)$. We assume that regularity conditions are fulfilled, which guarantee that these level sets are surfaces. For display, such level sets are usually approximated by triangulated *isosurfaces* and then displayed, or they are raycasted directly. Surfaces of this kind have been used for the segmentation of important anatomic structures of the human brain [22, 24].

DTI data, like all measured data, is affected by errors and uncertainty. This means that the *true values* of measured and derived quantities are unknown and the data can be safely interpreted only if the uncertainties are considered. The impact of noise and uncertainty on the results of several data acquisition schemes and processing methods in DT-MRI has been thoroughly studied (see Sect. 2). However, until now authors have only investigated the uncertainty of the resulting values and have not considered uncertainty propagation during thresholding and isosurface extraction.

In this chapter we study the propagation of errors and uncertainty from the initial tensor field through the anisotropy measures to the isosurfaces of these measures. For this we estimate the amplification or attenuation of uncertainty by the condition numbers of the numerical problem (Sect. 3). We also address the question whether one measure is more immune to uncertainty than the other. The spatial distribution of uncertain isosurfaces is described by level crossing probabilities and visualized using combined surface and volume rendering. We apply our methods to phantom and brain DTI data (Sect. 5) and discuss the results (Sect. 5).

2 Related Work

There are some general approaches to the analysis of uncertainties in DTI. Pajevic and Basser introduced a non-parametric statistical method, the DT-MRI bootstrap, and used it to confirm that the tensor components are usually normally distributed due to thermal noise [17]. They also estimated probability distributions for various other tensor-derived quantities. Koay et al. presented a framework to analyze error propagation in DTI for different diffusion tensor representations [13]. Considering objective functions for nonlinear least square optimizations they formulate error propagation equations that relate tensor-derived quantities to the diffusion-weighted MRI data. However, this method is restricted to the propagation of variances and does not directly yield the resulting probability distributions.

Another area of research, indirectly related to our work, is the investigation of uncertainty of fiber tracks in the brain computed from DTI data. Jones used bootstrapping to determine confidence intervals for fiber orientations (cones of uncertainty) [12]. Anderson investigated the effects of noise in DTI data of human brains to fiber tracking [1], while Lazar et al. focused on tractography in synthetic tensor fields [14]. The sensitivity of fiber tracking results to parameter changes was investigated by Brecheisen et al. [4]. Friman et al. proposed a Bayesian approach to generate distributions of fiber tracks [9]. The advantage of the latter approach is that prior knowledge about the fiber tracks can be incorporated using a fully probabilistic framework.

Several papers analyze the impact of noise and uncertainty on scalar DTI indices. Pierpaoli and Basser statistically compared rotationally variant and invariant anisotropy indices [20]. They show that for in vivo measurements the invariants are, in general, superior to the rotationally variant indices. Papadakis et al. studied the signal to noise ratios (SNR) of different anisotropy measures using data from simulations and in vivo experiments [18]. Chang et al. used matrix perturbation theory to estimate the uncertainty of several DTI-derived parameters including FA, RA and the direction of the principal eigenvectors [5]. Compared to bootstrap approaches this method requires significantly fewer diffusion weighted images. The work of Hasan et al. focused on the question whether FA is more robust to noise than RA [10]. For that they derived an analytical expression that directly relates RA and FA and that can be evaluated using Monte Carlo simulation. References [5, 10, 18] all state that, in general, *FA is superior to RA* regarding to noise immunity and uncertainty propagation. One of the aims of our work is to consider a further processing step and to assess the robustness of anisotropy isosurfaces, i.e. thresholding, of FA and RA.

Isosurfaces in anisotropy scalar fields generated from DTI data have been used to create segmentations of the ventricles, the corpus callosum, and the internal capsule of the human brain [24]. Large connected components of isosurfaces of FA have been used as segmentations of major brain structures by Schultz et al. [22]; they used additional information in the tensor field to automatically detect the specific brain region being represented by the isosurface segment. In a clinical study

Snook et al. used anisotropy isosurfaces for the comparison of different stages of neurodevelopment [23]. Pöthkow and Hege proposed methods for modeling and visualization of uncertain isocontours [21]; these methods are briefly recapitulated in the next section.

3 Methods

The uncertainty of the input data is modeled as a random field. To assess the propagation of uncertainty we compute condition numbers for both computation of anisotropy measures and isosurface extraction. The spatial distribution of uncertain isosurfaces is quantified using level crossing probabilities; these can be computed in real time.

3.1 Uncertainty Model

We model the uncertainty of the tensor field's eigenvalues using a discrete random field $\{\lambda_i(\mathbf{x}_j)\}$, where $\mathbf{x}_j$ runs through the vertices of the sampling grid. The values are distorted by additive measurement errors, i.e. an observation λ_i is given by

$$\lambda_i = \lambda_i^0 + \tilde{\lambda}_i \tag{3}$$

where λ_i^0 is the true but unknown quantity. We assume that each $\tilde{\lambda}_i$ is a zero-mean random variable. This means, we assume that the systematic errors have been minimized and can be neglected. A measure for the uncertainty of λ_i is the standard deviation $\sigma_{\lambda i}$ or its square, the variance $\sigma_{\lambda i}^2$ of $\tilde{\lambda}_i$. The variance can be estimated analytically from the variances of the diffusion weighted MR images [13]. The specific probability distributions can be estimated using parametric and non-parametric statistical methods. For example, it has been shown that the DT eigenvalues are affected by additive Gaussian noise [17]. Assuming that noise is the result of a combination of many sources of measurement errors, e.g. thermal noise, vibrations and background radiation, the presence of Gaussian noise is explained by the Central Limit Theorem which states that the distribution of a sum (or mean) of n random variables converges to a normal distribution for sufficiently large n.

3.2 Signal to Noise Ratio

In previous work the signal to noise ratio (SNR) was used to compare the noise immunity of *RA* and *FA* [10, 18]. For this chapter two different definitions of SNR are relevant. Let y be the function of interest (e.g. an image or a signal). When referring to complete datasets or images we use the average intensity μ_y and the standard deviation σ_y of the noise to define the *global* $\widehat{SNR}_y = \frac{\mu_y}{\sigma_y}$, assuming σ_y is constant for the whole image.

A *local* SNR can be defined for all points $\mathbf{x}$ in a dataset as $SNR(\mathbf{x})_y = \frac{y(\mathbf{x})}{\sigma_y(\mathbf{x})}$, where $\sigma_y(\mathbf{x})$ is the specific standard deviation at location $\mathbf{x}$, see [16, pp. 299–300].

3.3 Condition Numbers

For a given function ρ that maps some input value α to some output value $\rho(\alpha)$, the absolute normwise condition κ_{abs} provides an upper bound for the amplification (or attenuation) of perturbations ε by the mapping ρ:

$$\kappa_{abs} \geq \frac{\|\rho(\alpha) - \rho(\alpha + \varepsilon)\|}{\|\varepsilon\|} \tag{4}$$

If ρ is differentiable, the condition can be computed by means of the gradient

$$\kappa_{abs} = \|\nabla \rho(\alpha)\|. \tag{5}$$

The relative condition $\kappa_{rel} = \frac{\|\alpha\|}{\|\rho(\alpha)\|}\kappa_{abs}$ describes the propagation of relative errors. A problem is said to be *well-conditioned* if the condition number is low and it is said to be *ill-conditioned* if the condition number is high. The exact meaning of 'low' and 'high' depends on the problem at hand, see, e.g. [6] for more details.

3.3.1 Condition Numbers of AI Computation

The absolute normwise condition for FA is given by

$$\kappa_{FA}^{abs} = \|\nabla_\lambda FA\| = \left\| \left(\frac{\partial FA}{\partial \lambda_1}, \frac{\partial FA}{\partial \lambda_2}, \frac{\partial FA}{\partial \lambda_3} \right)^T \right\| = \sqrt{\frac{1}{2}} \frac{|\lambda_1 + \lambda_2 + \lambda_3|}{(\lambda_1^2 + \lambda_2^2 + \lambda_3^2)} \tag{6}$$

and describes the propagation of absolute errors. The relative normwise condition is given by $\kappa_{FA}^{rel} = \frac{\|(\lambda_1,\lambda_2,\lambda_3)\|}{FA}\kappa_{FA}^{abs}$ and describes the propagation of relative errors. Similiarly, the absolute normwise condition for RA is given by

$$\kappa_{RA}^{abs} = \|\nabla_\lambda RA\| = \left\| \left(\frac{\partial RA}{\partial \lambda_1}, \frac{\partial RA}{\partial \lambda_2}, \frac{\partial RA}{\partial \lambda_3} \right)^T \right\| = \frac{3\sqrt{\lambda_1^2 + \lambda_2^2 + \lambda_3^2}}{\sqrt{2}(\lambda_1 + \lambda_2 + \lambda_3)^2} \tag{7}$$

and the relative normwise condition is given by $\kappa_{RA}^{rel} = \frac{\|(\lambda_1,\lambda_2,\lambda_3)\|}{RA}\kappa_{RA}^{abs}$. Figure 1a–c show the *FA*, *RA* and their condition numbers for a 1D tensor field varying between isotropy and linear anisotropy, i.e. λ_1 increasing (linearly) in x direction while $\lambda_2 = \lambda_3 = 1$ are kept constant.

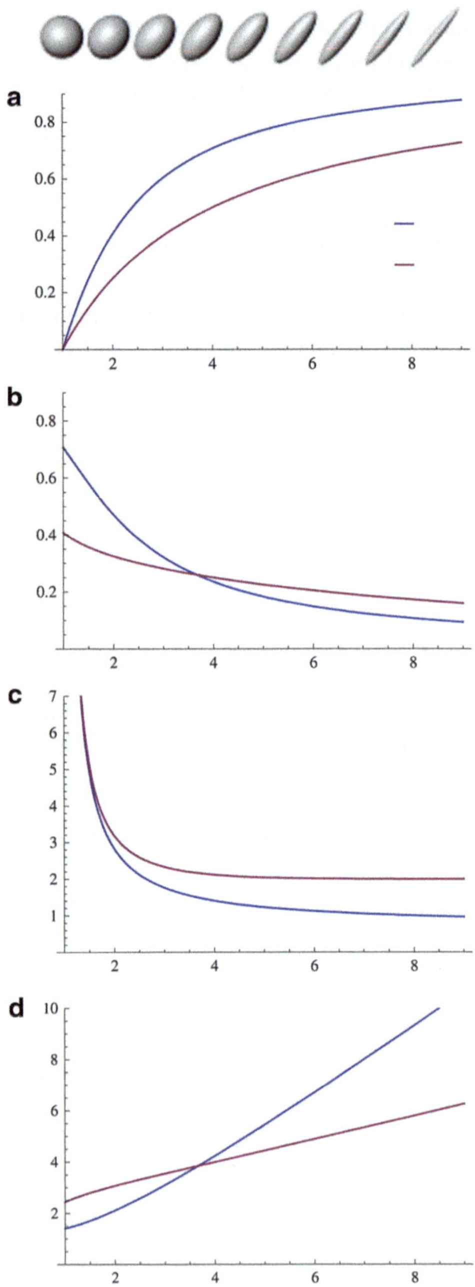

Fig. 1 In (**a**)–(**c**) the *FA*, *RA* and their condition numbers are shown for a 1D tensor field varying between isotropy and linear anisotropy, i.e. λ_1 increases linearly in x direction while $\lambda_2 = \lambda_3 = 1$ are constant. The condition numbers for isosurface extraction are shown in (**d**)

3.3.2 Condition Numbers of the Isosurface Problem

To extract isocontours from a scalar field y we have to find the locations $\mathbf{x}$ where level crossing occur by solving $y(\mathbf{x}) = \vartheta$. If the gradient $\nabla y(\mathbf{x})$ is invertible, then, according to the inverse function theorem, we can write $y^{-1}(\vartheta) = \mathbf{x}$. The derivative satisfies

$$\|(y^{-1})'(\vartheta)\| = \|\nabla y(\mathbf{x})\|^{-1}. \tag{8}$$

Thus, the absolute normwise condition of the problem (y^{-1}, ϑ) is

$$\kappa^{abs}_{y^{-1}(\vartheta)}(\mathbf{x}) = \|\nabla y(\mathbf{x})\|^{-1}. \tag{9}$$

We denote the condition numbers for isosurface extraction from anisotropy fields by $\kappa^{abs}_{FA^{-1}(\vartheta)}$ and $\kappa^{abs}_{RA^{-1}(\vartheta)}$. They are shown in Fig. 1d for the 1D tensor field.

3.4 *Uncertainty Propagation*

Using the condition numbers we can estimate the propagation of uncertainty. Let $\tilde{\lambda} = \|(\tilde{\lambda}_1, \tilde{\lambda}_2, \tilde{\lambda}_3)\|$ be a (random) perturbation of the eigenvalues. Then first order estimations of the perturbation of the results for *FA* and *RA* are given by

$$\widetilde{FA} = \kappa^{abs}_{FA}\,\tilde{\lambda} \quad \text{and} \quad \widetilde{RA} = \kappa^{abs}_{RA}\,\tilde{\lambda}. \tag{10}$$

Note that these are rough estimates because the Taylor series is truncated after the first term, i.e., covariances between the eigenvalues and higher derivatives are not considered. Analogously, the error propagation for isosurface extraction is estimated by

$$\widetilde{FA^{-1}(\vartheta)} = \widetilde{FA}\,\kappa^{abs}_{FA^{-1}(\vartheta)} \quad \text{and} \quad \widetilde{RA^{-1}(\vartheta)} = \widetilde{RA}\,\kappa^{abs}_{RA^{-1}(\vartheta)}, \tag{11}$$

where $\widetilde{FA^{-1}(\vartheta)}$ and $\widetilde{RA^{-1}(\vartheta)}$ are perturbations of the isosurface point positions.

Obviously these two steps can be integrated into one, resulting in a single measure for error propagation that we refer to as the *combined condition numbers*

$$\kappa^{abs}_{FA,FA^{-1}(\vartheta)} = \kappa^{abs}_{FA}\,\kappa^{abs}_{FA^{-1}(\vartheta)} \quad \text{and} \quad \kappa^{abs}_{RA,RA^{-1}(\vartheta)} = \kappa^{abs}_{RA}\,\kappa^{abs}_{RA^{-1}(\vartheta)}, \tag{12}$$

which relate the perturbations of the eigenvalues to the perturbations of the isosurfaces, i.e.

$$\widetilde{FA^{-1}(\vartheta)} = \tilde{\lambda}\kappa^{abs}_{FA,FA^{-1}(\vartheta)} \quad \text{and} \quad \widetilde{RA^{-1}(\vartheta)} = \tilde{\lambda}\kappa^{abs}_{RA,RA^{-1}(\vartheta)}. \tag{13}$$

We also use the condition numbers to approximate the standard deviation (or standard error) of the *FA* and *RA*:

$$\sigma_{FA} = \sigma_\lambda \kappa_{FA}^{abs} \quad \text{and} \quad \sigma_{RA} = \sigma_\lambda \kappa_{RA}^{abs}. \tag{14}$$

3.5 Uncertain Isosurfaces

Now we are interested in the spatial distribution of uncertain isosurfaces that results from the probability distributions of the input data. Pöthkow and Hege [21] introduced the level crossing probability (LCP) for quantification and visualization of uncertain isosurfaces.

Let $f_{\mathbf{x}}(t)$ be a probability density function (PDF) describing a probability distribution of a random variable Y at position $\mathbf{x} \in \mathbb{R}^d$, $F(t)$ the corresponding cumulative distribution function, and ϑ a threshold (isovalue). The PDF $f_{\mathbf{x}}(t)$ can be generated for all points $\mathbf{x}$ in a continuous domain by interpolating the expected values μ and the m-th roots of the central moments $\nu_m = \mathrm{E}((X - \mu)^m)^{\frac{1}{m}}$ in between the sample points and inserting these interpolated values in the PDFs. The interpolation method is described in detail in [21]. The LCP, defined by

$$\mathrm{P}_\vartheta = 2\, F_{\mathbf{x}}(\vartheta)\,(1 - F_{\mathbf{x}}(\vartheta)), \tag{15}$$

gives the probability that for two independent realizations y_a and y_b of $Y_{\mathbf{x}}$, distributed according to $f_{\mathbf{x}}(t)$, one of them is greater or equal to ϑ, while one is less or equal to ϑ. We use this probability as a spatially dependent measure for the presence of a ϑ-isosurface at some given point $\mathbf{x}$. The field of such probabilities indicates the spatial distribution of uncertain isosurfaces. See reference [21] for the derivation and further discussion of this function.

3.6 Visualization

To display the quantities *RA* and *FA*, as well as the related condition numbers and SNR, we show textured slices with color-mapped scalar values. See Fig. 3 for an illustration. There the spatial distributions of uncertain isosurfaces are visualized using GPU-based volume rendering of P_ϑ.

It has been shown in the literature, that under the influence of Gaussian noise the squares of anisotropy measures are χ^2 distributed [17]. This implies that RA and FA conform to a χ distribution which can be approximated by a (half-)Gaussian, see [19]. Thus, as input for for the calculation of P_ϑ we consider normal distributions

$$Y \sim \mathcal{N}(FA, \sigma_{FA}) \quad \text{or} \quad Y \sim \mathcal{N}(RA, \sigma_{RA}) \tag{16}$$

that are restricted to the interval [0, 1]. Equation 15 is evaluated in real time and the result is used for ray casting with a simple emission absorption model. The volume rendering is combined with crisp isosurfaces $FA^{-1}(\vartheta)$ or $RA^{-1}(\vartheta)$, respectively. These surfaces can be considered as the "mean" or "most probable" surfaces [21].

4 Results

We compute the condition numbers and estimate the propagation of uncertainty to the anisotropy indices and the related isosurfaces for a synthetic spiral and a brain DTI dataset.

4.1 Synthetic DTI Data

We employ a synthetic spiral DTI dataset generated as described in [3] that is visualized using ellipsoid glyphs in Fig. 2 with *FA* mapped to glyph color. The *FA*, *RA*, κ_{FA}^{abs}, κ_{RA}^{abs} and *SNR* are shown in Fig. 3. Note that the *FA* values are higher than *RA* as well as the κ_{FA}^{abs} values are higher than κ_{RA}^{abs}, i.e. absolute errors are amplified more for *FA* than for *RA*. Nevertheless the higher *FA* values lead to a higher SNR relative to *RA*. We exemplarily assume a constant $\widehat{SNR} = 20$ for all eigenvalues.

In Fig. 4 the condition numbers $\kappa_{FA^{-1}(\vartheta)}^{abs}$, $\kappa_{RA^{-1}(\vartheta)}^{abs}$, $\kappa_{FA,FA^{-1}(\vartheta)}^{abs}$ and the relative differences between $\kappa_{FA,FA^{-1}(\vartheta)}^{abs}$ and $\kappa_{RA,RA^{-1}(\vartheta)}^{abs}$ are shown along with two corresponding uncertain isosurfaces. The values of $\kappa_{FA^{-1}(\vartheta)}^{abs}$ are lower than $\kappa_{RA^{-1}(\vartheta)}^{abs}$, while the relative differences between $\kappa_{FA,FA^{-1}(\vartheta)}^{abs}$ and $\kappa_{RA,RA^{-1}(\vartheta)}^{abs}$ are smaller than 1 %. The uncertain isosurfaces in Fig. 4c and f are depicted by volume renderings of P_ϑ combined with crisp isosurfaces $FA^{-1}(\vartheta)$ and $RA^{-1}(\vartheta)$.

4.2 Brain DTI Data

To apply our methods to real world data we consider a brain dataset consisting of 148 × 190 × 160 DTs. The eigenvalue fields are smoothed using a 3D Gaussian kernel with a standard deviation of 1.2 voxel widths to estimate the mean values from the noisy data. The *FA* and *RA* as well as κ_{FA}^{abs}, $\kappa_{FA^{-1}(\vartheta)}^{abs}$, κ_{RA}^{abs} and $\kappa_{RA^{-1}(\vartheta)}^{abs}$ are shown in Fig. 5.

From the *FA* and *RA* fields and the corresponding uncertainty estimations for $\widehat{SNR}_\lambda = 10$ and $\widehat{SNR}_\lambda = 20$ we generated the uncertain isosurfaces shown in Fig. 6. We chose the threshold $\vartheta = 0.5$ for *FA* that was used previously for

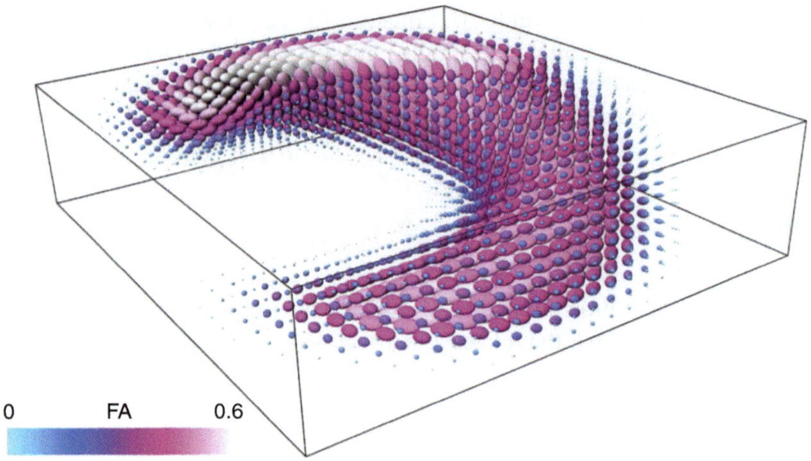

Fig. 2 The synthetic spiral DT dataset is visualized by ellipsoid glyphs with *FA* mapped to glyph color

the segmentation of brain structures using isosurfaces [22]. The threshold for the corresponding *RA* isosurface is $\vartheta \approx 0.32$. Again, the spatial distributions of the isosurfaces are indicated by volume renderings of P_ϑ that surround the mean (crisp) surfaces.

5 Discussion and Conclusions

Our outcomes reproduce the previous result [5, 10, 18] that *FA* yields higher *SNR* than *RA*, see Fig 3. This finding has led to the conclusion that *FA* is more immune to noise and uncertainty. This statement is true, but applicable only if the intended final result is an immediate depiction of diffusion anisotropy.

However, in some applications isosurface extraction or thresholding are subsequent steps in visual analysis of DTI anisotropy. If this step is included too, the uncertainty propagation from the eigenvalues to the spatial position of the isosurface has also to be considered. For sensitivity analysis of isosurface extraction not only the scalar field but also the gradient magnitude has to be taken into account.

The graphs in Fig. 1a, computed for a simple 1D tensor field, show the nonlinear nature of *FA* and *RA*. For small values of λ_1 the curves have a steep slope, while for increasing values the slope gets more flat. This means that the sensitivity of the functions depends on the actual values of all eigenvalues. On the left side of the plot in Fig. 1a small changes of λ_1 lead to large changes of *FA* and *RA*, i.e. perturbations are *amplified*. On the right side of the plot perturbations are *attenuated*. Both effects are stronger for *FA* than for *RA*.

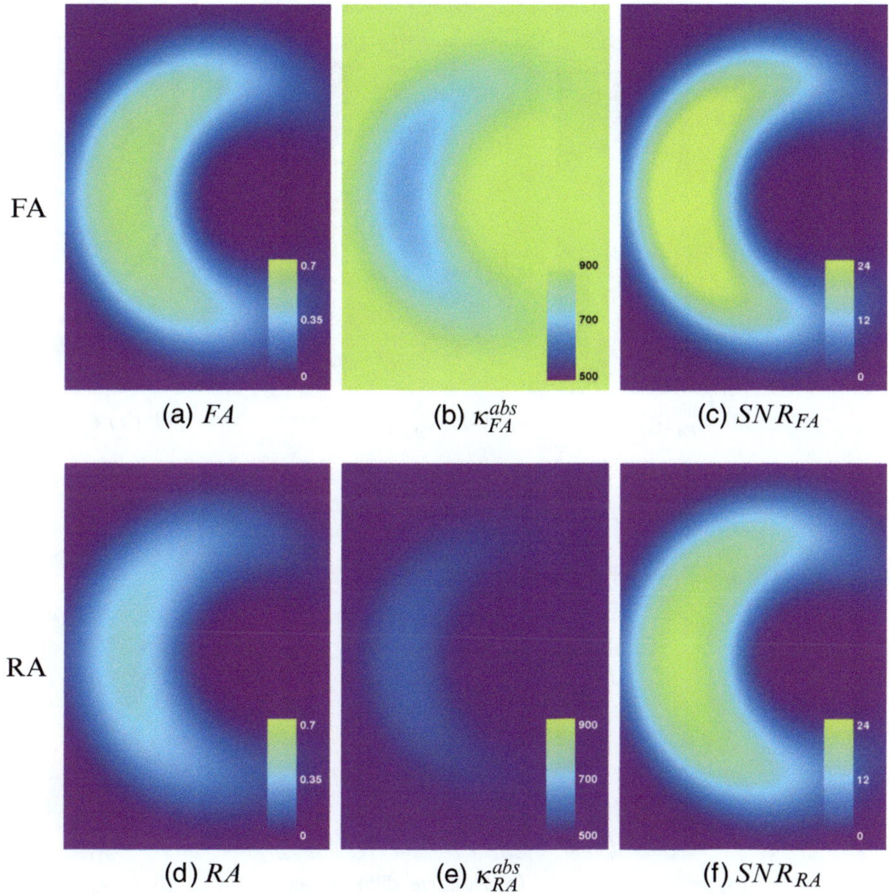

Fig. 3 Textured slices in the spiral dataset showing from *left* to *right*: anisotropy measure, absolute condition and *SNR* (assuming constant $\widehat{SNR}_\lambda = 20$ for all eigenvalues). First row: *FA*, second row *RA*

We can observe these properties in the plots of the condition numbers in Fig. 1b. On the left side of the plot κ_{FA}^{abs} is larger than κ_{RA}^{abs} and vice versa on the right side. On the other hand the condition numbers in Fig. 1b for isosurface extraction show a different behavior. On the left side of the plot $\kappa_{FA^{-1}(\vartheta)}^{abs}$ is smaller than $\kappa_{FA^{-1}(\vartheta)}^{abs}$ and vice versa on the right side. This corresponds to right side of the graph for *FA* in Fig. 1a which is closer to a plateau than that of *RA*, i.e. the isosurface extraction is more ill-conditioned.

If we compare Fig. 1b with Fig. 1d we see that where $\kappa_{FA}^{abs} < \kappa_{RA}^{abs}$ holds also $\kappa_{FA^{-1}(\vartheta)}^{abs} > \kappa_{RA^{-1}(\vartheta)}^{abs}$ holds, and vice versa. Indeed the combined condition numbers are equal:

$$\kappa_{FA,FA^{-1}(\vartheta)}^{abs} = \kappa_{RA,RA^{-1}(\vartheta)}^{abs}. \tag{17}$$

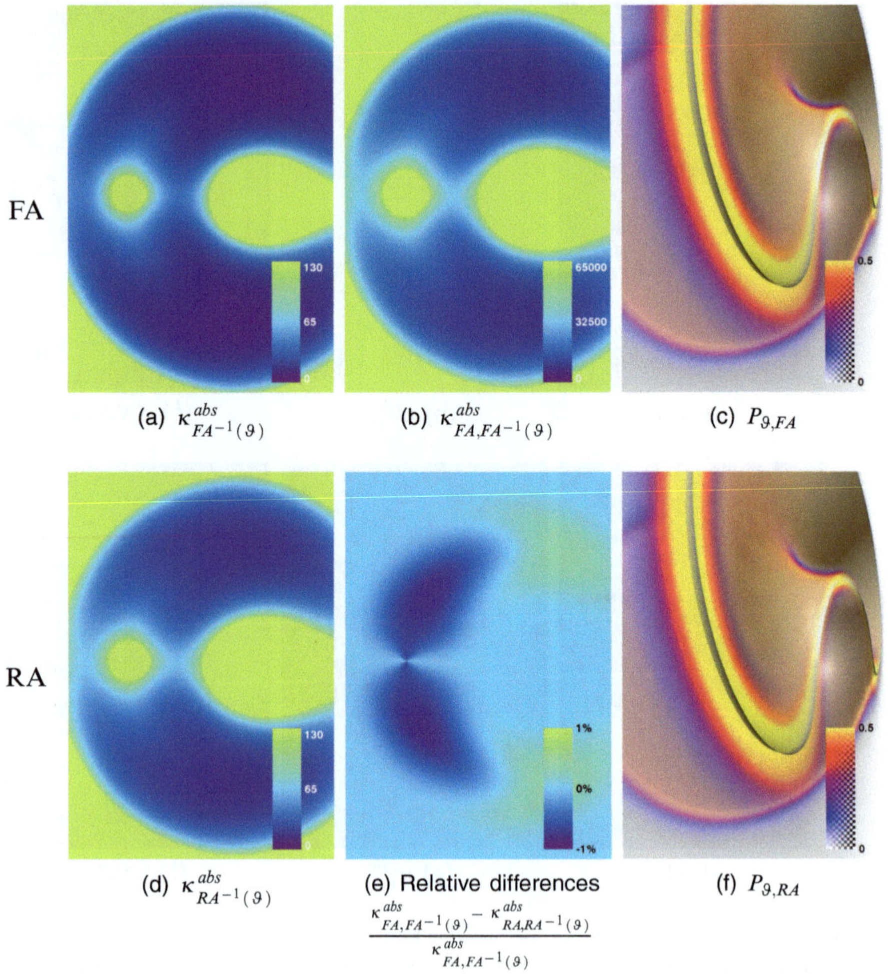

Fig. 4 Textured slices in the spiral dataset: condition number for isosurface extraction for FA (**a**) and RA (**d**), combined condition for FA (**b**) and relative differences between the combined condition numbers (**e**). In the *right* column uncertain isosurfaces (assuming constant SNR = 20 for all eigenvalues) are depicted

This means that—in a first order approximation—the *propagation of uncertainty for isosurface extraction* in FA and RA fields is *equal*.

This can also be verified visually by comparing Figs. 4c–f, 6a–c and 6b–d. So, in this context there is no clear superiority of *FA* compared to *RA*. In the Appendix we show the equality in Eq. (17) for the 2D case whose analytical treatment is still readily comprehensible. For the 3D datasets the numerical gradient estimation

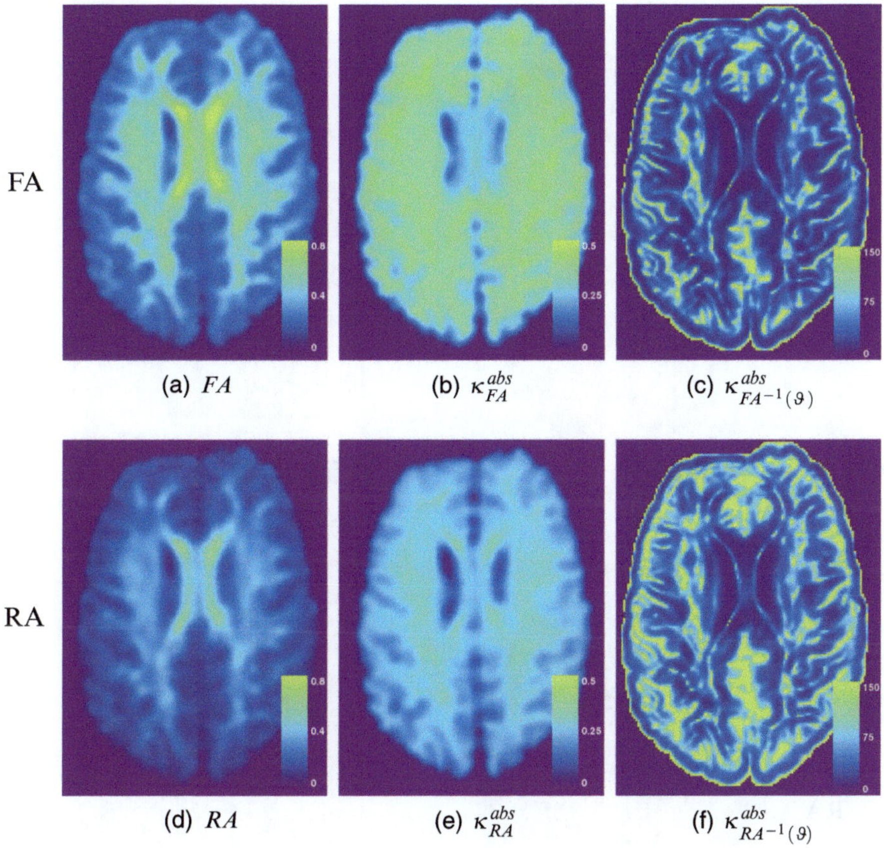

Fig. 5 Textured slices in the brain dataset showing from *left* to *right*: anisotropy measures, condition numbers for the anisotropy measures and condition numbers for isosurface extraction. First row: *FA*, second row *RA*

introduces additional errors, but the differences between the combined condition numbers for FA and RA are below 1 %.

Our results also have implications for fiber tracking. Many fiber tracking algorithms use thresholds of *FA* to restrict the resulting tracks to anisotropic areas of the brain. The uncertainty of the shape of these areas leads to uncertainties in the resulting fiber tracks. This is related to the sensitivity of fiber tracking results to variations of the anisotropy threshold that was investigated by Brecheisen et al. [4]. An area for further research is the more refined investigation of the spatial uncertainties of isosurfaces by means of higher order analytical expressions or numerical evaluation.

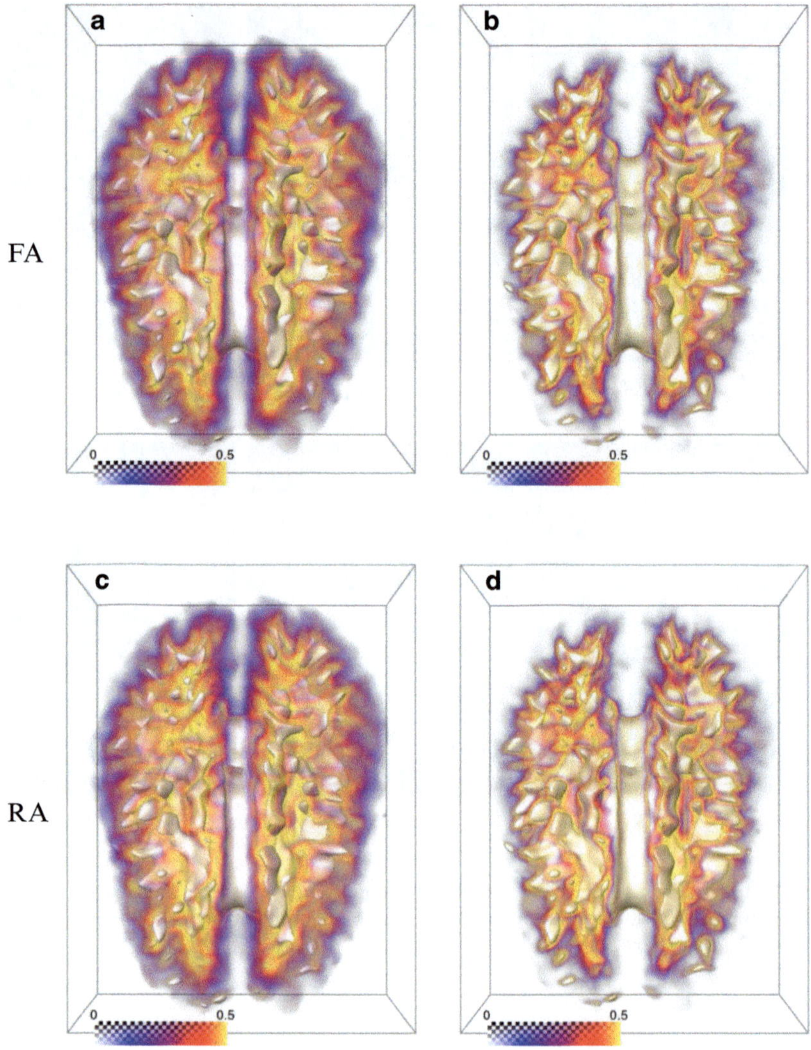

Fig. 6 Uncertain isosurfaces for $\widehat{SNR}_\lambda = 10$ (*left*) and $\widehat{SNR}_\lambda = 20$ (*right*) using *FA* (*top*) and *RA* (*bottom*). The mean crisp isosurfaces are shown in *white* while the level crossing probabilities are mapped to color for volume rendering. The threshold is $\vartheta = 0.5$ for FA and $\vartheta = 0.32$ for RA

Appendix

Let $D(x, y)$ be a diffusion tensor field in $\mathbb{R}^2$ where each tensor is described by its eigenvalues λ_i and eigenvectors $\mathbf{e}_i$. The fractional anisotropy is given by

$$FA = \sqrt{\frac{1}{2}}\sqrt{\frac{(\lambda_1 - \lambda_2)^2}{\lambda_1^2 + \lambda_2^2}} \tag{18}$$

and the relative anisotropy by

$$RA = \sqrt{\frac{1}{2}} \frac{\sqrt{(\lambda_1 - \lambda_2)^2}}{\lambda_1 + \lambda_2}. \tag{19}$$

With the derivatives

$$\frac{\partial FA}{\partial \lambda_1} = \frac{\lambda_2 \left(\lambda_1^2 - \lambda_2^2\right)}{|\lambda_1 - \lambda_2| \left(\lambda_1^2 + \lambda_2^2\right)^{3/2}} \tag{20}$$

and

$$\frac{\partial FA}{\partial \lambda_2} = \frac{\lambda_1 \lambda_2^2 - \lambda_1^3}{|\lambda_1 - \lambda_2| \left(\lambda_1^2 + \lambda_2^2\right)^{3/2}} \tag{21}$$

we can determine the absolute normwise condition for *FA* computation

$$\kappa_{FA}^{abs} = \left\| \left(\frac{\partial FA}{\partial \lambda_1}, \frac{\partial FA}{\partial \lambda_2} \right)^T \right\| = \frac{|\lambda_1 + \lambda_2|}{\lambda_1^2 + \lambda_2^2}. \tag{22}$$

Analogously, for *RA* we can write

$$\frac{\partial RA}{\partial \lambda_1} = \frac{2(\lambda_1 - \lambda_2)\lambda_2}{|\lambda_1 - \lambda_2| (\lambda_1 + \lambda_2)^2} \tag{23}$$

$$\frac{\partial RA}{\partial \lambda_2} = \frac{2\lambda_1 (\lambda_2 - \lambda_1)}{|\lambda_1 - \lambda_2| (\lambda_1 + \lambda_2)^2} \tag{24}$$

$$\kappa_{RA}^{abs} = \left\| \left(\frac{\partial RA}{\partial \lambda_1}, \frac{\partial RA}{\partial \lambda_2} \right)^T \right\| = \frac{2\sqrt{\lambda_1^2 + \lambda_2^2}}{(\lambda_1 + \lambda_2)^2}. \tag{25}$$

For an explicit formulation of Eq. (9) we use the gradient

$$\nabla FA = \left(\frac{\partial FA}{\partial x}, \frac{\partial FA}{\partial y} \right)^T \tag{26}$$

with

$$\frac{\partial FA}{\partial x} = -\frac{\left(\lambda_1^2 - \lambda_2^2\right) \left(\lambda_1 \frac{\partial \lambda_2}{\partial x} - \lambda_2 \frac{\partial \lambda_1}{\partial x} \right)}{|\lambda_1 - \lambda_2| \left(\lambda_1^2 + \lambda_2^2\right)^{3/2}} \tag{27}$$

and

$$\frac{\partial FA}{\partial y} = -\frac{\left(\lambda_1^2 - \lambda_2^2\right) \left(\lambda_1 \frac{\partial \lambda_2}{\partial y} - \lambda_2 \frac{\partial \lambda_1}{\partial y} \right)}{|\lambda_1 - \lambda_2| \left(\lambda_1^2 + \lambda_2^2\right)^{3/2}}, \tag{28}$$

leading to the condition number for isosurface extraction

$$\kappa^{abs}_{FA^{-1}(\vartheta)} = \sqrt{\frac{\left(\left|\frac{(\lambda_1^2-\lambda_2^2)\left(\lambda_1\frac{\partial\lambda_2}{\partial y}-\lambda_2\frac{\partial\lambda_1}{\partial y}\right)}{(\lambda_1^2+\lambda_2^2)^2}\right|^2+\left|\frac{(\lambda_1^2-\lambda_2^2)\left(\lambda_1\frac{\partial\lambda_2}{\partial x}-\lambda_2\frac{\partial\lambda_1}{\partial x}\right)}{(\lambda_1^2+\lambda_2^2)^2}\right|^2\right)(\lambda_1^2+\lambda_2^2)}{(\lambda_1-\lambda_2)^2}}. \tag{29}$$

Similarly for RA

$$\frac{\partial RA}{\partial x} = -\frac{2\,(\lambda_1-\lambda_2)\left(\lambda_1\frac{\partial\lambda_2}{\partial x}-\lambda_2\frac{\partial\lambda_1}{\partial x}\right)}{|\lambda_1-\lambda_2|\,(\lambda_1+\lambda_2)^2} \tag{30}$$

and

$$\frac{\partial RA}{\partial y} = -\frac{2\,(\lambda_1-\lambda_2)\left(\lambda_1\frac{\partial\lambda_2}{\partial y}-\lambda_2\frac{\partial\lambda_1}{\partial y}\right)}{|\lambda_1-\lambda_2|\,(\lambda_1+\lambda_2)^2} \tag{31}$$

give the condition number

$$\kappa^{abs}_{RA^{-1}(\vartheta)} = \frac{2\sqrt{\left|\lambda_2\frac{\partial\lambda_1}{\partial y}-\lambda_1\frac{\partial\lambda_2}{\partial y}\right|^2+\left|\lambda_2\frac{\partial\lambda_1}{\partial x}-\lambda_1\frac{\partial\lambda_2}{\partial x}\right|^2}}{(\lambda_1+\lambda_2)^2}. \tag{32}$$

By elementary algebra it can be shown that the relation

$$\kappa^{abs}_{FA}\kappa^{abs}_{FA^{-1}(\vartheta)} = \kappa^{abs}_{RA}\kappa^{abs}_{RA^{-1}(\vartheta)} \tag{33}$$

holds. This means that in a first order approximation the propagation of uncertainties from the eigenvalues to uncertainties of isocontour-positions is equal for FA and RA.

References

1. Anderson, A.W.: Theoretical analysis of the effects of noise on diffusion tensor imaging. Magn. Reson. Med. **46**(6), 1174–1188 (2001)
2. Special issue on Diffusion Imaging: Magn. Reson. Imaging Clin. N. Am. **19**(1) (2011)
3. Bergmann, O., Lundervold, A., Steihaug, T.: Generating a synthetic diffusion tensor dataset. In: Proceedings of 18th IEEE Symposium on Computer-Based Medical Systems, 2005, Dublin, pp. 277–281 (2005)
4. Brecheisen, R., Vilanova, A., Platel, B., ter Haar Romeny, B.: Parameter sensitivity visualization for DTI fiber tracking. IEEE Trans. Vis. Comput. Graph. **15**, 1441–1448 (2009)

5. Chang, L.C., Koay, C.G., Pierpaoli, C., Basser, P.J.: Variance of estimated DTI-derived parameters via first-order perturbation methods. Magn. Reson. Med. **57**(1), 141–149 (2007)
6. Deuflhard, P., Hohmann, A.: Numerical Analysis in Modern Scientific Computing: An Introduction, 2nd edn. Springer, New York (2003)
7. Ennis, D.B., Kindlmann, G.: Orthogonal tensor invariants and the analysis of diffusion tensor magnetic resonance images. Magn. Reson. Med. **55**(1), 136–146 (2006)
8. Filler, A.G.: MR neurography and diffusion tensor imaging: origins, history & clinical impact of the first 50,000 cases with an assessment of efficacy and utility in a prospective 5,000 patient study group. Neurosurgery **65**(4 Suppl), A29–A43 (2009)
9. Friman, O., Farneback, G., Westin, C.F.: A Bayesian approach for stochastic white matter tractography. TMI **25**(8), 965–978 (2006)
10. Hasan, K.M., Alexander, A.L., Narayana, P.A.: Does fractional anisotropy have better noise immunity characteristics than relative anisotropy in diffusion tensor MRI? An analytical approach. Magn. Reson. Med. **51**(2), 413–417 (2004)
11. Hotz, I., Sreevalsan-Nair, J., Hagen, H., Hamann, B.: Tensor field reconstruction based on eigenvector and eigenvalue interpolation. In: Hagen, H. (ed.) Scientific Visualization: Advanced Concepts, *Dagstuhl Follow-Ups*, vol. 1, pp. 110–123. Schloss Dagstuhl – Leibniz-Zentrum für Informatik, Germany (2010)
12. Jones, D.K.: Determining and visualizing uncertainty in estimates of fiber orientation from diffusion tensor MRI. Magn. Reson. Med. **49**, 7–12 (2003)
13. Koay, C.G., Chang, L.C., Pierpaoli, C., Basser, P.C.J.: Error propagation framework for diffusion tensor imaging via diffusion tensor representations. IEEE Trans. Med. Imaging **26**(8), 1017–1034 (2007)
14. Lazar, M., Alexander, A.L.: An error analysis of white matter tractography methods: synthetic diffusion tensor field simulations. NeuroImage **20**(2), 1140–1153 (2003)
15. Le Bihan, D., Mangin, J.F., Poupon, C., Clark, C.A., Pappata, S., Molko, N., Chabriat, H.: Diffusion tensor imaging: concepts and applications. J. Magn. Reson. Imaging **13**(4), 534–546 (2001)
16. Murphy, D.B.: Fundamentals of Light Microscopy and Electronic Imaging. Wiley-Liss, New York (2001)
17. Pajevic, S., Basser, P.J.: Parametric and non-parametric statistical analysis of DT-MRI data. J. Magn. Reson. **161**(1), 1–14 (2003)
18. Papadakis, N., Xing, D., Houston, G., Smith, J., Smith, M., James, M., Parsons, A., Huang, C.H., Hall, L., Carpenter, T.: A study of rotationally invariant and symmetric indices of diffusion anisotropy. Magn. Reson. Imaging **17**, 881–892(12) (1999)
19. Peizer, D.B., Pratt, J.W.: A normal approximation for the binomial, F, beta, and other common, related tail probabilities, I and II. J. Am. Stat. Assoc. **63**(324), 1416–1457 (1968)
20. Pierpaoli, C., Basser, P.J.: Toward a quantitative assessment of diffusion anisotropy. Magn. Reson. Med. **36**(6), 893–906 (1996)
21. Pöthkow, K., Hege, H.C.: Positional uncertainty of isocontours: condition analysis and probabilistic measures. IEEE Trans. Vis. Comput. Graph. **17**, 1393–1406 (2011) (Preprint)
22. Schultz, T., Theisel, H., Seidel, H.P.: Segmentation of DT-MRI anisotropy isosurfaces. In: Museth, K., Möller, T., Ynnerman, A. (eds.) EuroVis07: Joint Eurographics – IEEE VGTC Symposium on Visualization 2007, pp. 187–194. Eurographics, Norrköping (2007)
23. Snook, L., Plewes, C., Beaulieu, C.: Voxel based versus region of interest analysis in diffusion tensor imaging of neurodevelopment. NeuroImage **34**(1), 243–252 (2007)
24. Zhukov, L., Museth, K., Breen, D., Whitaker, R., Barr, A.H.: Level set modeling and segmentation of DT-MRI brain data. J. Electron. Imaging **12**, 125–133 (2003)

Part V
Beyond Second-Order Diffusion Tensor MRI

Classification Study of DTI and HARDI Anisotropy Measures for HARDI Data Simplification

Vesna Prčkovska, Maxime Descoteaux, Cyril Poupon, Bart M. ter Haar Romeny, and Anna Vilanova

Abstract High angular resolution diffusion imaging (HARDI) captures the angular diffusion pattern of water molecules more accurately than diffusion tensor imaging (DTI). This is of importance mainly in areas of complex intra-voxel fiber configurations. However, the extra complexity of HARDI models has many disadvantages that make it unattractive for clinical applications. One of the main drawbacks is the long post-processing time for calculating the diffusion models. Also intuitive and fast visualization is not possible, and the memory requirements are far from modest. Separating the data into anisotropic-Gaussian (i.e., modeled by DTI) and non-Gaussian areas can alleviate some of the above mentioned issues, by using complex HARDI models only when necessary. This work presents a study of DTI and HARDI anisotropy measures applied as classification criteria for detecting non-Gaussian diffusion profiles. We quantify the classification power of these measures using a statistical test of receiver operation characteristic (ROC) curves applied on *ex-vivo* ground truth crossing phantoms. We show that some of the existing DTI and HARDI measures in the literature can be successfully applied for data

V. Prčkovska (✉)
Center for Neuroimmunology, Department of Neurosciences, Institut Biomedical Research August Pi Sunyer (IDIBAPS), Hospital Clinic of Barcelona, Spain
e-mail: VPRCHKOV@clinic.ub.es

B.M. ter Haar Romeny · A. Vilanova
Department of Biomedical Engineering, Eindhoven University of Technology, Eindhoven, The Netherlands
e-mail: B.M.terHaarRomeny@tue.nl; A.Vilanova@tue.nl

C. Poupon
NeuroSpin, CEA Saclay, Gif-sur-Yvette Cedex, France
e-mail: cyril.poupon@cea.fr

M. Descoteaux
Computer science department, Université de Sherbrooke, Sherbrooke, QC, Canada
e-mail: m.descoteaux@usherbrooke.ca

D.H. Laidlaw and A. Vilanova (eds.), *New Developments in the Visualization and Processing of Tensor Fields*, Mathematics and Visualization,
DOI 10.1007/978-3-642-27343-8_12, © Springer-Verlag Berlin Heidelberg 2012

classification to the diffusion tensor or different HARDI models respectively. The chosen measures provide fast data classification that can enable data simplification. We also show that increasing the b-value and number of diffusion measurements above clinically accepted settings does not significantly improve the classification power of the measures. Moreover, we show that a denoising pre-processing step improves the classification. This denoising enables better quality classifications even with low b-values and low sampling schemes. Finally, the findings of this study are qualitatively illustrated on real diffusion data under different acquisition schemes.

1 Introduction

Diffusion tensor imaging (DTI) is a recent magnetic resonance imaging (MRI) technique that can map the orientation architecture of neural tissues in a completely non-invasive way by measuring the directional specificity (anisotropy) of local water diffusion [1]. The diffusion tensor model, however, has well known limitations in areas of complex intra-voxel heterogeneity with crossing fibers, where the diffusion process cannot be modeled as Gaussian. Nonetheless, DTI is still very popular and has many advantages such as fast and clinically feasible acquisition schemes (typically, number of gradients (NG) from 7 to 60, b-value of 1,000 s/mm^2 and total acquisition time of 3–5 min), fast post-processing of the data that allows interactivity in the data exploration, simple visualization techniques and modeling using well-developed tensor mathematics. To overcome the limitations of DTI, more sophisticated models were introduced using high angular resolution diffusion imaging (HARDI). For HARDI, significantly more diffusion gradients are acquired (from sixty to a several hundred) in order to reconstruct a spherical probability function (SPF) that either recovers the underlying fiber populations or depicts certain diffusion properties. Popular HARDI reconstruction techniques include apparent diffusion coefficient (ADC) modeling [2, 3], Q-Ball imaging [4], diffusion orientation transform (DOT) [5], spherical deconvolution (SD) [6, 7], and several other model-based methods. The output produced by the above techniques is always given in the form of a spherical function $\psi(\theta,\phi)$ that characterizes the local intra-voxel fiber structure. This function can be represented using a truncated spherical harmonics (SH) expansion

$$\psi(\theta,\phi) = \sum_{l=0}^{l_{max}} \sum_{m=-l}^{l} a_{lm} Y_{lm}(\theta,\phi) , \tag{1}$$

where Y_l^m represent the spherical harmonics of order l and phase m, and l_{max} is the truncation order of the SH series.

HARDI has obvious advantages over DTI in more composite fiber configurations, but has several drawbacks that accompany this complex modeling: longer processing time of the data (that can typically take a few hours up to a few days), inability to interactively explore the data because of over-cluttered and computationally heavy visualization as well as longer data acquisitions. Hence, one wonders if a complex high-order modeling of the data is always needed (i.e., at every

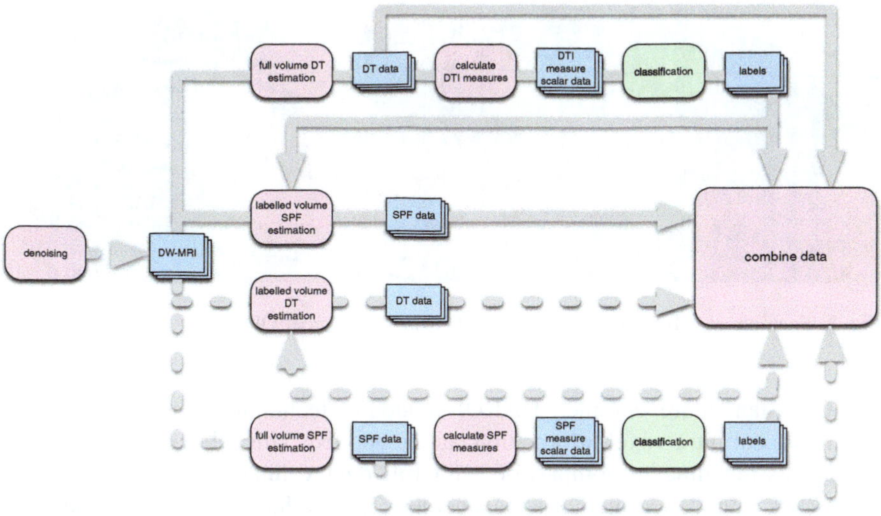

Fig. 1 Schematic diagram of the pipeline for DTI and HARDI combination. The *blue stacked rectangles* represent data volumes. The *pink round rectangles* represent data transformations. The main contribution of this chapter is focused in the classification part colored as *green*. The *arrows* capture the flow of the process, whereas *dashed arrows* depict optional scenarios

voxel) or merits its drawbacks? In crossing areas, it is certainly justified, but for a large part of the white matter, there is a significant number of single fiber voxels where high-order modeling might be redundant. Thus, it is important to be able to classify regions of single fiber (anisotropic-Gaussian) and crossing fibers (non-Gaussian) in white matter in a fast and reliable way. This can lead to reducing the modeling complexity in areas where it is not needed and enabling possibilities for data simplification. The advantages would be significant for further post-processing and visualization of the data, especially with respect to reducing computer memory requirements. This will undoubtedly make HARDI data easier to manipulate and interact with, making it more attractive for clinical applications.

One possibility for fast classification of the DW-MRI data is by identifying the type of anisotropy in each voxel by some of the anisotropy measures for DTI and HARDI. These measures are fast to calculate since they are scalar measures calculated on the eigenvalues of the diffusion tensor, **D**, in the first case or the SH coefficients in the latter.

Classification of the data by these measures in three compartments will allow masking of the data in the isotropic areas where the gray matter and the ventricles belong, using simple diffusion tensor model in the anisotropic-Gaussian regions and applying more sophisticated high order modelings in the non-Gaussian regions in a fast manner. There are two ways in which we can look at this problem for data combination and thus, simplification (see Fig. 1).

In the first scenario, from the DW-MRI data (that can be additionally denoised) we can calculate the DT model and apply scalar anisotropy measures that label the data into three compartments. Afterwards, in the compartment labeled as

Fig. 2 Example of hybrid visualization of CSD [6] and 2nd order ODFs

non-Gaussian, more complex HARDI modeling techniques that provide more accurate local intra-voxel information can be applied. This allows reducing of the postprocessing time in comparison with full data volume modeled by HARDI techniques and better memory management: one value in the area labeled as isotropic, a diffusion tensor for the anisotropic Gaussian areas and the rest modeled with high-order SPFs. However, this would require reliable classification by the DTI measures especially with respect to small number of false positives in the labeling, since this would underestimate the data.

In the second scenario, we model the data by a HARDI modeling technique (preferably by linear models like ADC or Q-Ball that are relatively fast to calculate) and then apply HARDI anisotropy measures. These measures also label the data into three compartments, and similarly the data can be modeled by DT or HARDI SPFs. In this scenario, to get better angular resolution, we can additionally choose to apply a non-linear (more time demanding) technique as constrained spherical deconvolution (CSD) [6, 7] for the non-Gaussian regions. Presenting the data by combining both of the data representation models would benefit in faster visualizations with better context especially since the data in the anisotropic Gaussian regions would be significantly simplified.

As an example, in Fig. 2 we show a hybrid visualization of the simplified data (labeling provided by generalized anisotropy (GA) classification) from an in-vivo dataset represented by 8th order CSD [6] in the non-Gaussian classified regions, and 2nd order ODFs in the anisotropic-Gaussian regions. The combining process follows the second scenario described previously. The difference in running time from the most naive implementation of CSD, is as follows. Computing CSD of order 8 for the whole brain in white matter mask: 540 min (36,601 voxels). Computing CSD of order 8 in labeled crossing : 120 min (8,164 voxels). Computing 2nd order ODFs in labeled linear : 19 s.[1] With hybrid data modeling, there is a gain of almost

[1]These times were calculated on a 1.66 GHz processor dual core Intel machine with 2 GB of RAM. CSD is a non-linear method that takes several iteration to perform the constrained regularization, which goes back and forth between at least 300 points on the sphere and the order 8 SH representation. This can be greedy and in our implementation takes approximately 0.5–1 s per

factor 5 in time for computation compared to modeling full brain data with the same high order model. Also interaction in the visualization pipeline becomes possible, even for a full brain slice.

In this work we examine and compare the classification power of different DTI and HARDI anisotropy measures. Additionally, we try to answer whether there is a significant gain from the measures calculated from the more complex HARDI models. We use pattern recognition schemes for investigating the classification potentials of these measures on a ground truth *ex-vivo* phantom under different b-values. Furthermore we show some qualitative real data results that corroborate some of the conclusions from the phantom data. To improve the quality of the real data scanned under clinically acceptable schemes, we use a denoising scheme [8] that recommends improvement of the coherence of the classified regions. We thus come to several conclusions suggesting that HARDI processing and data interaction are possible in a clinical setting.

2 Related Work

The work of Alexander et al. [3], is among the first that classifies the data in three compartments: isotropic, anisotropic-Gaussian, and non-Gaussian. This work uses ANOVA F-test based on the SH representation of the apparent diffusion coefficient (ADC) profile for several truncation orders l_{max}. Comparison of the measured ADC with the estimated one is required in each step, and therefore this process is slow and memory consuming. Furthermore it is necessary to calculate a critical value for the F-test to achieve stopping criteria, and this threshold is difficult to find given that the whole process is not interactive. Behrens et al. [9] use automatic relevance determination integrated in a Bayesian modeling framework to simplify the problem of tracking in a multi-orientation field. Hosey et al. [10] use an extension of a Markov chain method that infers the probability density function of up to two intra-voxel fiber orientations. However, both of the mentioned techniques are computationally intensive to implement. This is mainly due to the Bayesian estimation of the parameters making these methods iterative, with lack of simple user interaction. Interactive data classification can be of great importance for simplification of the HARDI data, especially in the case of clinical applications. It is also valuable for immediate identification of uncertainty regions in the DTI-based fiber tracking that has already been used widely. The speed in accurate identification of anisotropic-Gaussian and non-Gaussian regions in the data can accelerate the whole postprocessing pipeline for the complex HARDI data. Wide range of anisotropy measures has been proposed in literature [2, 4, 11–15]. Several authors [3, 12, 13] have attempted to use some of these measures to classify non-Gaussian profiles, but all these attempts have been made on the apparent diffusion

voxel. This time can obviously be improved by parallelizing the code and changing the parameters of CSD regularization (less iteration and faster stopping criteria).

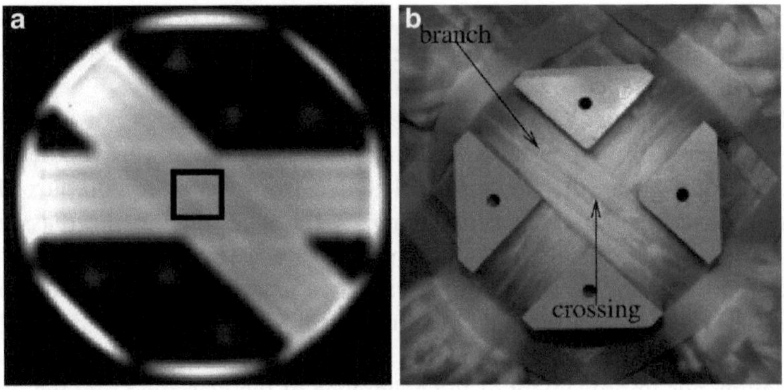

Fig. 3 (**a**) Fast spin-echo map and a region of interest for the crossing voxels of the 45° *ex-vivo* phantom and (**b**) Picture of the 90° *ex-vivo* phantom

coefficient (ADC) profiles and without convincing real data results. In our previous work [16], we applied these HARDI anisotropy measures not only to ADC profiles but to different spherical probability functions like Q-Ball and the DOT. In this chapter, however, we extend the previous ad-hoc analysis by thorough examination of the classification power of the measures by ROC curves, histograms and scatter plots. We additionally extend the analysis to DTI anisotropy measures and compare their classification power with the measures derived from HARDI data.

3 Diffusion Data Acquisition

The details of the *ex-vivo* phantom data and in-vivo human data used in this study are explained below.

Ex-vivo phantom: To test our classification measures, we use two *ex-vivo* phantoms with fibre bundles crossing at 45° and 90° [17] (Fig. 3). These datasets serve as ground truth, where the number of crossing voxels is known. The phantom data was acquired on a 1.5T Signa MR system (GE Healthcare), TE/TR = 130 ms/4.5 s,12.0 s (45° and 90° phantom, respectively), BW = 200 KHz. We analyze the data acquired at two b-values of $b = 2{,}000$ and $b = 8{,}000$ s/mm^2, along 200 uniform directions.

Human: Diffusion acquisitions were performed using a twice focused spin-echo echo-planar imaging sequence on a Siemens Allegra 3T scanner, with FOV 208 × 208 mm, isotropic voxels of 2 mm. Ten horizontal slices were positioned through the body of the *corpus callosum* and *centrum semiovale*. Uniform gradient direction schemes with 49 and 121 directions were generated with the electrostatic repulsion algorithm [18] and the diffusion-weighted volumes were interleaved with b_0 volumes every 12th scanned gradient direction. Datasets were acquired at b-values of 1,000, 1,500, 2,000, 3,000, 4,000 s/mm^2 and in the same session, two anatomical data sets (192 slices, isotropic 1 mm voxels) were acquired using the ADNI sequence for registration. Finally, before HARDI reconstruction of the ADC, Q-Ball and DOT, we applied a denoising preprocessing step [8], available online,[2] to correct for the Rician noise bias in the datasets.

4 Methods

4.1 HARDI Measures

We implemented several HARDI anisotropy measures from the literature: generalized anisotropy (GA) [14], generalized fractional anisotropy (GFA) [4], the cumulative residual entropy (CRE) [11, 12], as well as fractional multifiber index (FMI) [2], and R_0, R_2, R_i [13] (see Table 1). Most of the measures are calculated directly from the SH coefficients of the corresponding spherical probability functions and therefore are extremely fast to calculate.

These measures were applied on the ADC profiles [2, 3], analytical Q-ball [19] and the DOT [5]. Since the DOT was originally proposed in complex SH basis, we adapted it to real SH, and solve the spherical harmonic transform in the parametric DOT by least-square fit. This way we obtain a probability density function (PDF) represented in real spherical harmonic coefficients, and all the anisotropy measures can be applied to it. The DOT generally produces much sharper glyph profiles for high radius Rad_0, at the cost of more noisy profiles with spurious peaks. Finding the *best* Rad_0 in real data is difficult and often done by visual observation [5]. Hence, to avoid this Rad_0 selection problem and inspired by definitions of the ODF from Q-ball imaging [4] and the marginal ODF (mODF) from diffusion spectrum imaging (DSI) [20], we propose similar ODFs computed from the DOT as:

$$\begin{aligned} \psi_{\text{DOT-ODF}}(\theta,\phi) &= \int_0^{Rad_{0max}} P(r,\theta,\phi)dr, \\ \psi_{\text{DOT-mODF}}(\theta,\phi) &= \int_0^{Rad_{0max}} P(r,\theta,\phi)r^2dr, \end{aligned} \tag{2}$$

where $P(r,\theta,\phi)$ is the PDF computed from DOT [5], and Rad_0 is set to a conservatively high value.

As a discrete binary measure for the classification, we propose the number of maxima (NM). NM uses the number of local maxima of the min-max normalized SPFs profiles, where the discrete spherical function surpasses a certain threshold (here, we use 0.6) from points on a fine discrete mesh (5th order of tessellation of icosahedron), using a finite difference search on the mesh points. Moreover, for

[2]http://www.irisa.fr/visages/benchmarks/

Table 1 Scalar measures for HARDI. a_{lm} are spherical harmonics coefficients of order l and phase m

Name	Abbrev.	equation
Generalized anisotropy [14]	GA	$= 1 - \frac{1}{1 + (250\,\mathrm{V})^{e(V)}}$ where $e(V) = 1 + \frac{1}{1 + 5{,}000\,\mathrm{V}}$, $V = \frac{1}{9a_{00}^2} \sum_{l=2}^{l_{max}} \sum_{m=-l}^{l} \lvert a_{lm} \rvert$
Generalized fractional anisotropy [4]	GFA	$= \sqrt{1 - \frac{a_{00}^2}{\sum_{l=0}^{l_{max}} \sum_{m=-l}^{l} \lvert a_{lm} \rvert}}$
Cumulative residual entropy [11,12]	CRE	$= -\sum_{i=2}^{M} P(\psi_n > \lambda_i) \log P(\psi_n > \lambda_i) \Delta\lambda_i$, where $\lambda_1 < \ldots < \lambda_M$, $\psi_n = \psi_{norm}(\theta, \phi)$
Fractional multi-fiber index [2]	FMI	$= \frac{\sum_{l=4}^{l_{max}} \sum_{m=-l}^{l} \lvert a_{lm} \rvert^2}{\sum_{l=2} \sum_{m=-l}^{l} \lvert a_{lm} \rvert^2}$
Isotropic ratio [13]	R_0	$= \frac{\lvert a_{00} \rvert}{\sum_{l=0}^{l_{max}} \sum_{m=-l}^{l} \lvert a_{lm} \rvert}$
Linear ratio [13]	R_2	$= \frac{\sum_{l=2}^{l_{max}} \sum_{m=-l}^{l} \lvert a_{lm} \rvert}{\sum_{l=0}^{l_{max}} \sum_{m=-l}^{l} \lvert a_{lm} \rvert}$
Multi-fiber ratio [13]	R_i	$= \frac{\sum_{l=4}^{l_{max}} \sum_{m=-l}^{l} \lvert a_{lm} \rvert}{\sum_{l=0}^{l_{max}} \sum_{m=-l}^{l} \lvert a_{lm} \rvert}$

better visual perception, in our figures we generate min-max normalized RGB color coded glyphs, although one must keep in mind that this normalization enhances angular contrast of glyphs in the white matter but also deforms isotropic glyphs considerably.

Table 2 Scalar measures for DTI. $\lambda_1 > \lambda_2 > \lambda_3$ are the corresponding eigenvalues of the diffusion tensor **D**

Name	Abbrev. equation
Mean diffusivity [22]	$MD = \mathrm{tr}(\mathbf{D})/3 = (\lambda_1 + \lambda_2 + \lambda_3)/3$
Fractional anisotropy [22]	$FA = \frac{\sqrt{(\lambda_1-\lambda_2)^2+(\lambda_2-\lambda_3)^2+(\lambda_1-\lambda_3)^2}}{\sqrt{2(\lambda_1^2+\lambda_2^2+\lambda_3^2)}}$
Relative anisotropy [22]	$RA = \frac{\sqrt{(\lambda_1-\lambda_2)^2+(\lambda_2-\lambda_3)^2+(\lambda_1-\lambda_3)^2}}{\sqrt{2}(\lambda_1+\lambda_2+\lambda_3)}$
Linear anisotropy [21]	$C_l = (\lambda_1 - \lambda_2)/(\lambda_1 + \lambda_2 + \lambda_3)$
Planar anisotropy [21]	$C_p = 2(\lambda_2 - \lambda_3)/(\lambda_1 + \lambda_2 + \lambda_3)$
Isotropy [21]	$C_s = 3\lambda_3/(\lambda_1 + \lambda_2 + \lambda_3)$

4.2 DTI Measures

We implemented DTI anisotropy measures: linear anisotropy C_l, planar anisotropy C_p and isotropy C_s [21] as well as the well-known fractional anisotropy (FA) and mean diffusivity (MD) [22] (see Table 2).

These measures were applied on the diffusion tensors estimated from the same DW-MRI data as used for the HARDI modeling.

4.3 Analysis of Measures

To quantify the classification power of the DTI and HARDI measures, we use the statistical test of receiver operation characteristic (ROC) curves [23] for the hardware phantom data. The ROC curves describe the classification power of each of the measures for separation of the data into three distinct compartments: isotropic, non-Gaussian and anisotropic-Gaussian by using two thresholds. We apply this analysis only to the phantom data, since only there we know the ground truth for crossing voxels. Furthermore, we explore the distributions of the values from each measure in different phantom data configurations by histograms. At the end we suggest some interesting combinations of different measures for improving the classification power of the individual measures. In addition, we discuss the differences between DTI and HARDI measures. We describe each of these analysis in details below.

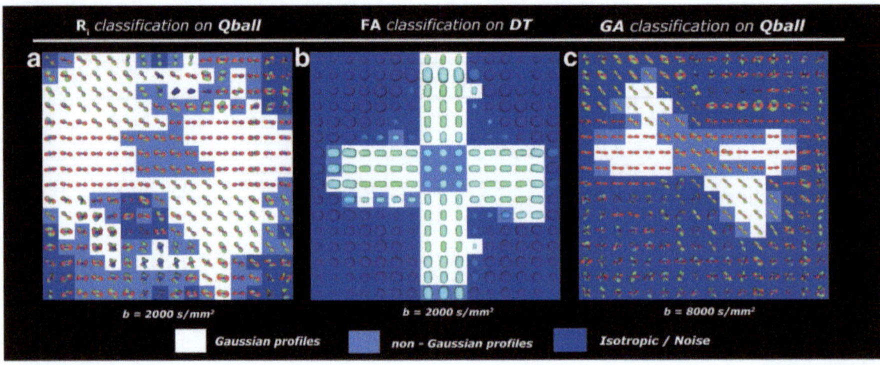

Fig. 4 Classification results from the phantom data

4.3.1 Different Phantom Configurations

Before explaining the data analysis we would like to review the *ex-vivo* phantom data and its configurations. We will derive our conclusions based on three different phantom crossing datasets varying either in configuration or b-value (see Fig. 4). We have the 45° of crossing phantom data, with exactly 12 voxels of ground truth crossings. The rest of the $16 \times 16 = 256$ voxels belong to single fiber voxels (estimated to 64 voxels) and isotropic voxels considerably deformed by the MRI noise. The 45° angle of crossing is a challenging angle where most of the HARDI techniques fail to recover multiple fiber populations, especially at low b-values. Therefore we investigate this configuration under two different b-values of 2,000 and 8,000 s/mm^2. In addition we analyze the phantom data of 90° only at b-value of 2,000 s/mm^2 (b-value of 8,000 s/mm^2 is not interesting to analyze for this angular configuration). Here, we have exactly nine voxels of ground truth crossings and the rest belong to single fiber population and noise. There are a few points that we need to keep in mind. Due to partial volume effect, some of the linear voxels might exhibit non-Gaussian diffusion properties. Since this is to be expected in real data as well, we simply need to keep it in mind when analyzing the data. Due to the high SNR value of the *ex-vivo* phantom data, we do not need a denoising phase.

4.3.2 ROC and Histogram Analysis of the Phantom Data

For the DTI and HARDI anisotropy measures we can quantitatively describe their classification power using binary classification statistical test. First, the measures must be thresholded to obtain the classification, and this process is sensitive. Two thresholds are needed to separate the interval of anisotropy values into three distinct compartments: isotropic, non-Gaussian and anisotropic-Gaussian. Afterwards we can calculate the ROC curves that graphically represent the relationship between

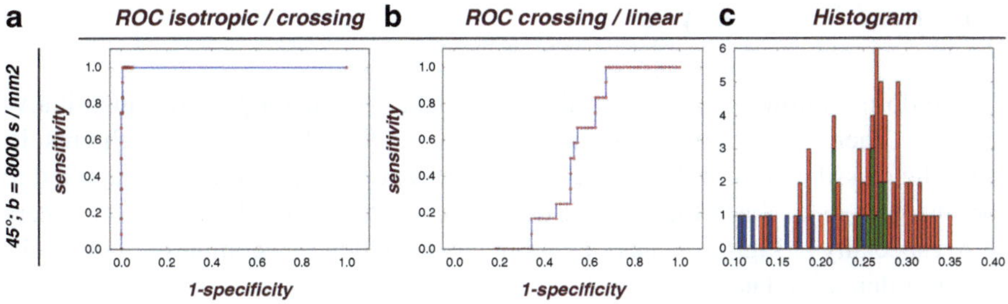

Fig. 5 ROC curves for isotropic/crossing and crossing/linear scenario and histogram for FA applied to the 45° *ex-vivo* phantom data at b-value of 8,000 s/mm^2. The color coding of the histogram: *blue*—isotropic, *green*—crossings, *red*—linear voxels

specificity and sensitivity of the voxels classified as: isotropic (from the noise), crossing (non-Gaussian) and linear (anisotropic-Gaussian) [23]. The sensitivity measures the proportion of actual positives which are correctly identified as such, and the specificity measures the proportion of negatives which are correctly identified.

$$\begin{aligned} sensitivity &= \frac{number\ of\ True\ Positives}{number\ of\ True\ Positives + number\ of\ False\ Negatives} \\ specificity &= \frac{number\ of\ True\ Negatives}{number\ of\ True\ Negatives + number\ of\ False\ Positives} \end{aligned} \quad (3)$$

Since separation into three distinct compartments is desirable, we need to calculate two ROC curves per measure, one of which represents the classification power between isotropic and non-Gaussian profiles (Fig. 5a), and the other between non-Gaussian and anisotropic Gaussian (Fig. 5b). Calculating two ROC curves is possible due to the distribution of the anisotropy as low in the isotropic parts, medium in the non-Gaussian regions and high for anisotropic Gaussian. To quantify the accuracy of the measures we calculate and report the area under the ROC curves (see appendix). The larger the area under the curve, the better the separation of the profiles by the examined measure.

For illustrating the distribution of anisotropy values, we use histograms (Fig. 5c). However, as expected many DTI measures do not have this smooth transition, whereas most of the HARDI measures exhibit more desirable properties.

4.3.3 Scatter Plot Analysis for Combination of Measures

To investigate the possibilities of combining and thus increasing the classification power of the measures we do some preliminary experiments with scatter plots, where we combine different DTI and HARDI measures together and look at the distribution of the combined measure values (see Fig. 8).

4.4 *Real Data Analysis*

As we do not know the ground truth in real data, we cannot perform quantitative analysis. Therefore we limit our analysis to qualitative observations on the results from the classification of the data done by the same measures that were applied to the phantom data. Since the SNR of our clinically obtained DW-MRI data is very low (especially at high b-values), experiments were done by comparing the results with pre-denoised data.

4.4.1 Denoising of the Real Data

We use a non-local mean filter with Rician noise correction to denoise the DW data before HARDI reconstruction. This method was shown [8] to have the desired effect of correcting for the noise bias without blurring-out figure crossing information. Hence, it improves scalar measures extracted from DTI and HARDI and does not reduce the angular profiles of HARDI glyphs. The computation time of this filter depends on spatial resolution and the number of maxima. For example, on our real dataset of $104 \times 104 \times 10 \times 121$, the denoising takes 16 min when computed over four processors 3 GHz and 8 GB RAM.

5 Results

In this section, we present the quantitative analysis of the *ex-vivo* phantoms and qualitative results from the real data.

5.1 *Phantom Results*

The 45° is a challenging angle where most of the HARDI techniques struggle to detect multiple maxima, especially at low b-values. We will first analyze the results from the maxima detection. As pointed out in the work of Prčkovska et al. [24], DOT has the potential of recovering small angles regardless of the b-value, which we show in the table of Fig. 6. Only the DOT (and its derivations) manages to recover two fiber populations in the crossing regions at low combination of crossing angle and b-value (in this case 45° and 2,000 s/mm^2). In the table we report the success at recovering two maxima in the crossing voxels by all of the examined SPFs for SH orders 4, 6 and 8. We additionally report the first Rad_0 for the DOT and its derivations in which the success is greater than 50 %. Even more interesting, we observe that the derivations of the DOT discussed in Sect. 4, with its ODFs (DOT-

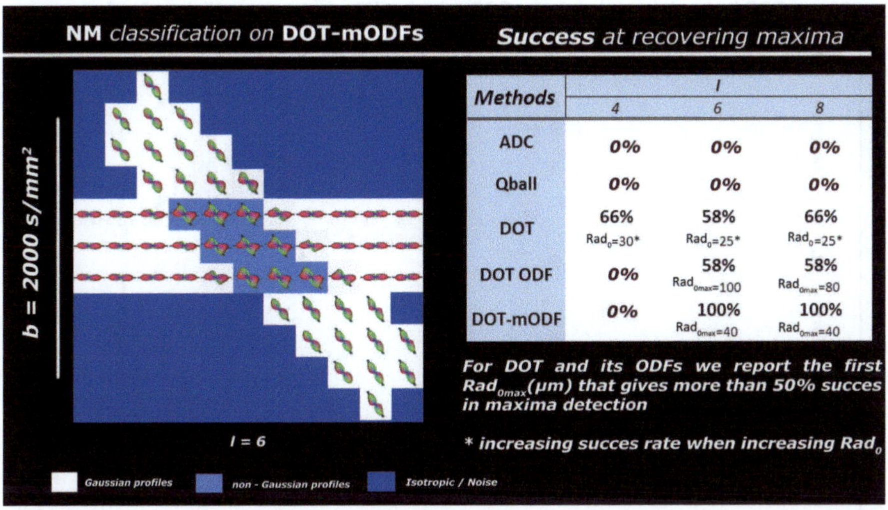

Fig. 6 Classification results from the 45° phantom data. The noise voxels were masked out by FA, since NM classification gives multiple maxima in these areas

ODF and DOT-mODF) manifest similar behavior to the DOT itself, which show a better angular resolution than Q-Ball and suggest a better choice of reconstruction algorithm for fiber tracking purpose. The results from the NM classification on the 90° phantom are omitted, due to the 100 % success in the classification of the non-Gaussian voxels demonstrated in all reconstruction methods. Increasing the b-value to 8,000 s/mm^2 improves the angular resolution of Q-Ball as expected, and crossings are starting to be observed in the 45° dataset. We presented the classification results from the real data to an anatomist who evaluated the accuracy of the classification, and suggested preferences over some of the results. In the following paragraphs, as we discuss the real data results we include the feedback from the anatomist.

For the DTI and HARDI anisotropy measures, we can quantitatively describe the classification power of the 45° and 90° phantoms by examining the shape of the corresponding ROC curves, and calculating the area under the ROC curves. Additionally we observe the distribution of the measure values in the histograms as described in Sect. 4.3.2. In the appendix, the area under all of the examined DTI and HARDI ROC curves is reported. From the observation of the ROC curves and the values of the area under the ROC curves, we draw several conclusions. Increasing the order of SH representation does not significantly improve the classification power of the measures. Therefore, for simplicity, in our phantom data results we report only an SH order of 4. For the DOT derivations (DOT-mODF and DOT ODF) the results become worse at high SH order, due to the very high and densely distributed anisotropy values. The DOT in every configuration and SH order gives

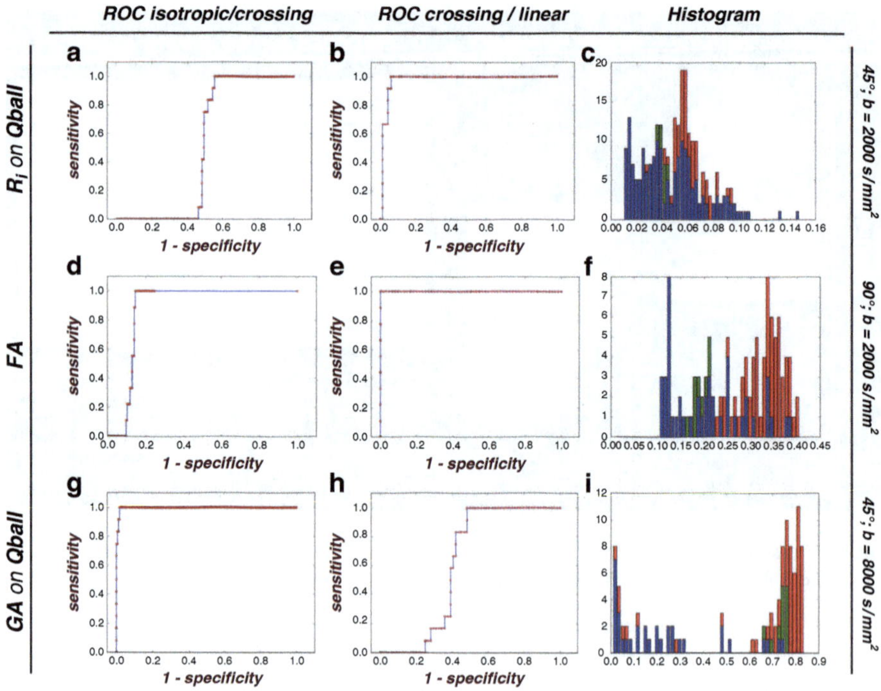

Fig. 7 ROC and histogram examples from the phantom data. The coloring in the histograms indicates: *blue*—isotropic, *green*—crossings, *red*—linear

bad results due to the reasons mentioned above, even though it produces sharper angular profiles. Also the measures applied to DOT derivations in general produce worse results than those applied to the rest of the HARDI models. This can be observed in the appendix Fig. A.1a colored with red stating bad classifier. Most of the measures (DTI and HARDI), are significantly better in separating the data between isotropic and crossing voxels (see Fig. 7d, g; Fig. A.1a, b) . The DTI measures even outperform the HARDI measures in many cases (see Fig. 7a, d); however, note that the presented results are for different angular configurations). For the separation of the crossing and linear areas, on the other hand, the situation is more complex. In general, many HARDI measures like CRE, GA, GFA and $R2$ on ADC and Q-Balls have medium classification power and are comparable to the DTI measures like C_l, FA and MD. Notably bad in many scenarios appear to be C_p . A measure that stands out for good classification, especially of the challenging 45° angle between crossing and linear combinations, is R_i applied on Q-Balls. This is to be expected, given the definition of the measure (see Fig. 7b).

From our *ex-vivo* phantom study we can conclude that the classification power between anisotropic Gaussian and non-Gaussian areas of the HARDI measures

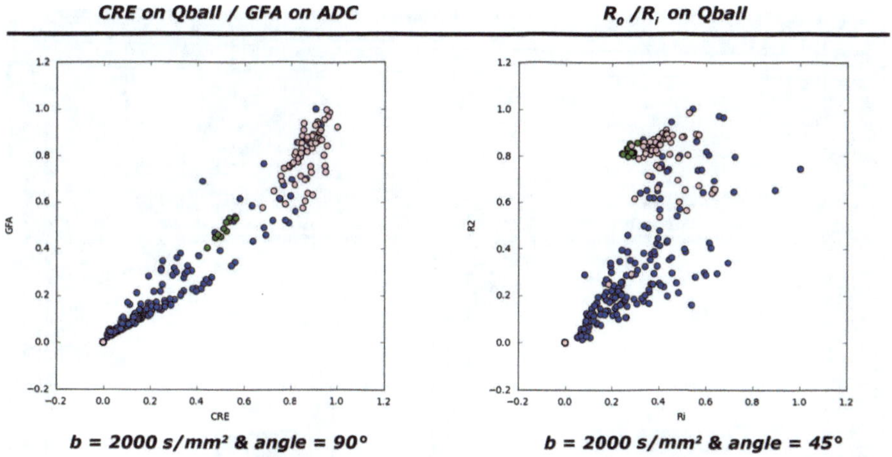

Fig. 8 Scatter plots from combination of different measures. *Blue* color stands for isotropic voxels, *green* crossing voxels and *red* single fiber voxels

in general is slightly better than that of the DTI measures. This difference is more prominent at smaller angles of crossings. For distinguishing isotropic from non-Gaussian regions, the DTI measures outperform HARDI measures. The distribution of the anisotropy values from low to high following the isotropic/non-Gaussian/anisotropic-Gaussian pattern, is more noticeable in the HARDI measures like GA, GFA and CRE (e.g., Fig. 7f, i). R_i on Q-Balls is a good classifier between non-Gaussian and anisotropic-Gaussian regions.

There is potential in combining measures to increase the separation between classes. Some measures show a better separation between isotropic and non-Gaussian, and others between non-Gaussian and anisotropic Gaussian. Furthermore, some measures perform better when applied to different SPFs. Preliminary results suggest that combination of CRE and GFA on Fig. 8 left and R_i and R_2 on Fig. 8 right can improve the classification power of the measures.

5.1.1 Human Data Results

The *centrum semiovale* was used to illustrate the qualitative analysis of the classification results. It is an interesting region for analysis, since fibers of the *corpus callosum* (CC), *corticospinal tract* (CST), and *superior longitudinal fasciculus* (SLF) form different two-fiber and three-fiber crossing configurations in that area. The region-of-interest (ROI) was defined on a coronal slice (see Fig. 9a). It is important to mention that all the real data results are from similar regions, since they are different DW-MRI scans from the same subject, and have not been registered. We presented the classification results to an anatomist who evaluated the accuracy

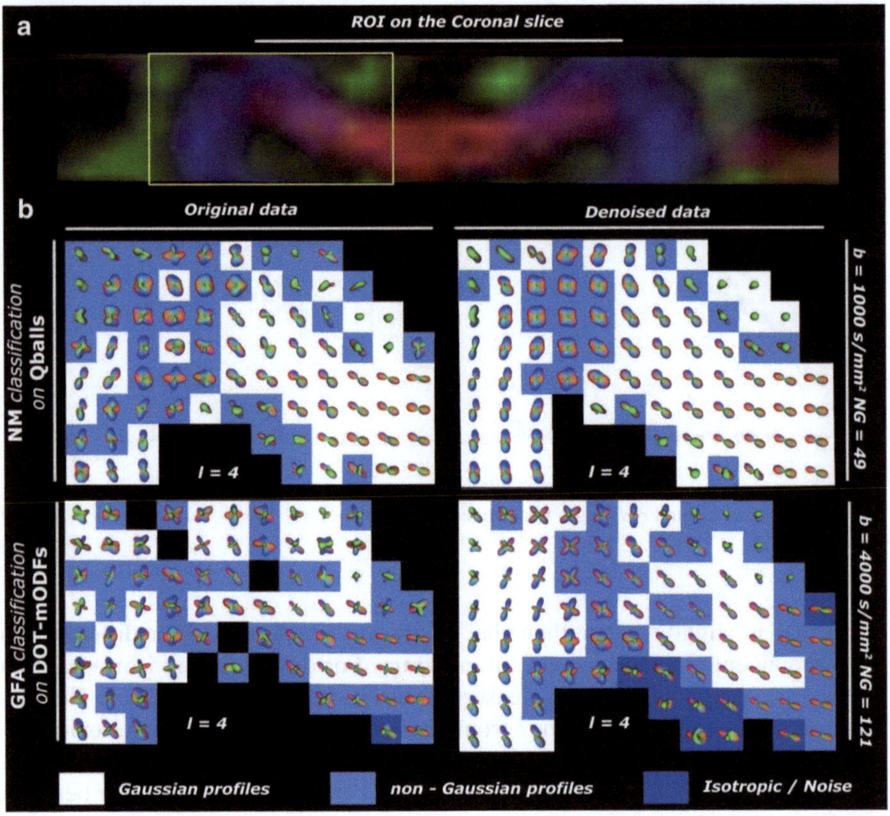

Fig. 9 The effect of denoising demonstrated on original versus denoised data in different acquisition schemes

of the classification, and suggested preferences over some of the results. In the following paragraphs, as we discuss the real data results we include the feedback from the anatomist. We applied the same classification measures as for the phantom study on the original and denoised data from our datasets. Denoising significantly improves the glyph profiles and the coherence of the non-Gaussian regions, as seen in Fig. 9. We also observe a decrease in the irregularities in the crossing profiles. Our results suggest that even at low b-value, low NG and low estimation SH order, there is success in recovering crossing diffusion patterns and identifying linear regions (see Fig. 10). The feedback from the anatomist followed the same rule. In general most of the classification done under b-value of 1,000 s/mm^2 and low sample of the gradients were found to be the best. For instance in Fig. 10 the classification from CRE applied to DOT-mODF at $b = 1{,}000$ s/mm^2 and NG $= 49$ was found to be the best due to the well spread crossing region.

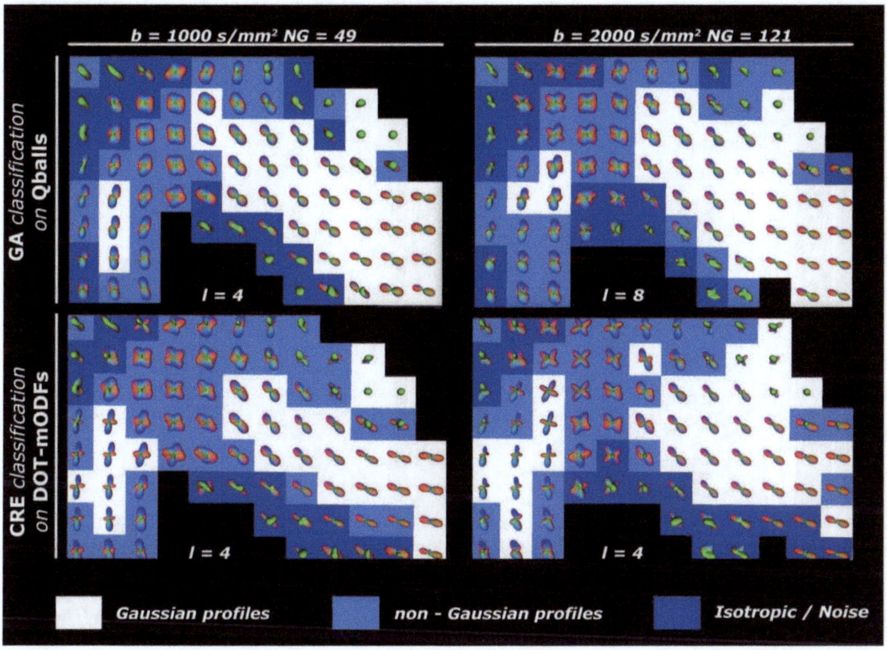

Fig. 10 Some examples from different classifications applied to in-vivo human data from the region of *centrum semiovale*

In contrast, going to very high b-values (i.e. $\geq$3,000 s/mm^2) and modeling the data with high SH order ($\geq$6) results in polluted glyphs regardless of whether denoising is performed. Comparing the results of the classification from different measures, we observe that increasing the b-value sharpens the HARDI profiles and benefits only for maxima extraction purposes. However, there is no significant gain in classification of non-Gaussian profiles, as observed in the phantom study. This is seen in Fig. 10, where we see sharper glyphs for DOT-mODF but similar classification power regardless of the measure or acquisition scheme. We also note that increasing the model order ($l > 4$) does not increase the classification power, which coincides with the conclusions from our phantom study. This leads to the conclusion that 49 directions is sufficient for recovering most of the crossings and non-Gaussian voxels, which means that the acquisition time can be significantly reduced (compared to a 121 NG acquisition).

Figure 11 demonstrates a comparison between FA and R_i. We observe that FA exhibits similar classification properties to the ones observed in 90° phantom, thus giving a nice contrast in the *centrum semiovale*. R_i on the other hand, has a bigger problem separating isotropic areas from non-Gaussian ones, which reference the poor performance in the isotropic/crossing ROC curve (Fig. 7a).

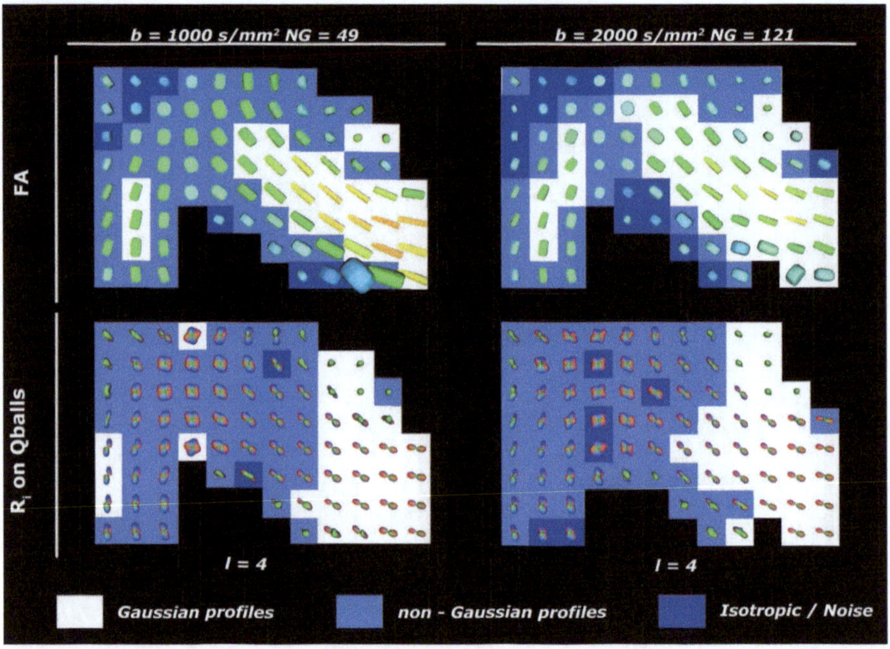

Fig. 11 Comparison DTI versus HARDI classification

The anatomist similarly found the classification from FA at $b = 1,000\,\text{s/mm}^2$ and $NG = 49$ to outperform the one from the higher b-value and denser gradient sampling. The classifications from R_i were found to have over-classified crossing area.

6 Discussion and Conclusions

Finding the correct threshold for classification in real data is important for accurate classification, and often depends on the b-value from the acquisition protocol and the angular configuration. In our study, the thresholds found in the 90° phantom configuration were very similar to the thresholds used to classify the real data at the same b-value. The thresholding process for the in-vivo data can be significantly improved by a semi-automatic algorithm for detection of the thresholds. The user can additionally give feedback by identifying regions with positive and negative examples.

There are a few important messages from this work. Denoising as a pre-processing step improves the coherence of the classification areas and enhances the

HARDI profiles, as reported by Descoteaux et al. [8]. ADC and Q-Ball demonstrate significant classification information, even though they sometimes lack sufficient angular resolution for small crossing-angle discrimination. The sharper and slightly more noisy profiles produced by DOT and its derivation (and we believe this would be the case for SD techniques [7] as well) find more accurate numbers of maxima and are better suited for fiber tracking applications. Increasing the acquisition parameters (b-value $> 2{,}000\,\mathrm{s/mm^2}$ and NG > 80) as well as the model order do not significantly improve the classification power. In contrary, high b-value acquisitions produce low SNR datasets that are worse for classification, and result in polluted HARDI profiles. This corroborates the observations from the anatomist, who found the classifications applied on data with $b = 1{,}000\,\mathrm{s/mm^2}$ and NG $= 49$ to be the best. It is even doubtful if, in practice, these higher b-value datasets improve fiber tracking. Further studies would be needed to be able to estimate the exact optimal acquisition parameters for classification. For example, it would be interesting to acquire *ex-vivo* ground truth crossing phantoms with a higher variety of acquisition parameters. The results of the study presented in this chapter indicate that the optimal acquisition will be possible in a clinical environment, since relatively low acquisition time will be needed for the preferred acquisition parameter setting. DTI anisotropy measures are comparable with the HARDI measures like GA, GFA and CRE and in classifying isotropic from crossing regions often outperform.

In this work, we investigated a broad range of different DTI and HARDI anisotropy measures proposed in the literature and applied them as classification criteria for discriminating different fiber configurations within the white matter. All the measures were applied on the HARDI reconstructions and all, except for CRE and NM, are measures directly implemented on SH representation of the model or DT that can be calculated and thresholded in real time. Some of the measures such as GA, GFA and CRE applied on Q-balls and ADC behave in a similar fashion and are relatively good classification criteria. However, their power is comparable to DTI measures such as FA and MD. R_i exhibits strong classification power for separating crossings from linear areas even at low angle. However, due to the poor isotropic/crossing performance, it is recommended combining it with other measures. The NM measure belongs to a different category of measures because it does not need a thresholding process for classification. However, it is dependent on the HARDI profiles and can produce many false positives in the presence of noise.

A strong message that comes out of this work, is that the measures can be applied on different SPFs and still have the same classification power (especially in the case of ADC and Q-Ball). This means that the users can use any existing HARDI modeling technique and apply classification measures to distinguish between anisotropic-Gaussian and non-Gaussian profiles. If the non-Gaussian voxels are correctly classified in a first step, one can ignore all the other single fiber voxels and properly focus on the modeling and more accurate reconstruction of these voxels. Hence, as a second step, one can use a complex modeling approach, such as CSD and PAS-MRI [25] that take long computation time. In clinical setting, the simplification of the data into anisotropic-Gaussian and non-Gaussian areas can be useful and presents a new contrast as such, even though complex structures are oversimplified as non-Gaussian. It can lead to new ways to study the white matter, especially by enabling the possibility for interactive visualization and inspection of the data.

Future work will address combination of different measures for better reliability of the classified regions. Comparison of our simple and fast classification with some of the existing classification schemes (as in the work of Schnell et al. [26]) is addressed as future work. These methods use support vector machines or learning approach such as boosting on the entire set of measures to statistically determine the discriminative strength of each feature and therefore cannot be calculated at interactive speed. Thus, the comparison should be done for validation purposes of the methods only.

In this chapter, we are not dealing with the behavior of the classification in partial volume effect regions. If there is partial volume effect between different classes, it is expected that the selected model will correspond to the one that can represent the most complex configuration. For example, in the case of partial volume effect between isotropic and anisotropic-Gaussian, the voxel will be classified as anisotropic-Gaussian. Further studies would be necessary to determine the validity of this assumption.

Nonetheless, in this work we have shown that possible classification of anisotropic-Gaussian and non-Gaussian profiles can be done with some of the existing measures including scalar indices calculated from DTI data. For the DTI indices, however, we need to be careful as the distribution of the anisotropy values does not always follow the isotropic/non-Gaussian/anisotropic-Gaussian pattern. The data can therefore be simplified into linear, crossing and isotropic voxels. This means that more sophisticated hybrid methods, which are more time consuming can be applied only in the non-Gaussian areas, whereas in the anisotropic-Gaussian the profiles can be modeled with a simple 2nd order ODFs (see Fig. 2) and the isotropic profiles masked out. This gives considerable potential for the employment of the HARDI techniques in a clinical setting due to the moderate post-processing time. Another application of the classification information can be in visualizing uncertainties in fiber tracking algorithms by, for example, attributing transparency to the unreliable fiber tracts.

Acknowledgements We thank Alard Roebroeck from Maastricht Brain Imaging Center, Department of Cognitive Neuroscience, Faculty of Psychology, Maastricht University, The Netherlands and Pim Pullens from Brain Innovation B.V., Maastricht, The Netherlands for providing us with in-vivo datasets. This study was financially supported by the VENI program of the Netherlands Organization for Scientific Research NWO (Anna Vilanova) and by the Netherlands Organization for Scientific Research (NWO), project number 643.100.503 MFMV (Vesna Prčkovska).

Appendix

a

Measure	Profile	Cross/Linear			Average	Noise/Cross			Average
		b=2000s/mm^2		b=8000s/mm^2		b=2000s/mm^2		b=8000s/mm^2	
		45°	90°	45°		45°	90°	45°	
CRE	*adc*	0,671	1,000	0,409	0,693	0,938	0,864	0,999	0,934
	dot	0,310	0,256	0,346	0,304	0,656	0,614	0,413	0,561
	dotodf	0,408	0,776	0,376	0,520	0,908	0,127	0,979	0,671
	mdotodf	0,411	0,850	0,421	0,561	0,907	0,736	0,981	0,875
	qball	0,755	1,000	0,622	0,793	0,895	0,856	0,997	0,916
FMI	*adc*	0,551	0,107	0,660	0,439	0,091	0,070	0,008	0,056
	dot	0,785	0,761	0,754	0,767	0,317	0,228	0,748	0,431
	dotodf	0,633	0,663	0,608	0,635	0,132	0,157	0,485	0,258
	mdotodf	0,589	0,333	0,471	0,464	0,091	0,514	0,249	0,284
	qball	0,678	0,720	0,691	0,697	0,044	0,041	0,037	0,041
GA	*adc*	0,680	1,000	0,530	0,737	0,934	0,875	0,998	0,936
	dot	0,000	0,000	0,000	0,000	0,000	0,000	0,000	0,000
	dotodf	0,394	0,496	0,365	0,418	0,915	0,702	0,986	0,868
	mdotodf	0,902	0,618	0,994	0,838	0,025	0,445	0,000	0,157
	qball	0,749	1,000	0,611	0,786	0,898	0,873	0,997	0,923
GFA	*adc*	0,680	1,000	0,530	0,737	0,934	0,875	0,998	0,936
	dot	0,000	0,000	0,000	0,000	0,000	0,000	0,000	0,000
	dotodf	0,393	0,494	0,363	0,417	0,915	0,702	0,986	0,868
	mdotodf	0,902	0,617	0,995	0,838	0,025	0,444	0,000	0,157
	qball	0,749	1,000	0,611	0,786	0,898	0,873	0,997	0,923
R0	*adc*	0,475	0,000	0,605	0,360	0,074	0,187	0,001	0,087
	dot	0,801	0,778	0,824	0,801	0,318	0,228	0,756	0,434
	dotodf	0,602	0,524	0,633	0,586	0,089	0,234	0,017	0,113
	mdotodf	0,456	0,185	0,171	0,270	0,300	0,740	0,843	0,628
	qball	0,216	0,000	0,439	0,218	0,142	0,214	0,002	0,119
R2	*adc*	0,521	1,000	0,133	0,551	0,944	0,838	0,999	0,927
	dot	0,823	1,000	0,758	0,860	0,878	0,698	0,973	0,850
	dotodf	0,411	0,493	0,371	0,425	0,907	0,915	0,979	0,934
	mdotodf	0,405	0,826	0,368	0,533	0,911	0,044	0,982	0,646
	qball	0,570	1,000	0,152	0,574	0,940	0,828	0,999	0,923
Ri	*adc*	0,430	0,124	0,771	0,442	0,316	0,605	0,981	0,634
	dot	0,000	0,000	0,000	0,000	0,000	0,000	0,000	0,000
	dotodf	0,543	0,500	0,633	0,559	0,114	0,078	0,029	0,074
	mdotodf	0,598	0,204	0,632	0,478	0,089	0,952	0,017	0,353
	qball	0,973	1,000	0,760	0,911	0,501	0,529	0,980	0,670

b

Measure	Cross/Linear			Average	Noise/Cross			Average
	b=2000s/mm^2		b=8000s/mm^2		b=2000s/mm^2		b=8000s/mm^2	
	45°	90°	45°		45°	90°	45°	
Ca	0,482	0,672	0,496	0,550	0,220	0,208	0,049	0,159
Cl	0,724	1,000	0,521	0,748	0,910	0,772	0,998	0,894
Cp	0,038	0,000	0,039	0,026	0,944	1,000	0,999	0,981
FA	0,676	1,000	0,470	0,715	0,934	0,874	0,999	0,935
MD	0,846	0,878	0,595	0,773	0,840	0,859	0,992	0,897
RA	0,676	1,000	0,470	0,715	0,934	0,874	0,999	0,935

Fig. A.1 Areas under the ROC curves for (**a**) HARDI and (**b**) DTI models. The higher the value the better the measure is for classification of the data. The color stand for *green*—good classifier; *red*—bad classifier

References

1. Basser, P.J., Mattiello, J., Lebihan, D.: MR diffusion tensor spectroscopy and imaging. Biophys. J. **66**(1), 259–267 (1994)
2. Frank, L.R.: Characterization of anisotropy in high angular resolution diffusion-weighted mri. Magn. Reson. Med. **47**(6), 1083–99, (2002)

3. Alexander, D.C., Barker, G.J., Arridge, S.R.: Detection and modeling of non-gaussian apparent diffusion coefficient profiles in human brain data. Magn. Reson. Med. **48**(2), 331–40, (2002)
4. Tuch, D.: Q-ball imaging. Magn. Reson. Med. **52**, 1358–1372 (2004)
5. Özarslan, E., Shepherd, T.M., Vemuri, B.C., Blackband, S.J., Mareci, T.H.: Resolution of complex tissue microarchitecture using the diffusion orientation transform (DOT). NeuroImage **36**(3), 1086–1103 (2006)
6. Tournier, J.D., Calamante, F., Connelly, A.: Robust determination of the fibre orientation distribution in diffusion MRI: non-negativity constrained super-resolved spherical deconvolution. NeuroImage **35**(4), 1459–1472 (2007)
7. Jian, B., Vemuri, B.C.: A unified computational framework for deconvolution to reconstruct multiple fibers from Diffusion Weighted MRI. IEEE Trans. Med. Imaging **26**(11), 1464–1471 (2007)
8. Descoteaux, M., Wiest-Daesslé, N., Prima, S., Barillot, C., Deriche, R.: Impact of Rician Adapted Non-Local Means Filtering on HARDI. In: MICCAI, Berlin/New York, vol. 5242, pp. 122–130. Springer, Berlin (2008)
9. Behrens, T.E., Johansen-Berg, H., Jbabdi, S., Rushworth, M.F., Woolrich, M.W.: Probabilistic diffusion tractography with multiple fibre orientations: What can we gain? NeuroImage **34**(1), 144–55 (2007)
10. Hosey, T., Williams, G., Ansorge, R.: Inference of multiple fiber orientations in high angular resolution diffusion imaging. Magn. Reson. Med. **54**, 1480–1489 (2005)
11. Rao, M., Chen, Y., Vemuri, B.C., Wang, F.: Cumulative residual entropy: a new. measure of information. IEEE Trans. Inf. Theory **50**(6), 1220–1228 (2004)
12. Chen, Y., Guo, W., Zeng, Q., Yan, X., Rao, M., Liu, Y.: Apparent diffusion coefficient approximation and diffusion anisotropy characterization in DWI. In: Information Processing in Medical Imaging, Glenwood Springs, MICCAI, Berlin/New York, pp. 246–257. Springer, Berlin/New York (2005)
13. Descoteaux, M., Angelino, E., Fitzgibbons, S., Deriche, R.: Apparent diffusion coefficients from high angular resolution diffusion imaging: estimation and applications. Magn. Reson. Med. **56**, 395–410 (2006)
14. Özarslan, E., Vemuri, B.C., Mareci, T.H.: Generalized scalar measures for diffusion MRI using trace, variance, and entropy. Magn. Reson. Med. **53**(4), 866–76 (2005)
15. Leow, A., Zhu, S., Zhan, L., McMahon, K., de Zubicaray, G., Meredith, M., Wright, M., Thompson, P.: A study of information gain in high angular resolution diffusion imaging (HARDI). In: Computational Diffusion MRI Workshop, MICCAI, Berling/New York, pp. 97–105 (2008). http://picsl.upenn.edu/cdmri08/proceedings.pdf
16. Prčkovska, V., Vilanova, A., Poupon, C., Haar Romeny, B.M., Descoteaux, M.: Fast classification scheme for hardi data simplification. In: Davcev, D., Gómez, J.M. (eds.) ICT Innovations 2009, Ohrid, pp. 345–355. Springer, Berlin/Heidelberg (2010)
17. Poupon, C., Rieul, B., Kezele, I., Perrin, M., Poupon, F., Mangin, J.F.: New diffusion phantoms dedicated to the study and validation of HARDI models. Magn. Reson. Med. **60** (2008) 1276–1283
18. Jones, D., Horsfield, M., Simmons, A.: Optimal strategies for measuring diffusion in anisotropic systems by magnetic resonance imaging. Magn. Reson. Med. **42**, 515–525 (1999)
19. Descoteaux, M., Angelino, E., Fitzgibbons, S., Deriche, R.: Regularized, fast and robust analytical q-ball imaging. Magn. Reson. Med. **58**, 497–510 (2007)
20. Wedeen, V.J., Hagmann, P., Tseng, W.Y., Reese, T.G., Weisskoff, R.M.: Mapping complex tissue architecture with diffusion spectrum magnetic resonance imaging. Magn. Reson. Med. **54**(6), 1377–1386 (2005)
21. Westin, C.F., Peled, S., Gudbjartsson, H., Kikinis, R., Jolesz, F.A.: Geometrical diffusion measures for MRI from tensor basis analysis. In: ISMRM '97, Vancouver, p. 1742 (1997)
22. Vilanova, A., Zhang, S., Kindlmann, G., Laidlaw, D.: An introduction to visualization of diffusion tensor imaging and its applications. In: Weickert, J., Hagen, H. (eds.) Visualization and Processing of Tensor Fields. Mathematics and Visualization, pp. 121–153 Springer, Berlin/Heidelberg/New York (2005)

23. Fawcett, T.: An introduction to roc analysis. Pattern Recognit. Lett. **27**(8), 861–874 (2006). ROC Analysis in Pattern Recognition
24. Prčkovska, V., Roebroeck, A.F., Pullens, W., Vilanova, A., ter Haar Romeny, B.M.: Optimal acquisition schemes in high angular resolution diffusion weighted imaging. In: MICCAI, Berlin/New York. Lecture Notes in Computer Science, vol. 5242, pp. 9–17. Springer, Berlin/New York (2008)
25. Jansons, K.M., Alexander, D.: Persistent angular structure: new insights from diffusion magnetic resonance imaging data. Inverse Probl. **19**, 1031–1046 (2003)
26. Schnell, S., Saur, D., Kreher, B., Hennig, J., Burkhardt, H., Kiselev, V.: Fully automated classification of HARDI in vivo data using a support vector machine. NeuroImage **46**(3), 642–651 (2009)

Towards Resolving Fiber Crossings with Higher Order Tensor Inpainting

Thomas Schultz

Abstract The use of second-order tensors for the modeling of data from Diffusion Weighted Magnetic Resonance Imaging (DW-MRI) is limited by their inability to represent more than one dominant direction in cases of crossing fiber bundles or partial voluming. Higher-order tensors have been used in High Angular Resolution Diffusion Imaging (HARDI) to overcome these problems, but their larger number of parameters leads to longer measurement times for data acquisition. In this work, we demonstrate that higher-order tensors that indicate likely fiber directions can be estimated from a small number of diffusion-weighted measurements by taking into account information from local neighborhoods. To this end, we generalize tensor voting, a method from computer vision, to higher-order tensors. We demonstrate that the resulting even-order tensor fields facilitate fiber reconstruction at crossings both in synthetic and in real DW-MRI data, and that the odd-order fields differentiate crossings from junctions.

1 Introduction

Diffusion Weighted Magnetic Resonance Imaging (DW-MRI) is a medical imaging modality that allows for a non-invasive investigation of fibrous tissue, such as the nerve fiber bundles in the human brain [1]. In cases where a clear principal fiber direction exists, it is generally well-aligned with the main diffusion direction captured by the second-order diffusion tensor (DT-MRI) model [2]. However, DT-MRI provides insufficient information in cases of partial voluming and crossing or spreading fiber bundles. High Angular Resolution Diffusion Imaging (HARDI) uses

T. Schultz (✉)
University of Chicago, Computation Institute, 5735 S. Ellis Ave #209A, Chicago, IL, 60637, USA
e-mail: tschultz@tuebingen.mpg.de

D.H. Laidlaw and A. Vilanova (eds.), *New Developments in the Visualization and Processing of Tensor Fields*, Mathematics and Visualization,
DOI 10.1007/978-3-642-27343-8_13, © Springer-Verlag Berlin Heidelberg 2012

more complex models like higher-order tensors [3, 4], but requires a larger number of measurements, which are usually too time-consuming in a clinical context.

Inpainting is a process by which a damaged image is restored, or an object is removed from an image in an unobtrusive way. Different strategies have been developed to fill in the missing information in such cases in an automated manner. They can be classified roughly into structure-based methods, which try to continue the surrounding image geometry into the missing region [5] and texture synthesis approaches, that generate patches from statistical models [6] or by copying pixels from example images [7].

Our work is motivated by the observation that the image inpainting problem resembles the problem of complex fiber configurations in DT-MRI data: In both cases, information is missing in certain regions of a dataset. Our goal is to use ideas from inpainting in a preprocess that facilitates fiber tracking through such complex regions. Our method belongs to the class of techniques that fill in small holes by continuing existing structures in the neighborhood. It is based on tensor voting, an approach from computer vision that can be used to infer likely continuations of lines, and that we extend to higher-order tensors in order to preserve directional information at crossings.

This chapter is structured as follows: After reviewing related work in Sect. 2, we will introduce the concept of higher-order tensor voting in Sect. 3. In Sect. 4, we explain the application of this method to the problem of estimating tensors that can be used for fiber tracking. Finally, results are presented in Sect. 5 and the paper is concluded in Sect. 6.

2 Related Work

Tensor voting [8] is a framework for perceptual organization. Based on principles of human perception, it tries to group tokens (like points, lines, or surface segments) into structures that appear natural to a human observer. The method was first proposed for the automated detection of perceptual contours [9], but has been extended to numerous other applications, including the inpainting problem [10].

Our work differs from this existing use of tensor voting for inpainting in that our input is a three-dimensional second-order tensor field rather than a two-dimensional color image. The formulation of texture synthesis as a tensor voting problem, which is one of the main contributions of [10], does not apply to our problem. Instead, we extend tensor voting by using higher-order tensors to represent directional information at crossings.

Inpainting small image regions is closely related to image interpolation. Weickert and Welk [11] have developed a PDE-based method for interpolation of second-order tensor fields. As part of our experiments, we successfully applied their approach to higher-order tensor fields. However, we found that tensor voting makes it easier to re-orient the propagated tensors and, in our particular application, allows for a simpler, non-iterative, and relatively fast implementation. A comparison between tensor voting and PDE-based methods in a different context is presented by Moreno et al. in Chap. 9 of this book.

The higher-order tensors generated by our method are similar to the "tractosemas" proposed by Barmpoutis et al. [12] and the "extrapolated spherical diffusion functions" by Prčkovska [13] in that they do not have an immediate physical meaning, but are designed to indicate likely fiber directions. However, the main purpose of [12] is to create distributions from HARDI data which are no longer antipodally symmetric. Even though Sect. 5.3 will demonstrate that asymmetries can also be detected by tensor voting, our focus is to infer information about crossings from data with low angular resolution.

A study by Caan et al. [14] has pursued a similar goal, but under different conditions: They acquire low angular resolution data from different subjects, and estimate a high angular resolution atlas from the coregistered results. In contrast, our method takes its information from spatial neighborhoods rather than a cohort of subjects and thus works on individual datasets.

Previous work that integrates neighborhood information into the fiber tracking process includes the tensorlines algorithm by Weinstein et al. [15]. However, it only uses information from the previous tracking step, while our voting process takes the full neighborhood into account.

The spin glass model by Mangin et al. [16] aims at balancing local diffusion directions with global curvature constraints by simulating coupled compass needles in a magnetic field which is defined from the diffusion tensors. Even though a simple synthetic crossing has been resolved using a two-compass variant [17], it has not been applied to curved tracts, and results on real data have only been presented based on a single-compass model that does not support crossings [16].

An alternative voting-based approach to tractography, inspired by the Hough transform, was recently presented by Aganj et al. [18]. They identify the most plausible fiber trajectories by voting on a large number of possible curves, while our method votes on local fiber directions and does a tractography only in a subsequent step. Since [18] is only an extended abstract, it does not provide enough detail for a reproduction and side-by-side comparison of results.

3 Higher-Order Tensor Voting

3.1 Basics of Tensor Voting

The input of the tensor voting algorithm is a set of tokens. Even though tensor voting allows for different types of tokens, like unoriented points and curve or surface elements, we are specifically interested in reconstructing the trajectories of major nerve fiber bundles, so we only make use of the part of the framework that deals with curves.

In tensor voting, each curve element generates hypotheses about likely continuations and votes for them. Individual votes encode the direction of the proposed

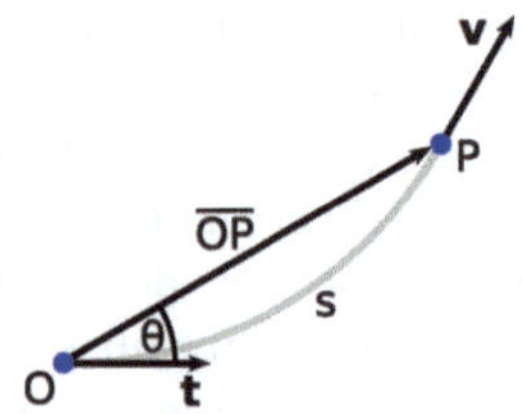

Fig. 1 Given two points O and P along with a tangent **t** at O, tensor voting considers the osculating circle (*gray*) to be their most likely connection, leading to the inferred tangent **v** at P

curve, and a scalar saliency, indicating the confidence that the voter has in it. In a subsequent step, the accumulated votes are analyzed to identify the curve for which all input tokens together provide the strongest evidence. Accumulated votes reflect the overall saliency, the average direction, weighted by individual saliencies, and the spread around it.

For a curve element given by a point O and a tangent direction **t**, tensor voting assumes that the most likely continuation that includes a second point P is an arc of their osculating circle, i.e., the circle that passes through both O and P and shares the tangent **t** at O (cf. Fig. 1). Consequently, the direction of the vote cast from O to P is the tangent **v** of the osculating circle at that point. The saliency of the vote decreases both with arc length s and with the curvature κ of the circle. If the angle θ between **t** and the line $\overline{\text{OP}}$ exceeds 45°, the saliency is set to zero. A detailed justification of these choices is given in [19].

3.2 *Introducing Higher-Order Tensors*

In the original framework, votes are represented by second-order tensors with sorted eigenvalues $\lambda_1 \geq \lambda_2 \geq \lambda_3 \geq 0$. The relative magnitudes of the eigenvalues reflect the type of a structure, the eigenvector directions describe its orientation.

Our implementation deviates from the established tensor voting algorithm in two ways: First, curve elements are traditionally encoded as planar tensors ($\lambda_1 = \lambda_2 \gg \lambda_3$) whose larger eigenvector pair spans the normal plane. In contrast, we let the major eigenvector represent the tangent direction. This agrees with the role of the principal eigenvector in DT-MRI, which is assumed to be tangential to the fiber trajectory, and it allows us to use simple stick votes for curves, which are the only relevant elements in our application.

Second, traditional tensor voting represents crossings and junctions as isotropic tensors ($\lambda_1 = \lambda_2 = \lambda_3$), which do not possess any directional information. Since it is the main motivation of our work to overcome the inability of DT-MRI to resolve the involved fiber directions at crossings, we replace the second-order tensors with a higher-order tensor representation, which retains directional information even when averaging differently oriented stick tensors. This is done in analogy to the higher order structure tensors in [20] and is illustrated in Fig. 2.

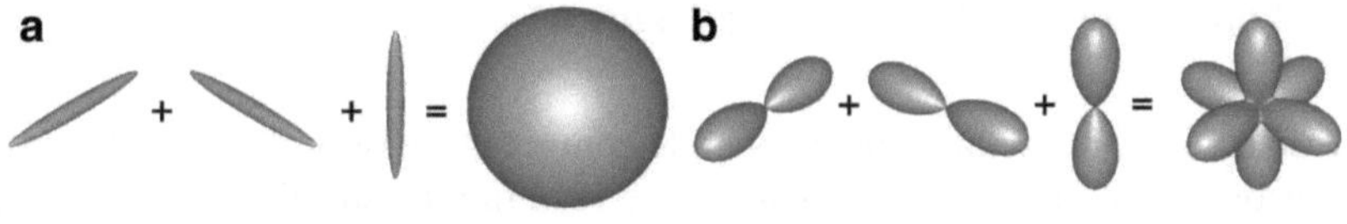

Fig. 2 When adding orthogonal second-order tensors, the result has no directional information (**a**). In the sixth-order case, the orientations of the individual terms is reflected in the sum (**b**). Therefore, we use higher-order tensors in our voting

3.3 Formalizing the Voting Process

The votes in our algorithm are created in three steps: First, the correct direction is determined and represented as a unit vector $\mathbf{v}$. Then, the scalar saliency ψ is computed. Finally, a higher-order tensor vote $\mathcal{V}$ is formed by a repeated outer product of $\mathbf{v}$ with itself, scaled with saliency ψ.

The first two steps are analogous to the traditional creation of stick votes [19]: Let θ be the angle between the tangent $\mathbf{t}$ at O and the vector $\overline{\mathrm{OP}}$ (Fig. 1). Then, $\mathbf{v}$ is the tangent of the osculating circle at P, and is created by rotating $\mathbf{t}$ by the angle 2θ around the axis given by $\mathbf{t} \times \overline{\mathrm{OP}}$. Thus, $\mathbf{t}$, $\mathbf{v}$, and $\overline{\mathrm{OP}}$ lie in a common plane.

Saliency decreases with arc length s and curvature κ:

$$s = \frac{\theta \, \|\overline{\mathrm{OP}}\|}{\sin\theta} \qquad \kappa = \frac{2\sin\theta}{\|\overline{\mathrm{OP}}\|} \tag{1}$$

The voting has two main parameters: σ determines how rapidly saliency decays with distance, ρ controls the decay with curvature. If $\theta > 45°$, the saliency ψ is set to zero. Otherwise, it is given as

$$\psi(s,\kappa) = e^{-\frac{s^2}{\sigma^2} - \frac{\kappa^2}{\rho^2}}. \tag{2}$$

In the original tensor voting approach, ρ is set as a function of σ, based on the assumption that it is equally plausible to join two orthogonal curve elements with a smooth curve or a sharp corner [9]. In fiber tracking, it is commonly assumed that fiber bundles do not bend sharply [21]. Therefore, we penalize curvature more than regular tensor voting would. This is achieved by fixing $\sigma = 2$ and $\rho = 0.3$.

With these ingredients, the final vote is computed as $\mathcal{V} = \psi(s,\kappa)\mathbf{v}^{\otimes l}$, where $\mathbf{v}^{\otimes l}$ denotes taking the outer product of $\mathbf{v}$ with itself l times. In component notation,

$$[\mathcal{V}]_{i_1 i_2 \ldots i_l} = \psi(s,\kappa) v_{i_1} v_{i_2} \cdots v_{i_l}. \tag{3}$$

Since $\mathcal{V}$ is invariant under arbitrary index permutations, it is sufficient to store a small number of non-redundant components (cf. [20]). All presented experiments use tensor order $l = 6$, which has been found sufficient to resolve intersections of three fiber bundles [22].

3.4 Analyzing the Accumulated Votes

Individual votes are accumulated by simple component-wise addition of the respective tensors. In traditional tensor voting, the resulting tensor $\mathbf{T}$ is analyzed via its spectral decomposition into eigenvalues $\lambda_1 \geq \lambda_2 \geq \lambda_3$ and eigenvectors $\mathbf{e}_1$, $\mathbf{e}_2$, $\mathbf{e}_3$:

$$\mathbf{T} = \sum_{i=1}^{3} \lambda_i \mathbf{e}_i \otimes \mathbf{e}_i \tag{4}$$

From this, a stick, plate, and a ball component are detected with saliencies

$$\psi_s = \lambda_1 - \lambda_2, \quad \psi_p = \lambda_2 - \lambda_3, \quad \psi_b = \lambda_3 \tag{5}$$

respectively. No exact equivalent of the spectral decomposition exists for the higher-order case, but a previous work [22] proposed an algorithm to approximate a given order-l tensor $\mathscr{T}$ with a sum of symmetric rank-1 terms:

$$\mathscr{T} \approx \sum_{i=1}^{r} \lambda_i \mathbf{e}_i^{\otimes l} \tag{6}$$

Here, the vectors $\mathbf{e}_i$ are still unit-length, but no longer pairwise orthogonal. The rank-1 terms $\mathbf{e}_i^{\otimes l}$ correspond to stick components, and represent curve elements in our modified tensor voting framework. Since equally large $\lambda_1 \approx \lambda_2$ can now indicate two salient stick components in different directions, we can no longer rely on the difference $\lambda_1 - \lambda_2$ to define saliency, as in Eq. (5). Rather, we assign high saliency to a direction if the tensor's homogeneous form

$$T(\mathbf{v}) = \mathscr{T} \cdot^l \mathbf{v} = \sum_{i_1, i_2, \ldots, i_l} [T]_{i_1 i_2 \ldots i_l} v_{i_1} v_{i_2} \cdots v_{i_l}, \tag{7}$$

as it is plotted in Fig. 2b, has large convex curvature in that direction.

Restrict $T(\mathbf{v})$ to the unit sphere and let $\mathbf{H}_i$ be the symmetric 2×2 Hessian matrix of second derivatives of $T(\mathbf{v})$ on the sphere, evaluated at the curve direction $\mathbf{e}_i$ that was estimated in Eq. (6). This Hessian can be obtained either numerically [23] or analytically, by expressing $T(\mathbf{v})$ in spherical coordinates [24]. Let $\mu_1 \geq \mu_2$ be the sorted eigenvalues of $\mathbf{H}_i$. If $\mathbf{e}_i$ is a salient stick component, we expect $T(\mathbf{v})$ to be strongly convex ($\mu_1 \ll 0$), so we measure saliency ψ as

$$\psi = -\frac{\mu_1}{l}. \tag{8}$$

The normalization by tensor order l ensures that the maximum saliency of component $\mathbf{e}_i^{\otimes l}$ is given by $\psi = \lambda_i$. In case of the second-order tensors used in traditional tensor voting, $\mu_1 = 2(\lambda_2 - \lambda_1)$. Therefore, the definition of ψ in Eq. (8) is equivalent to ψ_s in Eq. (5) in this case.

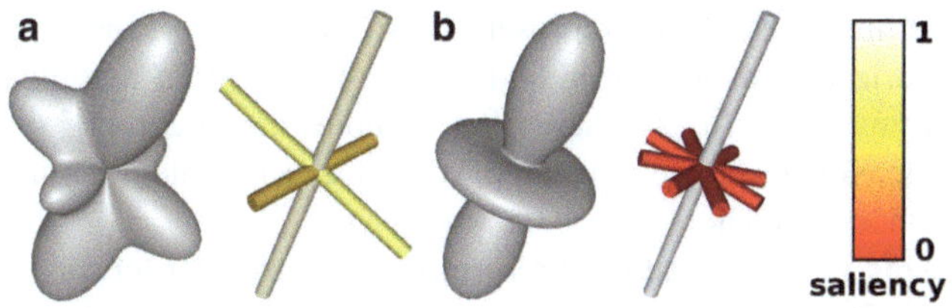

Fig. 3 Accumulated higher-order votes are approximated with a sum of rank-1 terms that represent stick components. In Subfigure (**a**), all three contributions have significant saliency, as computed according to Eq. (8) from the convex shape in these directions. In (**b**), the planar contribution orthogonal to the main direction is decomposed into three sticks with zero saliency

Figure 3a illustrates that a single accumulated higher-order tensor vote can contain multiple stick components, with different angles and saliencies. The plate component in Fig. 3b can be subdivided further into four rank-1 terms, but they have zero saliency as stick components. This is completely analogous to second-order plate tensors, which can be written as $\mathbf{e}_1 \otimes \mathbf{e}_1 + \mathbf{e}_2 \otimes \mathbf{e}_2$ for any two orthonormal vectors $\mathbf{e}_1$ and $\mathbf{e}_2$ that span the plane.

4 Inpainting as a Preprocess for Tractography

Fiber tracking [25], also known as tractography [21], infers likely fiber trajectories by computing integral curves which are everywhere tangential to the major eigenvector of the diffusion tensor field. Tracking is stopped when the difference $\lambda_1 - \lambda_2$ between the larger two eigenvalues becomes too small, since this typically indicates that the diffusion tensor no longer represents a single homogeneous fiber bundle. A common criterion is the linearity measure of Westin et al. [26],

$$c_l = \frac{\lambda_1 - \lambda_2}{\lambda_1}. \tag{9}$$

In our experiments, we stop when $c_l < 0.4$.

In order to use tensor voting as a preprocess for fiber tracking, we need to define a set of tokens that generate the votes, and to mark the region in which the inpainting should happen. Tokens are placed in all voxels where c_l is large enough for fiber tracking. Their direction is given by the principal eigenvector of the diffusion tensor. In analogy to streamline tractography, which makes a hard binary decision about whether or not to follow an eigenvector direction, our method assigns the same saliency to all input tokens. The inpainting mask is given by the voxels in which c_l is too low for fiber tracking. Information from the neighborhood is propagated into the inpainting region by the tensor voting process described in Sect. 3.

Outside the mask, tokens are represented as rank-1 tensors, as given by Eq. (3). To keep the norm of tensors within and outside the mask in a comparable range, we scale all votes by $2/\iiint \psi$, where $\iiint \psi$ denotes the integral of ψ from Eq. (2) over three-dimensional space. In practice, we evaluate the integral numerically over a spatial neighborhood where $\psi > 0.01$.

Fiber tracking is performed on the resulting higher-order tensor field using the algorithm from [22]. Voxels outside the brain are marked a priori based on their low MR signal, and are not taken into account at any point. Setting up the inpainting does not involve any new parameters, since the threshold on c_l is taken from the algorithm for DT-MRI tractography. However, the higher-order tracking process needs a termination criterion; in our examples, we stop when no rank-1 contribution with $\psi > 0.01$ is found within 20° of the current tracking direction.

5 Results

In order to validate our method, we have applied it to three synthetic datasets. They were created by modeling fiber crossings as a mixture of second-order diffusion tensors with fractional anisotropy $\mathrm{FA} = 0.87$ [27], by simulating diffusion-weighted images (DWIs) from them, and estimating a DT-MRI model through a linearized least squares fit [2]. DWIs were simulated in 12 evenly distributed directions, with a b-value of $b = 1{,}000\,\mathrm{s/mm}^2$.

We also used our method for the reconstruction of a major fiber bundle in a real dataset of a healthy human brain. Like the synthetic data, it consisted of 12 diffusion-weighted images at $b = 1{,}000\,\mathrm{s/mm}^2$, plus one non-weighted image, and the same fitting procedure was used.

5.1 *Results on Synthetic Data*

Our first example is a crossing of two orthogonal fiber bundles. Figure 4 shows a superquadric glyph visualization [28] of the synthetic DT-MRI data (a), the inpainted higher-order tensor field (b), and a tractography based on the higher-order tensors (c). Our inpainting reconstructs this simple configuration perfectly.

As a more challenging test case, we created a second dataset in which a straight fiber bundle is intersected by a parabolic one. In Fig. 5a, a ground truth tractography is presented. It is based on a spherical deconvolution model [29] that has been computed from simulated HARDI measurements (60 directions, $b = 1{,}000\,\mathrm{s/mm}^2$). In all examples, the tracking is seeded at the top and at the left side of the image.

Since the parabolic shape violates the constant curvature assumption made by the tensor voting approach, the curved bundle is not reconstructed perfectly in this case; however, the correct connectivity is still inferred in a large part of the bundle (b). For comparison, Subfigure (c) demonstrates that the tensorlines algorithm [15], a previous approach that integrates neighborhood information in the fiber tracking process, fails completely to reconstruct the crossing.

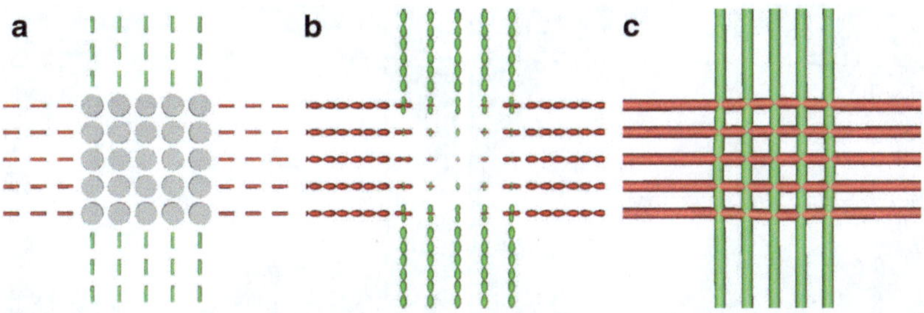

Fig. 4 At a 90° crossing, second-order tensors become planar, preventing fiber tracking (**a**). Our inpainting process infers higher-order tensors (**b**) that allow for a reconstruction of the crossing (**c**)

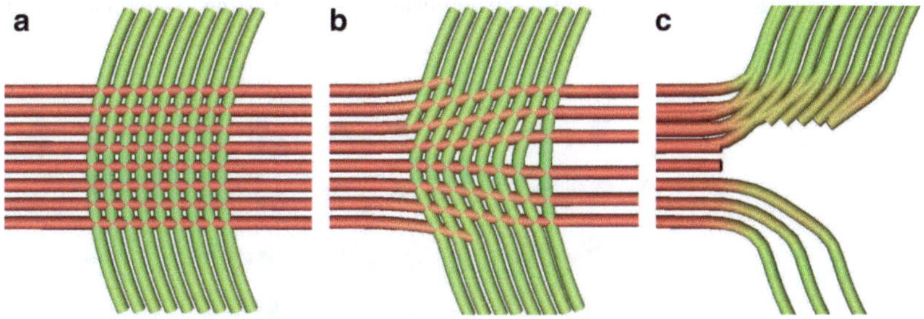

Fig. 5 Since the parabolic (*green*) fibers in (**a**) violate the constant curvature assumption of tensor voting, which was illustrated in Fig. 1, our inpainting produces imperfect results, but is still able to infer the correct connectivity (**b**). The tensorlines algorithm is unable to resolve the crossing (**c**)

5.2 Result on Real Data

In the real dataset, we aimed at tracking the left pyramidal tract by seeding in the internal capsule. Due to its high importance to motor function, reconstructions of this tract have been used repeatedly for surgical planning [30–32].

With second-order diffusion tensors alone, it is difficult to capture the bundle in its entirety, since c_l drops when the tract crosses the transcallosal fibers that run through the corpus callosum (Fig. 6a, red ellipse). As shown in Fig. 6b, higher-order tensor inpainting successfully bridges this gap, and allowed us to continue the tracking towards the cortical surface.

5.3 Distinguishing Crossings from Junctions

Recently, Barmpoutis et al. [12] have employed a diffusion process to create a field of asymmetric spherical functions that differentiate between X-shaped crossings and Y-shaped junctions. To illustrate that the same distinction can also be made via a tensor voting, we have created a synthetic junction, shown in Fig. 7a.

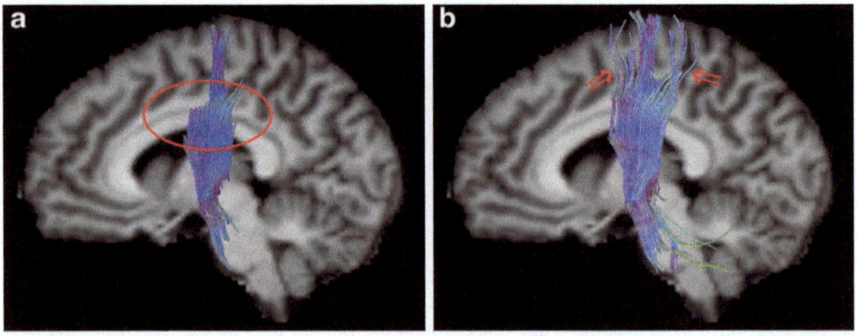

Fig. 6 Due to crossing transcallosal fibers, tracking of the pyramidal tract in this DT-MRI dataset ended prematurely (**a**). Higher-order tensor inpainting allowed the tracking to continue towards the cortex (**b**)

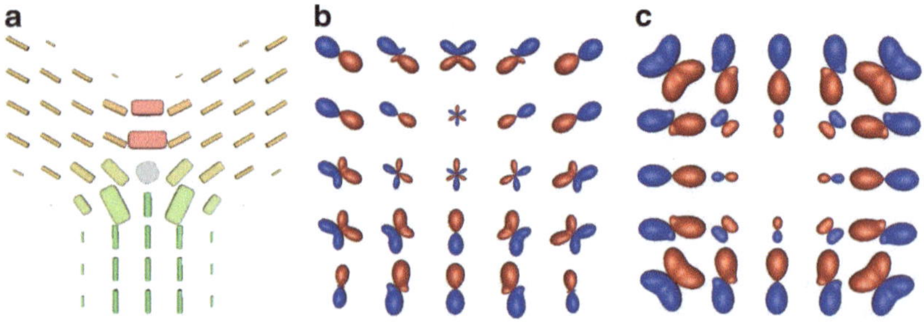

Fig. 7 The direction and polarity of fibers meeting in a 120° junction (**a**) is clearly shown by third-order tensor votes (**b**, closeup), whose homogeneous forms are antisymmetric (*blue* indicates negative values, *red* positive). At the center of crossings, odd-order tensor votes cancel out (**c**), effectively distinguishing crossings from junctions

In order to find the location and polarity of boundaries, Tong et al. [33] extend the traditional second-order tensor voting framework towards vectors ("first-order tensors"): Regions in the center of a structure will receive vector votes from both sides, which cancel out due to their opposite orientation. At structure boundaries, however, vector votes accumulate.

This approach does not carry over to our problem directly, since vectors cancel out both at crossings and at regular (e.g., 120° triple) junctions. However, we observe that tensors of odd order $l \geq 3$ still cancel out at crossings, while they have non-zero norm and indicate the directions of the involved bundles at junctions.

This is because the homogeneous form $T(\mathbf{v})$ that characterizes a symmetric tensor (Eq. 7) is antipodally antisymmetric $T(-\mathbf{v}) = -T(\mathbf{v})$ for odd l, but can become multimodal for $l \geq 3$. Odd-order votes are generated in complete analogy to even-order ones (cf. Sect. 3.3).

Figure 7b, c plot the homogeneous forms of accumulated third-order tensor votes; blue indicates positive values, red negative ones. The glyph at the center of the junction (b) clearly indicates the directions and polarities of the three joining bundles. At the center of crossings (c), odd-order votes cancel out; the directions of the involved bundles is instead given by the even-order votes (Fig. 4b).

Taken together, even-order (antipodally symmetric) and odd-order (antipodally antisymmetric) tensors hold all the information that is present in general asymmetric functions on the sphere, as they are used in [12].

6 Conclusion and Future Work

In this chapter, we have made two contributions. First, we have introduced higher-order tensor voting, an extension of the standard tensor voting method that allows one to preserve directional information at crossings and junctions. We expect that higher-order tensor voting will also prove useful in other applications. However, to leverage the full framework, rather than being restricted to inferring curves, more research is required on decompositions of totally symmetric higher-order tensors.

Second, we have approached the problem of missing information in DT-MRI due to partial voluming from a new perspective, by treating it in analogy to image inpainting. For this, we have relied on generic rules of what constitutes a good continuation of a curve. They are encoded in the tensor voting algorithm and include proximity, similarity, and simplicity [19]. In the future, one might consider exploiting more specific prior knowledge about likely continuations of fiber bundles, which could be given in the form of a HARDI brain atlas. While we have concentrated on structure-inferring inpainting in this work, such an approach could take inspiration from example-based texture synthesis [7].

It is not the goal of our research to establish inpainting as an alternative to HARDI. However, our results indicate that in cases where the acquisition of high angular resolution data cannot be afforded, inpainting can help with the extraction of clinically relevant fiber tracts.

As part of our future work, we would like to employ tensor voting techniques similar to the one described in this chapter to stabilize the tracking process in cases where HARDI data is available, and exploit the antisymmetric information provided by odd-order tensor votes.

Acknowledgements I would like to thank Alfred Anwander (MPI CBS, Leipzig, Germany) for providing the DW-MRI dataset that was used to create Fig. 6. This work was supported by a fellowship within the Postdoc Program of the German Academic Exchange Service (DAAD).

References

1. Le Bihan, D., Breton, E., Lallemand, D., Grenier, P., Cabanis, E., Laval-Jeantet, M.: MR imaging of intravoxel incoherent motions: application to diffusion and perfusion in neurologic disorders. Radiology **161**, 401–407 (1986)
2. Basser, P.J., Mattiello, J., Le Bihan, D.: Estimation of the effective self-diffusion tensor from the NMR spin echo. J. Magn. Reson. B **103**, 247–254 (1994)

3. Özarslan, E., Mareci, T.: Generalized diffusion tensor imaging and analytical relationships between diffusion tensor imaging and high angular resolution diffusion imaging. Magn. Reson. Med. **50**, 955–965 (2003)
4. Liu, C., Bammer, R., Acar, B., Moseley, M.E.: Characterizing non-gaussian diffusion by using generalized diffusion tensors. Magn. Reson. Med. **51**, 924–937 (2004)
5. Bertalmio, M., Sapiro, G., Caselles, V., Ballester, C.: Image inpainting. In: Proceedings of ACM SIGGRAPH, New Orleans, pp. 417–424. ACM, New York (2000)
6. Portilla, J., Simoncelli, E.P.: A parametric texture model based on joint statistics of complex wavelet coefficients. Int. J. Comput. Vis. **40**, 49–71 (2000)
7. Efros, A.A., Leung, T.K.: Texture synthesis by non-parametric sampling. In: Proceedings of IEEE International Conference on Computer Vision (ICCV), Kerkyra, pp. 1033–1038. IEEE, Los Alamitos (1999)
8. Medioni, G., Lee, M.S., Tang, C.K.: A Computational Framework for Segmentation and Grouping. Elsevier, Amsterdam/New York (2000)
9. Guy, G., Medioni, G.: Inferring global perceptual contours from local features. Int. J. Comput. Vis. **20**, 113–133 (1996)
10. Jia, J., Tang, C.K.: Image repairing: robust image synthesis by adaptive ND tensor voting. In: IEEE Conference on Computer Vision and Pattern Recognition (CVPR), Madison, vol. 2, pp. 643–650. IEEE, Los Alamitos (2003)
11. Weickert, J., Welk, M.: Tensor field interpolation with PDEs. In: Weickert, J., Hagen, H. (eds.) Visualization and Processing of Tensor Fields, pp. 315–325. Springer, Berlin/Heidelberg (2006)
12. Barmpoutis, A., Vemuri, B.C., Howland, D., Forder, J.R.: Extracting tractosemas from a displacement probability field for tractography in DW-MRI. In: Metaxas D. et al. (eds.) Proceedings of Medical Image Computing and Computer-Assisted Intervention (MICCAI), New York. Lecture Notes in Computer Science, vol. 5241, pp. 9–16. Springer, Berlin/Heidelberg (2008)
13. Prčkovska, V.: High angular resolution diffusion imaging – processing & visualization. Ph.D. thesis, Technische Universiteit Eindhoven (2010)
14. Caan, M., Sage, C., van der Graaf, M., Grimbergen, C., Sunaert, S., van Vliet, L., Vos, F.: Dual tensor atlas generation based on a cohort of coregistered non-HARDI datasets. In: Yang, G.Z. et al. (eds.) Proceedings of Medical Image Computing and Computer-Assisted Intervention (MICCAI), London. Lecture Notes in Computer Science, vol. 5761, pp. 869–876. Springer, Berlin/Heidelberg (2009)
15. Weinstein, D., Kindlmann, G., Lundberg, E.: Tensorlines: advection-diffusion based propagation through diffusion tensor fields. In: Proceedings of IEEE Visualization, IEEE Computer Society Press, San Francisco, pp. 249–253 (1999)
16. Mangin, J.F., Poupon, C., Cointepas, Y., Rivière, D., Papadopoulos-Orfanos, D., Clark, C.A., Régis, J., Le Bihan, D.: A framework based on spin glass models for the inference of anatomical connectivity from diffusion-weighted MR data – a technical review. NMR Biomed. **15**, 481–492 (2002)
17. Cointepas, Y., Poupon, C., Le Bihan, D., Mangin, J.F.: A spin glass based framework to untangle fiber crossing in MR diffusion based tracking. In: Dohi, T., Kikinis, R. (eds.) Proceedings of Medical Image Computing and Computer-Assisted Intervention (MICCAI), Tokyo. Lecture Notes in Computer Science, vol. 2488, pp. 475–482. Springer, Berlin/Heidelberg (2002)
18. Aganj, I., Lenglet, C., Keriven, R., Sapiro, G., Harel, N., Thompson, P.: A Hough transform global approach to diffusion MRI tractography. In: Proceedings of International Society of Magnetic Resonance in Medicine (ISMRM), Honolulu, vol. 17, p. 854. The Society, Berkeley (2009)
19. Mordohai, P., Medioni, G.: Tensor voting: a perceptual organization approach to computer vision and machine learning. Morgan & Claypool, San Rafael (2007)
20. Schultz, T., Weickert, J., Seidel, H.P.: A higher-order structure tensor. In: Laidlaw, D.H., Weickert, J. (eds.) Visualization and Processing of Tensor Fields – Advances and Perspectives, pp. 263–280. Springer, Berlin/Heidelberg (2009)

21. Basser, P.J., Pajevic, S., Pierpaoli, C., Duda, J., Aldroubi, A.: In vivo fiber tractography using DT-MRI data. Magn. Reson. Med. **44**, 625–632 (2000)
22. Schultz, T., Seidel, H.P.: Estimating crossing fibers: a tensor decomposition approach. IEEE Trans. Vis. Comput. Graph. (Proc. IEEE Visualization) **14**, 1635–1642 (2008)
23. Seunarine, K.K., Cook, P.A., Hall, M.G., Embleton, K.V., Parker, G.J.M., Alexander, D.C.: Exploiting peak anisotropy for tracking through complex structures. In: Proceedings of IEEE Workshop Mathematical Methods in Biomedical Image Analysis (MMBIA), IEEE, Rio de Janeiro, Brazil, pp. 1–8 (2007)
24. Schultz, T., Kindlmann, G.: A maximum enhancing higher-order tensor glyph. Comput. Graph. Forum (Proc. EuroVis) **29**, 1143–1152 (2010)
25. Mori, S., Crain, B.J., Chacko, V.P., van Zijl, P.C.M.: Three-dimensional tracking of axonal projections in the brain by magnetic resonance imaging. Ann. Neurol. **45**, 265–269 (1999)
26. Westin, C.F., Maier, S., Khidhir, B., Everett, P., Jolesz, F., Kikinis, R.: Image processing for diffusion tensor magnetic resonance imaging. In: Proceedings of Medical Image Computing and Computer-Assisted Intervention (MICCAI), Cambridge. Lecture Notes in Computer Science, vol. 1679, pp. 441–452. Springer, Berlin/New York (1999)
27. Basser, P.J., Pierpaoli, C.: Microstructural and physiological features of tissues elucidated by quantitative-diffusion-tensor MRI. J. Magn. Reson. B **111**, 209–219 (1996)
28. Kindlmann, G.: Superquadric tensor glyphs. In: EG/IEEE Symposium on Visualization (SymVis), Eurographics Association, Konstanz, Germany, pp. 147–154 (2004)
29. Tournier, J.D., Calamante, F., Gadian, D.G., Connelly, A.: Direct estimation of the fiber orientation density function from diffusion-weighted MRI data using spherical deconvolution. NeuroImage **23**, 1176–1185 (2004)
30. Niizuma, K., Fujimura, M., Kumabe, T., Higano, S., Tominaga, T.: Surgical treatment of paraventricular cavernous angioma: fibre tracking for visualizing the corticospinal tract and determining surgical approach. J. Clin. Neurosci. **13**, 1028–1032 (2006)
31. Nimsky, C., Ganslandt, O., Enders, F., Merhof, D., Hammen, T., Buchfelder, M.: Visualization strategies for major white matter tracts for intraoperative use. Int. J. Comput. Assist. Radiol. Surg. **1**, 13–22 (2006)
32. Qazi, A.A., Radmanesh, A., O'Donnell, L., Kindlmann, G., Peled, S., Westin, C.F., Golby, A.J.: Resolving crossings in the corticospinal tract by two-tensor streamline tractography: method and clinical assessment using fMRI. NeuroImage **47**, T98–T106 (2009)
33. Tong, W.S., Tang, C.K., Mordohai, P., Medioni, G.: First order augmentation to tensor voting for boundary inference and multiscale analysis in 3D. IEEE Trans. Pattern Anal. Mach. Intell. **26**, 594–611 (2004)

Representation and Estimation of Tensor-Pairs

Carl-Fredrik Westin and Hans Knutsson

Abstract Over the years, several powerful models have been developed to represent specific elementary signal patterns, e.g. locally linear and planar structures. However, in real world problems there is often a need for handling more than one elementary pattern simultaneously. The straightforward approach of adaptive model selection has proven to be difficult and fragile. At the core of this problem is the vicious intractable search space created by having to simultaneously select models and corresponding samples. This calls for higher order models where multiple patterns are represented as one more complex pattern. In this work, we illustrate the advantages of this approach on data that has bi-modal tensor-valued distributions.

The method uses first and second order invariants as a representation, and an eigenvector based solution for recovering the elementary tensor components. We show that this method allows estimation of the two tensors that best represent a given tensor distribution. This distribution can for example be samples from a local neighborhood. A bi-modal distribution will produce the two tensors corresponding to the peaks of the distribution. In addition, numbers indicating the amount of samples belonging to each sub distribution are produced. We demonstrate the potential of the approach by processing a number of simple tensor image examples. The results clearly show that new valuable information regarding the local tensor structure is revealed.

C.-F. Westin
Laboratory of Mathematics in Imaging, Brigham and Women's Hospital,
Harvard Medical School, Boston
e-mail: westin@bwh.harvard.edu

H. Knutsson
Department of Biomedical Engineering, Linköping University, Sweden and Center for Medical Image Science and Visualization (CMIV), Linköping, Sweden
e-mail: knutte@imt.liu.se

D.H. Laidlaw and A. Vilanova (eds.), *New Developments in the Visualization and Processing of Tensor Fields*, Mathematics and Visualization,
DOI 10.1007/978-3-642-27343-8_14,

1 Introduction

This chapter discusses how un-ordered pairs of tensors can be represented to enable operations such as consistent averaging of two distinct, simultaneously present, unimodal tensor fields. For a single tensor field, spatial averaging is straightforward, but for a field consisting of un-ordered pairs of tensors, standard averaging cannot be applied. The reason is that since the pairs can not be ordered a priori, the two-tensor field cannot easily be processed separately as two consistent single tensor fields.

For readability we use standard algebraic notation to describe the theory in this chapter. We perform "vectorization" of the tensors so that the standard vector outer product can express outer products of tensors. The vectorization implies a lineup operation which arranges the elements of a multi dimensional array into a lexicographic ordered column vector (where the order is according to the tensor indices). For example, the outer product of two second order tensors is a fourth order tensor. This operation can be performed using the standard vector outer product if the components of the tensors are first rearranged into a vector array. If needed, the resulting outer product matrix can be rearranged to a fourth order tensor.

We propose a new continuous representation of vector- or tensor-pairs using two quantities **s** and **P**. This representation has the important property of enabling consistent definitions of statistical operations. In this chapter we describe the theory of the representation and present results from some examples of how tensor-pair mean-values can be estimated from distributions of pairs. We continue to show how the theory can be extended to handles bi-modal single-tensor distributions.

2 Tensor-Pair Representation

Below, we derive the results using vectors, with the implicit notion that these vectors may represent tensors that have been "vectorized". This implies no loss of generality since the vectors contain all the components of the corresponding tensor and can thus can easily be reinterpreted as a tensor.

2.1 Representing an Un-Ordered Pair of Vectors

To represent un-ordered pairs, we need to find a representation that is invariant to the order of them. The only first order invariant is

$$\mathbf{s} = \mathbf{u} + \mathbf{v} \tag{1}$$

and the possible second order invariants are represented by

$$\mathbf{P} = \mathbf{u}\mathbf{v}^T + \mathbf{v}\mathbf{u}^T \tag{2}$$

$$\mathbf{Q} = \mathbf{u}\mathbf{u}^T + \mathbf{v}\mathbf{v}^T \tag{3}$$

If we have $\mathbf{s}$, the second order invariants can be related to each other since

$$\mathbf{s}\mathbf{s}^T = (\mathbf{u}+\mathbf{v})(\mathbf{u}+\mathbf{v})^T \tag{4}$$

$$= \left(\mathbf{u}\mathbf{u}^T + \mathbf{v}\mathbf{v}^T\right) + \left(\mathbf{u}\mathbf{v}^T + \mathbf{v}\mathbf{u}^T\right) \tag{5}$$

$$= \mathbf{Q} + \mathbf{P} \tag{6}$$

Thus we only need two of the three invariants for our representation. To retain the sign of $\mathbf{u}$ and $\mathbf{v}$, we need the first order invariant $\mathbf{s}$, and we choose to use $\mathbf{P}$ as the second order invariant. In related work, Herberthson et al. presented a solution to this problem for representation of two vectors in 2D [1]. His method is based on symmetrized outer products, which relates to $\mathbf{P}$ in our representation. We will extend this work to tensors and present a solution that is valid for arbitrary dimensions of the field.

2.2 *Reconstruction of the Two Vectors from the Representation*

It turns out that two vectors $\mathbf{u}$ and $\mathbf{v}$ can be reconstructed using a closed-form expression of their first and second order invariants. Figure 1 shows that the vectors $\mathbf{u}$ and $\mathbf{v}$ can be expressed by two orthogonal vectors $\mathbf{a}$ and $\mathbf{b}$ that are appropriately scaled by a value α:

$$\mathbf{u} = \alpha\ (\mathbf{a}+\mathbf{b}) \tag{7}$$

$$\mathbf{v} = 1/\alpha\ (\mathbf{a}-\mathbf{b}) \tag{8}$$

Inserting this into the expression of $\mathbf{P}$ gives

$$\mathbf{P} = \mathbf{u}\mathbf{v}^T + \mathbf{v}\mathbf{u}^T \tag{9}$$

$$= (\mathbf{a}+\mathbf{b})(\mathbf{a}-\mathbf{b})^T + (\mathbf{a}-\mathbf{b})(\mathbf{a}+\mathbf{b})^T$$

$$= 2\left(\mathbf{a}\mathbf{a}^T - \mathbf{b}\mathbf{b}^T\right) \tag{10}$$

Recalling that $\mathbf{a}$ and $\mathbf{b}$ are orthogonal, we can identify that $\mathbf{P}$ here is expressed in its eigenvector system (and ignoring the unimportant scaling factor of 2 in Eq. 10),

$$\mathbf{P} = \lambda_1\mathbf{e}_1\mathbf{e}_1^T + \lambda_2\mathbf{e}_2\mathbf{e}_2^T \tag{11}$$

Note that λ_2 is here negative to match the structure of Eq. 10. Note also that regardless of the dimensionality of $\mathbf{P}$ there will only be two non-zero eigenvalues. We can then identify the expressions for the eigenvectors and eigenvalues as

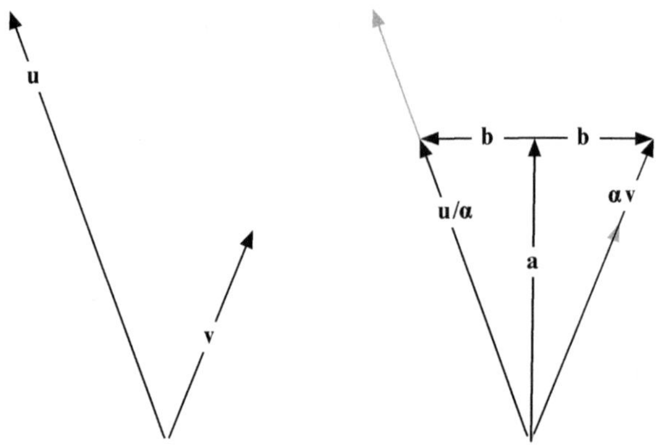

Fig. 1 The direction of the vectors **u** and **v** (*left*) can be described by the sum and difference of two orthogonal vectors **a** and **b** (*right*)

$$\mathbf{e}_1 = \hat{\mathbf{a}} \text{ and } \lambda_1 = \|\mathbf{a}\|^2 \quad (12)$$

$$\mathbf{e}_2 = \hat{\mathbf{b}} \text{ and } \lambda_2 = -\|\mathbf{b}\|^2 \quad (13)$$

which gives

$$\mathbf{a} = \sqrt{\lambda_1}\ \mathbf{e}_1 \quad (14)$$

$$\mathbf{b} = \sqrt{-\lambda_2}\ \mathbf{e}_2 \quad (15)$$

Via Eqs. 7–8 we then get explicit expressions for **u** and **v**:

$$\mathbf{u} = \alpha \left(\sqrt{\lambda_1}\ \mathbf{e}_1 + \sqrt{-\lambda_2}\ \mathbf{e}_2\right) \quad (16)$$

$$\mathbf{v} = 1/\alpha \left(\sqrt{\lambda_1}\ \mathbf{e}_1 - \sqrt{-\lambda_2}\ \mathbf{e}_2\right) \quad (17)$$

The scaling α factor cannot be determined from **P** since it only contains products of **u** and **v**. We will need the first order term for this ($\mathbf{s} = \mathbf{u} + \mathbf{v}$). Since we have found the directions of **u** and **v** we know that, regardless of the value of α, it is true that

$$\mathbf{s} = \mathbf{u} + \mathbf{v} \quad (18)$$

$$= \mu_1 \hat{\mathbf{u}} + \mu_2 \hat{\mathbf{v}} \quad (19)$$

$$= (\hat{\mathbf{u}}\ \hat{\mathbf{v}}) \begin{pmatrix} \mu_1 \\ \mu_2 \end{pmatrix} \quad (20)$$

$$= \mathbf{U}\boldsymbol{\mu}. \quad (21)$$

where $\hat{\mathbf{u}}$ and $\hat{\mathbf{v}}$ are the normalized **u** and **v** vectors, **U** is a matrix holding the components of $\hat{\mathbf{u}}$ and $\hat{\mathbf{v}}$ as two rows and $\boldsymbol{\mu}$ is a two component vector containing the scaling factors μ_1 and μ_2. The scaling factors can then be calculated using the

Moore-Penrose inverse,

$$\mu = \mathbf{U}^{+}\mathbf{s} \tag{22}$$

$$= \left(\mathbf{U}^T\mathbf{U}\right)^{-1}\mathbf{U}^T\mathbf{s} \tag{23}$$

and finally we have

$$\mathbf{u} = \mu_1\,\hat{\mathbf{u}} \tag{24}$$

$$\mathbf{v} = \mu_2\,\hat{\mathbf{v}} \tag{25}$$

For completeness we may note that by inserting Eqs. 24 and 25 into Eqs. 16 and 17 gives the relation to α: $\frac{\mu_1}{\mu_2} = \alpha^2$.

3 Mean Vector-Pair Estimation from Distributions

So far we have discussed how one set of vectors **u** and **v** can be recovered from the vector pairs' first and second order invariants **s** and **P**. In this section we will extend this notion to recovery of a single mean vector-pair from a distribution of vectors.

Two different distribution scenarios are discussed in this section. In the first case we have a set of un-ordered pairs of vectors. In the second case, the vectors do not come as pairs, but as one set containing all vectors.

3.1 Mean Vector-Pair Estimation from a Distribution of Vector-Pairs

In the first case of vector-pairs as data, the set of unordered pairs can expressed as:

$$\mathcal{S}_{uv} = \{\{\mathbf{u}_1, \mathbf{v}_1\}, \{\mathbf{u}_2, \mathbf{v}_2\}, \ldots \{\mathbf{u}_n, \mathbf{v}_n\}\} \tag{26}$$

where $\{\mathbf{u}, \mathbf{v}\}$ denotes an unordered pair.

The estimation of the mean **u** and **v** pair is calculated from sums of the first and second order invariants introduced in Eqs. 1–3. The expression for **s** and **P** from these sums of invariants is given by:

$$\mathbf{s} = \frac{1}{n}\sum_{i=1}^{n}\mathbf{u}_i + \mathbf{v}_i \tag{27}$$

$$\mathbf{P} = \frac{1}{n}\sum_{i=1}^{n}\mathbf{u}_i\mathbf{v}_i^T + \mathbf{v}_i\mathbf{u}_i^T \tag{28}$$

In contrast to the previous case with only one vector-pair (in Sect. 2.2), all eigenvalues of **P** will here in general be non-zero. In a least square sense **P** is best represented by the eigenvectors associated with eigenvalues with the largest magnitudes. To find representatives $\hat{\mathbf{u}}$ and $\hat{\mathbf{v}}$ we pick the eigenvectors, $\mathbf{e}_1$ and $\mathbf{e}_2$, corresponding to the two largest eigenvalue magnitudes. For distributions that can be reasonably represented by a mean vector-pair there will be only two eigenvalues with large magnitudes, $\lambda_1 > 0$, and $\lambda_2 < 0$. Thus, in order to estimate $\hat{\mathbf{u}}$ and $\hat{\mathbf{v}}$ from **s** and **P**, we insert the two eigenvalues and eigenvectors in Eqs. 16–17 and normalize.

$$\hat{\mathbf{u}} = \frac{\left(\sqrt{\lambda_1}\,\mathbf{e}_1 + \sqrt{-\lambda_2}\,\mathbf{e}_2\right)}{\sqrt{\lambda_1 - \lambda_2}} \tag{29}$$

$$\hat{\mathbf{v}} = \frac{\left(\sqrt{\lambda_1}\,\mathbf{e}_1 - \sqrt{-\lambda_2}\,\mathbf{e}_2\right)}{\sqrt{\lambda_1 - \lambda_2}} \tag{30}$$

Next, by minimizing the difference between **s** and a linear combination of the representative vectors $\hat{\mathbf{u}}$ and $\hat{\mathbf{v}}$ we can find **u** and **v**:

$$\min_{\mu_1,\mu_2} ||\mathbf{s} - (\mu_1\hat{\mathbf{u}} + \mu_2\hat{\mathbf{v}})\,||^2 \tag{31}$$

$$= \min_{\mu_1,\mu_2} ||\mathbf{s} - (\hat{\mathbf{u}}\ \hat{\mathbf{v}}) \begin{pmatrix}\mu_1\\ \mu_2\end{pmatrix} ||^2 \tag{32}$$

$$= \min_{\boldsymbol{\mu}} ||\mathbf{s} - \mathbf{U}\boldsymbol{\mu}||^2 \tag{33}$$

giving the solution

$$\boldsymbol{\mu} = \left(\mathbf{U}^{\mathrm{T}}\mathbf{U}\right)^{-1}\mathbf{U}^{\mathrm{T}}\mathbf{s} \tag{34}$$

and

$$\mathbf{u} = \mu_1\,\hat{\mathbf{u}} \tag{35}$$

$$\mathbf{v} = \mu_2\,\hat{\mathbf{v}} \tag{36}$$

3.2 *Mean Vector-Pair Estimation from a Distribution of Vectors*

In the second case, we pick pairs from one mixed set that contains data from two distributions.

$$\mathcal{S}_x = \{\mathbf{x}_1, \mathbf{x}_2, \ldots, \mathbf{x}_n\} \tag{37}$$

In practice we only need to pick the $n(n-1)/2$ distinct pairs, however to to simplify the formulas below we pick all n^2 possible pairs:

$$\mathcal{S}_{xx} = \{\{\mathbf{x}_1, \mathbf{x}_1\}, \{\mathbf{x}_1, \mathbf{x}_2\}, \ldots \{\mathbf{x}_2, \mathbf{x}_1\}, \{\mathbf{x}_2, \mathbf{x}_2\}, \ldots \{\mathbf{x}_n, \mathbf{x}_n\}\} \tag{38}$$

Some of the pairs will be chosen from the same distribution and we would like to minimize the influence of such pairs, since they will introduce errors in our model assuming there are two different distributions. It is likely that pairs from the same distribution are more similar than pairs from different distributions. For this reason, we introduce similarity-dependent weights in the sums defining **s** and **P**, where the weight w_{ij} is based on the sine of the angle between the vectors:

$$w_{ij} = \sqrt{1 - (\hat{\mathbf{x}}_i^T \hat{\mathbf{x}}_j)^2} = |\sin(\varphi_{ij})| \tag{39}$$

The expressions for our invariants as weighted sums are now given by:

$$\mathbf{s} = \sum_{i=1}^{n} \sum_{j=1}^{n} \frac{w_{ij}}{\mathrm{w}} (\mathbf{x}_i + \mathbf{x}_j) \tag{40}$$

$$\mathbf{P} = \sum_{i=1}^{n} \sum_{j=1}^{n} \frac{w_{ij}}{\mathrm{w}} \left(\mathbf{x}_i \mathbf{x}_j^T + \mathbf{x}_j \mathbf{x}_i^T\right) \tag{41}$$

$$\mathrm{w} = \sum_{i=1}^{n} \sum_{j=i}^{n} w_{ij} \tag{42}$$

Note that if all vectors are equal, all w_{ij} will be 0, however the weights based on the normalized sum $\frac{w_{ij}}{\mathrm{w}}$ in Eq. 40 will, using limit calculus, equal $1/n^2$. In practice w_{ij} is computed adding a small number ϵ in Eq. 39. There are other ways to define the dissimilarity weights, and the best way will likely be application dependent. On a side note, the expression in Eq. 41 is related to variance calculation with weights which has found applications in statistical classification. Potential connections between our current work and statistical classification are left for interesting future research.

To estimate the mean vector-pair **u** and **v**, from **s** and **P**, we follow the recipe above. Again, in a least square sense, **P** is best represented by the eigenvectors associated with eigenvalues with the largest magnitudes. We insert the eigenvectors, $\mathbf{e}_1$ and $\mathbf{e}_2$, and their corresponding eigenvalues in Eqs. 29–30, solve the least squares problem in Eqs. 31–33, and recover the scaling (μ_1, μ_2) from Eqs. 34 to 36:

$$\mathbf{u} = \mu_1 \frac{\left(\sqrt{\lambda_1}\,\mathbf{e}_1 + \sqrt{-\lambda_2}\,\mathbf{e}_2\right)}{\sqrt{\lambda_1 - \lambda_2}} \tag{43}$$

$$\mathbf{v} = \mu_2 \frac{\left(\sqrt{\lambda_1}\,\mathbf{e}_1 - \sqrt{-\lambda_2}\,\mathbf{e}_2\right)}{\sqrt{\lambda_1 - \lambda_2}} \tag{44}$$

3.3 Estimating the Amount of Samples from u and v

It is often important to estimate the number of samples that were taken from each distribution. Weights representing these amounts can be estimated in the following way. We start with the first order invariance sum,

$$\mathbf{s}_0 = \frac{1}{n^2}\sum_{i=1}^{n}\sum_{j=1}^{n}\mathbf{x}_i + \mathbf{x}_j \tag{45}$$

$$= \frac{2}{n}\sum_{i=1}^{n}\mathbf{x}_i \tag{46}$$

where the subscript of $\mathbf{s}_0$ denotes unweighted summation to distinguish it from the the weighted sum $\mathbf{s}$. Next, by minimizing the difference between $\mathbf{s}_0$ and a linear combination of the representative vectors we get an expression for the amount of samples m_1 and m_2 from the two classes $\mathbf{u}$ and $\mathbf{v}$ respectively.

$$\min_{m_1,m_2} ||\mathbf{s}_0 - (\mathbf{u}+\mathbf{v})||^2 \tag{47}$$

$$= \min_{m_1,m_2} ||\mathbf{s}_0 - (m_1\hat{\mathbf{u}} + m_2\hat{\mathbf{v}})\,||^2 \tag{48}$$

$$= \min_{m_1,m_2} ||\mathbf{s}_0 - (\hat{\mathbf{u}}\;\hat{\mathbf{v}})\begin{pmatrix} m_1 \\ m_2 \end{pmatrix}||^2 \tag{49}$$

$$= \min_{\boldsymbol{m}} ||\mathbf{s}_0 - \mathbf{U}\boldsymbol{m}||^2 \tag{50}$$

giving the solution

$$\boldsymbol{m} = \left(\mathbf{U}^{\mathbf{T}}\mathbf{U}\right)^{-\mathbf{1}}\mathbf{U}^{\mathbf{T}}\mathbf{s}_0 \tag{51}$$

In summary, we have now shown how to (1) estimate the mean vector pair, and (2) the amounts of the samples. The amounts of the samples is important for further processing as well as for visualizing the result.

4 Tensor Neighborhoods

As described above in Sect. 2, the theory that we have derived regarding vector pairs can be directly applied to tensors that have been "vectorized". This implies no loss of generality since the vectors contain all the components of the corresponding tensor and can thus can easily be reinterpreted as a tensor.

We will apply this theory to two distinct types of tensor fields, corresponding to the two types of vector distributions discussed above in Sect. 3. The first type of

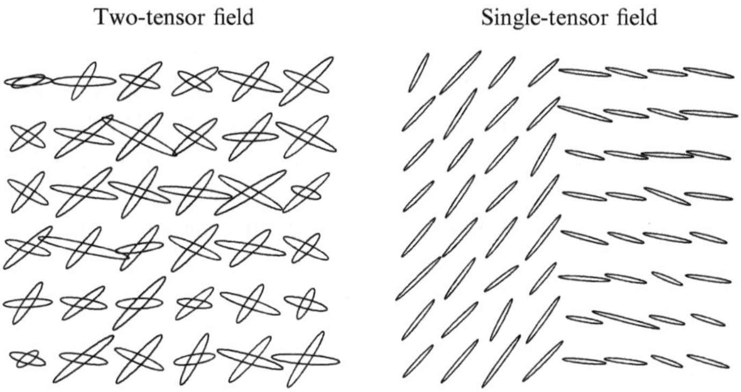

Fig. 2 Example of a two-tensor field (*left*) and a single-tensor field (*right*)

tensor field has two tensors (a tensor-pair) defined at each spatial location (Fig. 2, left). The second type of tensor field has one tensor defined in each location (Fig. 2, right), and tensor-pairs can be created by picking tensors from multiple locations.

Analogous to our estimation goal of finding the mean vector from a distribution, here we propose to do a local analysis to define the tensor distribution under a spatial lowpass kernel or mask a.

Estimation of the mean tensor is not straightforward in the two-tensor field case because there is no order defined, meaning that the field cannot easily be divided into two consistent single-tensor fields. In the experiments below, pairs of tensors are selected from a 3×3 spatial neighborhood defined by the lowpass mask a. The weights of the pairs are then calculated from the mask values:

$$w_i = a_i \tag{52}$$

For the single-tensor field case, all possible pairs of tensors are created from the set of tensors under the mask. Here the weights will be a combination of the dissimilarity-weight, and the weights from the mask a:

$$w_{ij} = \sqrt{a_i a_j \left(1 - \left(\hat{\mathbf{x}}_i^T \hat{\mathbf{x}}_j\right)^2\right)} = \sqrt{a_i a_j}\, |\sin(\varphi_{ij})| \tag{53}$$

where $\mathbf{x}_i$ and $\mathbf{x}_j$ are a pair of vectorized tensors under the mask.

5 Neighborhood Averaging

It is often desirable to estimate properties representing a larger neighborhood. Unfortunately the number of possible pairs increases quickly with the size of the neighborhood. Instead of working with large neighborhoods directly, we can take advantage of the fact that the proposed representation allows **s** and **P** to be averaged. For example in a 3D $n \times n \times n$ neighborhood we have $n^3(n^3 + 1)/2$ pairs, e.g. for $n = 7$ we have approximately 60,000 pairs. We can cover the same region by constructing the representation **P** from a 3×3×3 neighborhood (378 pairs), followed by spatial averaging of **P** with a 5×5×5 lowpass filter. Using such a 2-stage approach the computational burden can be reduced by orders of magnitude. Of course the two different operations are not equivalent, but in practice they often give comparable results.

5.1 *Normalized Convolution*

The first stage of the proposed 2-stage estimation scheme will, in general, produce estimates that should be given different weights in the second stage. The appropriate weight to use is the sum of the local neighborhood weights (w_{ij}, Eq. 53), i.e. the weight of an estimate in the first stage is at each position given by

$$\mathrm{w} = \sum_{ij} w_{ij} \tag{54}$$

When performing the neighborhood averaging operations, special care has to be taken in incorporating the effect of the weights. We have earlier designed a technique to handle this, normalized convolution [2,3]. A special case of normalized convolution is normalized averaging. When averaging the representation **s**, $\mathbf{s}_0$, and **P** using normalized averaging with a lowpass filter a, we also need to apply the filter to the weights of the estimates w. The result is then obtained by then normalized by the filtered weights:

$$\mathbf{s}_a = \frac{a * \mathbf{s}}{a * \mathrm{w}} \tag{55}$$

$$\mathbf{s}_{0a} = \frac{a * \mathbf{s}_0}{a * \mathrm{w}} \tag{56}$$

$$\mathbf{P}_a = a * \mathbf{P} \tag{57}$$

where $*$ denotes convolution. As above, there is no need to normalize **P**.

6 Experiments and Results

In this section we apply the presented theory to some examples to illuminate the strength of the approach. Examples are intended to display the potential performance in a few distinct situations.

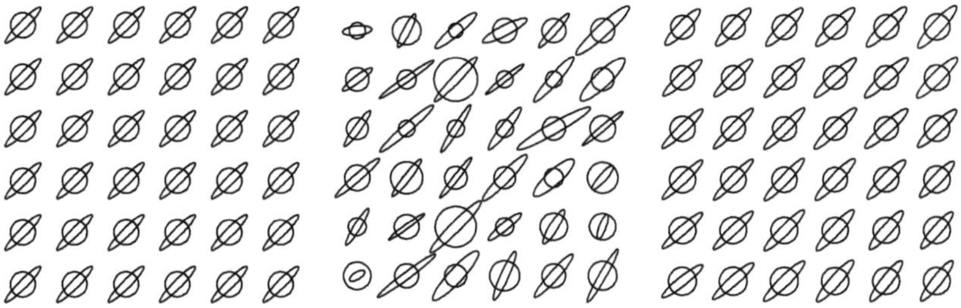

Fig. 3 Example with un-ordered two-tensor field (*left*) with added noise (*middle*). Result from estimating the two tensor representatives from the representation $\mathbf{s}_a$ and $\mathbf{P}_a$ (*right*)

6.1 Two-Tensor Field Neighborhoods

In the first example, the two-tensor field consists of tensor pairs that have one of the tensors isotropic, see Fig. 3 (left). A high level of noise was then added to produce the test data (middle). Then the representation ($\mathbf{s}_a$ and $\mathbf{P}_a$) was calculated using a 3×3 mask for defining the pairs, and a 3×3 mask for further spatial regularization. The reconstructed **u** and **v** field (right) is close to the original noise free field (left), clearly showing the robustness of the algorithm.

In the second example, three fields are presented with varying angles between the tensor pairs. Figure 4 shows the three fields where the tensor pairs have an average angle of 60° (top), 30° (middle), and 15° (lower). Again in this example a 3×3 mask was used for defining pairs, and a 3×3 mask was used for subsequent averaging of the invariants and the corresponding weight field. A slight reduction in anisotropy can be noticed after averaging, but as in the previous example, the results are very good.

6.2 Single-Tensor Field Neighborhoods

As mentioned above, for single-tensor field neighborhoods, the pairs of tensors are constructed by creating all possible pairs from the tensors under the spatial mask. When calculating the weights for this case, the weights are based on both the dissimilarity-weights and the spatial mask as described by Eq. 53.

In the third example, the tensor field consists of two distinct areas are shown Fig. 5 (top left), and with added noise (top right). The representation $\mathbf{s}_a$, $\mathbf{s}_{0a}$, and $\mathbf{P}_a$ were calculated using a 3×3 mask for defining the pairs, and a 3×3 mask for further spatial regularization.

The field $\mathbf{s}_{0a}$ (lower left), shows a low-pass version of the original tensor field. The result illustrates how standard spatial averaging would blur the edge of the data. This results is also used to estimate the weights of **u** and **v** as described by Eq. 51.

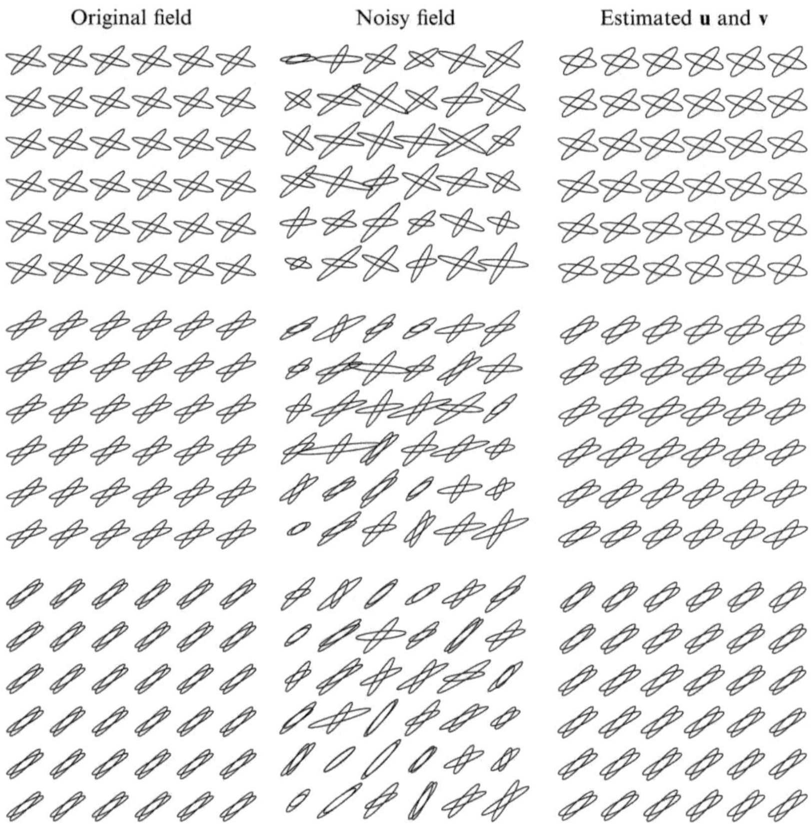

Fig. 4 Three examples with un-ordered two-tensor fields with different angles between the pairs (*left*) with added noise (*middle*). Results from estimating the two tensor representatives from the representation $\mathbf{s}_a$ and $\mathbf{P}_a$ (*right*)

By instead using our new representation, the averaging operation will not mix the two tensor classes. Instead two different averages are found, and relative weight of the two classes can be calculated. Equation 51 shows that these weights are in fact calculated from the standard tensor average field, $\mathbf{s}_{0a}$ (lower left). The strength of these weights are shown as the gray-level (darker corresponds to a higher weight) of the ellipses representing **u** and **v** (lower right).

7 Discussion and Conclusion

In this chapter we have discussed how un-ordered pairs of tensors can be represented. The representation is based on first and second order tensor-pair invariants. An eigenvector-based inverse transform is presented enabling an analytic

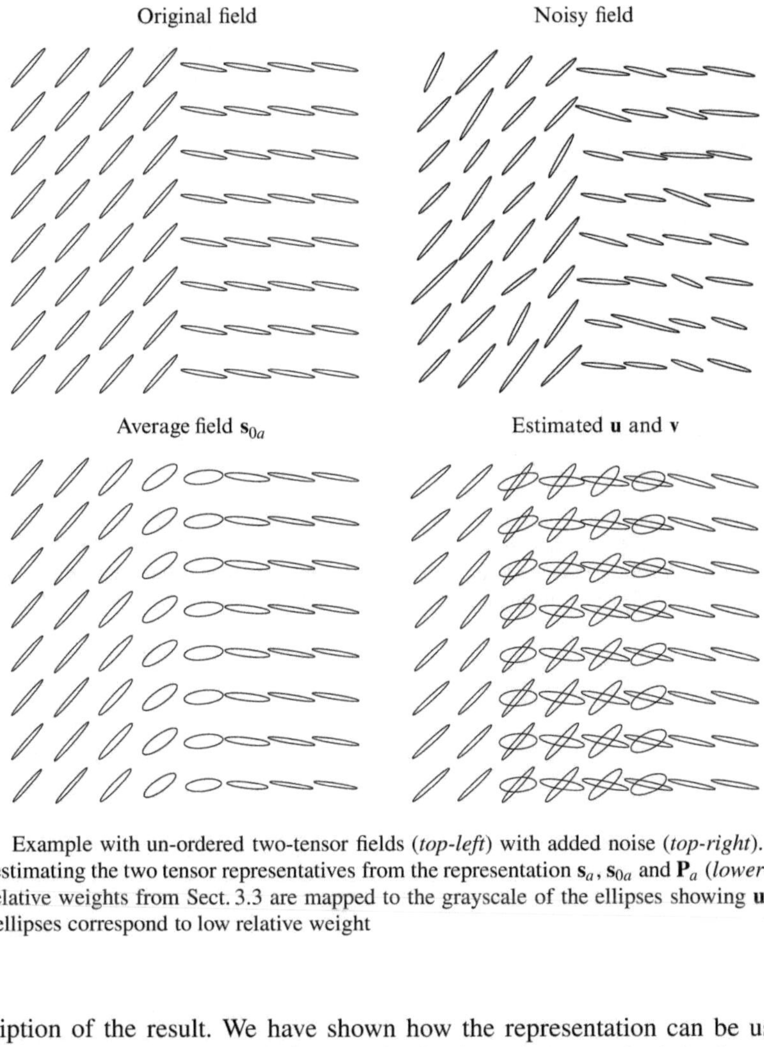

Fig. 5 Example with un-ordered two-tensor fields (*top-left*) with added noise (*top-right*). Result from estimating the two tensor representatives from the representation $\mathbf{s}_a$, $\mathbf{s}_{0a}$ and $\mathbf{P}_a$ (*lower-right*). The relative weights from Sect. 3.3 are mapped to the grayscale of the ellipses showing **u** and **v**. Faint ellipses correspond to low relative weight

description of the result. We have shown how the representation can be used to estimate an average tensor-pair from a distribution of tensors, both in the case of tensor-pair distributions and in the case of un-paired tensor distributions. Finally, we demonstrated the potential of the approach by processing a number of simple tensor image examples. The results show how the tensor-pair representation gives additional valuable information about the neighborhood and consistently promotes high robustness in the presence of noise.

Applications of the presented theory are likely to be found in areas where tensor fields are used to describe signals and features. One such area is in diffusion MRI (dMRI). In dMRI reconstruction of the white matter fiber architecture is complicated by the fact that fiber bundles are not spatially separated but often are crossing or interleaved. In the same measurement, there are then two classes of tensors, where the average tensor has no anatomical meaning. This implies that the standard tensor model used in dMRI is not adequate for disentangling multiple fiber orientation, but is designed for single orientation fiber bundles. Our proposed framework has the potential to handle two crossing fiber directions in a new and compact manner.

Acknowledgements The NIH grants R01MH074794, R01MH092862, P41EB015902 and P41RR013218, and grants from the Swedish Research Council, Swedish e-Science Research Center, Linköpings Universitet are gratefully acknowledged for supporting this work. We also thank Lauren O'Donnell for insightful comments on the manuscript.

References

1. Herberthson, M., Brun, A., Knutsson, H.: Representing pairs of orientations in the plane. In: Proceedings of the 15th Scandinavian conference on image analysis (SCIA'07). Aalborg. Lecture Notes in Computer Science, vol. 4522, pp. 661–670. Springer, Berlin/New York (2007)
2. Knutsson, H., Westin, C.F.: Normalized and differential convolution: methods for interpolation and filtering of incomplete and uncertain data. In: Proceedings of Computer Vision and Pattern Recognition (CVPR'93), New York City, pp. 515–523. IEEE, Los Alamitos (1993)
3. Westin, C.F.: A tensor framework for multidimensional signal processing. Ph.D. thesis, Linkoping University, S-581 83 Linkoping (1994). Dissertation No 348, ISBN 91-7871-421-4

Part VI
Tensor Metrics

On the Choice of a Tensor Distance for DTI White Matter Segmentation

Rodrigo de Luis-García, Carlos Alberola-López, and Carl-Fredrik Westin

Abstract The segmentation of anatomical structures within the white matter of the brain from DTI is an important task for white matter analysis, and has therefore received considerable attention in the literature during the last few years. Any segmentation method relies on the choice of a tensor dissimilarity measure, which should be small between tensors belonging to the same region and large between tensors belonging to different structures. Many different tensor distances have been proposed in the literature (Frobenius, Kullback-Leibler, Geodesic, Log-Euclidean, Hybrid...) for segmentation or other purposes, and there exist reasons (either theoretical or empirical) to justify the choice of any of them. Thus, determining which is the most appropriate tensor distance for a specific segmentation problem has become an extremely difficult decision. In this chapter we present a study on different tensor dissimilarity measures and their performance for white matter segmentation. The study is based on the use of two different DTI atlases of human brain, which provide a ground truth upon which the distances can be fairly compared. In order for the comparison to be independent of the segmentation method employed, it has been performed in terms of the separability of the different clases. Results show the Hybrid distance to perform better than other traditional tensor dissimilarity measures in terms of separability between classes, while the Frobenius, Kullback-Leibler, Geodesic and Log-Euclidean distances perform similarly.

R. de Luis-García (✉) · C.-F. Westin
Laboratory of Mathematics in Imaging, Harvard Medical School, Boston, MA, USA
e-mail: rodrigo@bwh.harvard.edu; westin@bwh.harvard.edu

C. Alberola-López
Laboratorio de Procesado de Imagen, Universidad de Valladolid, Valladolid, Spain
e-mail: caralb@tel.uva.es

D.H. Laidlaw and A. Vilanova (eds.), *New Developments in the Visualization and Processing of Tensor Fields*, Mathematics and Visualization,
DOI 10.1007/978-3-642-27343-8_15,

1 Introduction

Diffusion Tensor Magnetic Resonance Imaging (DT-MRI, also known as DTI, Diffusion Tensor Imaging) is a medical imaging modality based on Magnetic Resonance Imaging (MRI) which is able to quantify the anisotropic diffusion of water mollecules in highly structured biological tissues [6, 7]. The probability density function of the three-dimensional molecular motion is modeled, at each voxel, by a normal distribution of 0-mean and whose covariance matrix is given by the diffusion tensor. Currently, brain imaging is the most common application of diffusion MRI, as the brain has a complex structure of grey matter areas connected by white matter fibres. Diffusion MRI can be therefore employed for the visualization of the fibre tracts in the white matter of the brain.

DTI analysis of brain structures has shown to be relevant in a number of neurological pathologies, such as brain ischemia, multiple sclerosis or epilepsy, among others [18, 42, 45, 46]. In schizophrenia, alterations in the diffusion in several areas within the white matter have been found through a number of group studies [25]. The automatic segmentation of these structures from DTI has spurred significant research effort recently, due to its importance for these studies. Consequently, several authors have addressed this issue in the last years, as we briefly review in the next paragraphs.

Whatever the segmentation method employed may be, tensor field segmentation requires the use of a certain tensor distance. In [52], Zhukov et al. define an invariant anisotropy measure in order to drive the evolution of a level set and isolate strongly anisotropic regions of the brain. Rousson et al. [8, 39, 40] proposed to employ a Euclidean distance on the components of the diffusion tensor, which are arranged as a vector. The tensor components were also considered in the same way in [14]. In [16], the local structure tensor of the tensor data was employed for segmentation.

More recent approaches to tensor field segmentation employ intrinsic tensor dissimilarity measures in their approaches. Wiegell et al. proposed to use the Frobenius distance, combined with the Mahalanobis tensor distance with a K-means algorithm [51]. The Frobenius distance was also employed in [48, 53]. In [20], Jonasson et al. defined the Normalized Tensor Scalar Product (NTSP) to quantify tensor similarity with a great sensitivity to changes in orientation. The use of the symmetrized Kullback-Leibler distance for tensor fields was first proposed in [48], and later further explored in [12, 27–31, 50, 53]. Lenglet et al. introduced the Geodesic distance in [27–29], which was also employed in [12, 12, 29–31]. In [1, 2], Arsigny et al. proposed the Log-Euclidean framework for tensor computations, which was applied for DT-MRI segmentation in [5, 50]. Finally, an hybrid distance was proposed in [13] that considers the shape and the orientation of the tensor independently, and thus can adapt the relative importance of these two tensor properties. In Table 1, the most relevant tensor segmentation techniques in the literature are summarized, with special attention to the tensor distances that were employed.

The proposal of all these different tensor dissimilarity measures was motivated by both theoretical and empirical reasons that showed limitations in the different tensor distances. However, the proliferation of metrics and the increasing number

Table 1 Summary of tensor segmentation methods

Technique	Application	Tensor distance	Segmentation type
Peled et al. [37]	DTI (corpus callosum, internal capsule)	Linear, planar and spherical components	Quantitative analysis (not really segmentation)
Zhukov et al. [52]	DTI (regions with high diffusivity/ anisotropy)	Diffusivity measure I_1 Anisotrpy measure C_a	GAC (level sets)
Rousson et al. [8, 39–41]	LST (unsupervised segmentation)	Euclidean distance tensor components as vector	GAR (level sets)
Feddern et al. [16]	DTI	LST of the tensor data	GAC
Wiegell et al. [51]	DTI (talamic nuclei)	Combination of tensor distance Frobenius and Mahalanobis voxel distance	K-means
Jonasson et al. [20–22]	DTI (corpus callosum, corticospinal tract)	Normalized tensor scalar product (NTSP)	Level set, curve propagation based on tensor distance
Wang and Vemuri [49]	LST	Frobenius	Mumford-Shah functional (level sets)
Wang and Vemuri [48]	DTI LST	K-L distance	Chan and Vese model (level sets)
Lenglet et al. [26, 28]	DTI	K-L distance Geodesic distance	GAR on the distance to the mean tensor
Lenglet et al. [29–31]	DTI	Euclidean distance K-L distance Geodesic distance	GAR with Gaussian distribution over tensor fields
Ziyan et al. [53]	DTI (thalamus nuclei)	Frobenius distance K-L distance Eigenvectors angular diff.	Spectral clustering with Markovian relaxation
de Luis-Garcia and Alberola [12]	DTI (corpus callosum)	K-L distance Geodesic distance	GAR with mixtures of Gaussians on tensors
Weldeselasie and Hamarneh [50]	DTI (corpus callosum and cardiac)	Log-Euclidean K-L distance	Graph-cuts
Awate et al. [4, 5]	DTI (cingulum, corticospinal tract)	Log-Euclidean	Fuzzy C-means with nonparametric statistical models
de Luis-Garcia et al. [13]	DTI (corpus callosum, corona radiata, cingulum...)	Hybrid distance	GAR with Gaussian distributions over tensor fields

GAC geodesic active contours [9, 23], *GAR* geodesic active regions [33], *LST* local structure tensor [19, 24]

of works that claim the superiority of their respective choices make it very difficult to assess the relative performance of all these tensor distances and, consequently, make an appropriate decision for a specific segmentation task.

The study of the properties of different tensor metrics has been addressed in several works in the literature. In [35], the different measures proposed for DT-MRI analysis were reviewed and classified according to several important properties such as their sensitivities to changes in orientation or size, or the fulfillment of the conditions to be a Riemannian metric. From this classification, a tensor distance can be selected considering the requirements that may be relevant for a specific application. A review of tensor field segmentation techniques was presented in [13] with an important focus in the choice of the tensor distances. Rodrigues et al. devised a semiautomatic method to define a combination of tensor distances that better suits a given dataset. Finally, in [34] the authors investigate the properties of two tensor distances that define a Riemann manifold from a physical point of view.

In this chapter, we investigate the behaviour of the most important tensor distances that have been employed in the literature for DT-MRI segmentation, in terms of their ability to separate tensors belonging to different anatomical regions of interest in the white matter. To that end, we employ two different DT-MRI atlases of the human brain, as they provide a ground truth on which fair comparisons can be made.

No specific segmentation method has been employed in this chapter, as it is the aim of this work to compare different tensor distances independently of the chosen segmentation technique. However, the comparisons presented here are better suited towards region-based segmentation methods, whose success depends on the relative coherence among the tensors inside a region of interest with respect to those in the background. Many of the segmentation DT-MRI segmentation techniques described before are region oriented, although they have intrinsic limitations when dealing with curved structures where tensors slowly change their orientation to finally present a wide range of tensor values. Edge-based segmentation techniques, on the other hand, would act locally by looking for abrupt changes in the tensor values, independently of their homogeneity inside the region of interest. Distance comparisons adapted for this segmentation paradigm should incorporate the desired properties of a tensor dissimilarity measure in that context.

The remaining of the chapter is organized as follows: in the next section, we present and provide some background on the different tensor distances that will be studied in this work. Section 3 focuses on the techniques that will be employed to perform the comparisons between the different dissimilarity measures. These comparison results are provided in Section 4 together with a thorough discussion of their implications. Finally, some conclusions are drawn in Section 5.

2 Tensor Similarity Measures

2.1 Frobenius Distance

The Frobenius distance is the most natural metric for matrices, and is defined as the Frobenius norm of the difference of two matrices (diffusion tensors, in our case):

$$d(\mathbf{T_1}, \mathbf{T_2}) = \| \mathbf{T_1} - \mathbf{T_2} \|_F = \left[\text{Trace}\left((\mathbf{T_1} - \mathbf{T_2})^T (\mathbf{T_1} - \mathbf{T_2})\right)\right]^{1/2} \tag{1}$$

Wang and Vemuri employed the Frobenius norm of the difference of tensors as a tensor distance in their segmentation approach [49]. Indeed, this is equivalent to the consideration of the tensor as a vector with a Euclidean distance. Although the experimental results were good, the Frobenius norm of the difference of tensors has been considered not to be completely appropriate, as it uses the same weights for different components of the tensor and ignores the fact that tensors are symmetric positive-definite matrices.

2.2 *Kullback-Leibler Distance*

Wang and Vemuri introduced in [48] a new tensor dissimilarity measure that takes into account the positive-definiteness of tensors and is also affine invariant. Recalling that, in the context of DT-MRI, the displacement of water molecules over a time t follows a Gaussian distribution whose covariance matrix is the diffusion tensor, it is naturally justified to use the distance between Gaussian distributions to induce a distance between the tensors. The most frequently used information theoretic dissimilarity measure is the Kullback-Leibler (KL) divergence, which is defined as:

$$KL(p \parallel q) = \int p(\mathbf{x}) \log\left(\frac{p(\mathbf{x})}{q(\mathbf{x})}\right) d\mathbf{x} \tag{2}$$

As the KL divergence is not symmetric, it is commonly symmetrized as:

$$J(p, q) = \frac{1}{2}\left(KL(p \parallel q) + KL(q \parallel p)\right) \tag{3}$$

This symmetrized KL divergence is also called J-divergence. Based on it, Wang and Vemuri proposed in [48] a dissimilarity measure for SPD tensors, given by the square root of the J-divergence between two Gaussian distributions with zero mean and covariances $\mathbf{T_1}$ and $\mathbf{T_2}$:

$$d(\mathbf{T_1}, \mathbf{T_2}) = \sqrt{J(p(\mathbf{x}|\mathbf{T_1}, t), p(\mathbf{x}|\mathbf{T_2}, t))} \tag{4}$$

Taking the square root in Eq. 4 is justified because the KL divergence, and thus twice the J-divergence, is the square distance of two infinitesimally nearby points on a Riemannian manifold of parametrized distributions. It turns out that the J-divergence has a simple form for the Gaussian distribution considered [48]:

$$d(\mathbf{T_1}, \mathbf{T_2}) = \frac{1}{2}\sqrt{\text{trace}(\mathbf{T_1^{-1}}\mathbf{T_2} + \mathbf{T_2^{-1}}\mathbf{T_1}) - 2n} \tag{5}$$

where n is the size of the tensors.

2.3 Information Geodesic Distance

In order to define a new tensor dissimilarity measure, Lenglet et al. considered in [27, 28] the statistical manifold $\mathscr{M}$ representing the family of three-dimensional, zero-mean Gaussian PDFs described through the six free parameters of their covariance matrix.[1] A Riemannian metric can then be introduced in terms of the Fisher information matrix [38] that allows for the definition of geodesic distances on this manifold.

The geodesic distance induced by the Riemannian metric derived from the Fisher information matrix has been investigated for several parametric distributions. If we concentrate on the family of multivariate normal distributions with common mean vector ξ but different covariance matrices $\mathbf{T}$ or, equivalently, on the manifold $S^+(n, \mathbb{R})$, i.e., the set of $n \times n$ real SPD (symmetric positive definite) matrices, the information geodesic distance between any two elements $\mathbf{T_1}$ and $\mathbf{T_2}$ is given by (theorem by S. T. Jensen, 1976, originally proved in [3]):

$$d(\mathbf{T_1}, \mathbf{T_2}) = \sqrt{\frac{1}{2}\text{trace}\left(\log^2\left(\mathbf{T_1}^{-1/2}\mathbf{T_2}\mathbf{T_1}^{-1/2}\right)\right)} = \sqrt{\frac{1}{2}\sum_{i=1}^{m}\log^2(\lambda_i)} \tag{6}$$

where the λ_i are the m eigenvalues of the determinantal equation $|\lambda\mathbf{T_2} - \mathbf{T_1}| = 0$.

The information geodesic distance has been used for segmentation in a number of works in the literature [12, 26, 28]. For this purpose, the mean value of the tensor field over each region with respect to this new geodesic distance is needed, which is also known as the Riemannian barycenter (see [27] for details).

2.4 Log-Euclidean Metrics

Even though the information geodesic distance has shown very good results for tensor estimation and segmentation purposes, they yield in practice slow algorithms because of their complexity. In order to overcome this limitation, the Log-Euclidean metrics were proposed [1, 2] that produce similar results but using simpler and faster computations.

Euclidean distances (also referred to as Frobenius distances) are well adapted to general square matrices, but they are not designed for the tensor specific case. If Euclidean operations are performed on tensors, null or negative eigenvalues can appear, as these Euclidean operations are not convex [2]. Besides, Euclidean averaging of tensors produces what is called *tensor swelling effect* [10], meaning

[1] Throughout the theoretical derivation of the geodesic distance, 3×3 diffusion tensors are considered.

that the determinant of the Euclidean mean of several tensors can be larger than the determinants of the original tensors.[2]

With respect to information geodesic metrics, the computational burden is related to the intensive use of matrix inverses, square roots, logarithms and exponentials. Besides, there is not a closed form for the tensor mean, and the computation needs to be done in an iterative manner.

In order to define the Log-Euclidean metrics, the notions of matrix logarithm and exponential are first needed. For any matrix **M**, its exponential is given by:

$$\exp(\mathbf{M}) = \sum_{k=0}^{\infty} \frac{\mathbf{M}^k}{k!} \tag{7}$$

The matrix logarithm is defined as the inverse of the exponential. The existence and uniqueness of the logarithm is not guaranteed for a general invertible matrix, but it is well defined for a positive definite tensor, yielding a symmetric matrix. Conversely, the exponential of any symmetric matrix is a positive definite tensor.

The introduction of Log-Euclidean metrics is based on the idea of defining a novel vector space structure on tensors. It corresponds to Euclidean metrics in the domain of logarithms. Using the Euclidean norm $||\cdot||$ on symmetric matrices, the distance between two tensors can be written as:

$$d(\mathbf{T_1}, \mathbf{T_2}) = ||\log(\mathbf{T_1}) - \log(\mathbf{T_2})|| \tag{8}$$

Log-Euclidean metrics do not satisfy full affine-invariance as the Riemannian metric introduced in the previous section. However, a number of them are invariant by similarity, that is, orthogonal transformation and scaling. The simplest similarity-invariant Log-Euclidean metric is given by:

$$d(\mathbf{T_1}, \mathbf{T_2}) = \left[\mathrm{Trace}\left((\log(\mathbf{T_1}) - \log(\mathbf{T_2}))^2\right)\right]^{1/2} \tag{9}$$

Log-Euclidean metrics have been mainly employed for interpolation and regularization of tensor fields [2], but some authors have also used for segmentation purposes, as in [4, 5, 50].

2.5 Hybrid Distance

Although the different tensor distances introduced before are mathematically well founded, they present a major drawback. These distances cannot control the relative

[2]It has recently been shown that the swelling effect can actually be a desirable property. We refer the reader to the chapter of this book devoted to this issue: O. Pasternak, N. Sochen and P.J. Basser, *Metric Selection and Diffusion Tensor Swelling*.

importance of different tensor properties such as the shape and the orientation. To address this issue, in [13] it was proposed to treat the shape and the orientation of the tensor independently. In order to do so, let us first consider the feature vector $\mathbf{K}(\mathbf{x})$ composed of the three tensor invariants proposed in [15], $\mathbf{K} = [K_1\ K_2\ K_3]^T$, where

$$K_1 = \text{trace}(\mathbf{T}) \qquad K_2 = |\text{dev}(\mathbf{T})| \qquad K_3 = 3\sqrt{6}\,\det\left(\frac{\text{dev}(\mathbf{T})}{|\text{dev}(\mathbf{T})|}\right) \tag{10}$$

where $\text{dev}(\mathbf{T})$ is the deviatoric part of tensor $\mathbf{T}$:

$$\text{dev}(\mathbf{T}) = \mathbf{T} - \bar{\mathbf{T}} = \mathbf{T} - \frac{1}{3}\text{tr}(\mathbf{T})\mathbf{I} \tag{11}$$

where $\bar{\mathbf{T}}$ is the isotropic part of tensor $\mathbf{T}$. The three invariants, which are orthogonal, completely describe the tensor shape. K_1 represents the tensor size, while K_2 and K_3 describe the amount of anisotropy and the tensor mode, respectively (fractional anisotropy is not employed as K_2 because this measure is not orthogonal to trace). In order to describe differences between the invariants corresponding to two tensors, Euclidean distance can be easily employed.

To unambiguously represent the orientation of the main eigenvector $\mathbf{e}_1$ of the diffusion tensor, the outer product $\mathbf{U} = \mathbf{e}_1\mathbf{e}_1^T$ can be computed. This new tensor indeed describes the main orientation of the diffusion tensor, and differences can be computed using a Frobenius distance.[3]

Putting together the invariants and the orientation, the Hybrid distance between two diffusion tensors is:

$$d(\mathbf{T}_1, \mathbf{T}_2) = \alpha||\mathbf{K}(\mathbf{T}_1) - \mathbf{K}(\mathbf{T}_2)|| + (1-\alpha)||\mathbf{U}(\mathbf{T}_1) - \mathbf{U}(\mathbf{T}_2)||_F \tag{12}$$

As can be seen in the equation, the relative importance of the invariants and the orientation can be controlled by means of parameter α.

3 Tensor Distances Evaluation

3.1 DT-MRI White Matter Atlases

Two different DT-MRI atlases of the white matter of human brain were employed in this study:

[3]By definition, the outer product $\mathbf{U}$ has null eigenvalues, and therefore Kullback-Leibler or geodesic distances cannot be applied. Frobenius distance is thus chosen.

- **JHU MNI SS:** This is a single-subject atlas (a.k.a. *New Eve*) with a comprehensive white matter parcelation developed at Johns Hopkins University [32,47]. The subject is a 32 years old female, and the volume size is 181× 217×181 with 1 mm resolution. This atlas is B0-distortion corrected by non-linearly warping DTI data to a co-registered T2-weighted anatomical image. The atlas was linearly normalized to ICBM-152 template (MNI space). The white matter parcellation map was manually segmented based on FA and color (fiber orientation) information.
- **UCLA 81:** This second atlas, created at the University of California at Los Angeles, is a stereotaxic probabilistic white matter atlas that fuses DTI-based white matter information with an anatomical template (ICBM-152). The atlas is based on probabilistic tensor maps obtained from 81 normal subjects acquired under an initiative of the International Consortium of Brain Mapping (ICBM). The subjects were normal right-handed adults ranging from 18 to 59 years of age. A hand-segmented white matter parcellation map was created from this averaged map, where deep white matter regions were manually segmented into various anatomic structures based on fiber orientation information. All studies were obtained on 1.5T MR units (Siemens). DT imaging data were acquired by using a single-shot, echo-planar imaging sequence with sensitivity encoding and a parallel imaging factor of 2.0.

3.2 *Tensor Distances Evaluation*

In order to study the behaviour of the different tensor dissimilarity measures, we intend to asses how the different white matter structures are separated from their background when choosing different tensor distances. By employing the DTI white matter atlases described before as a ground truth, we are able to assess the suitability of each tensor distance to separate a region of interest from the background. We considered three regions of interest for this study: the corpus callosum, the corona radiata and the cingulum. The suitability of each tensor distance was studied using several analysis strategies, which we present next.

3.2.1 Kernel Target Alignment

Kernel matrices are symmetric distance matrices constructed by computing the distance between all pairs of objects in a set [43, 44]. For a tensor volume, let us define a labeled set $\mathbf{S} = (\mathbf{T_i}, l_i)$ of diffusion tensors $\mathbf{T_i}$ with a label l_i. If the tensor belongs to a region of interest, $l_i = 1$, whereas $l_i = -1$ for the background.

The kernel matrix $\mathbf{K}$ can be regarded as a matrix of pairwise similarity. For the set $\mathbf{S}$ on N tensors, we can define the feature vector

$$\mathbf{f_i} = [d(\mathbf{T_i}, \mathbf{T_1}), d(\mathbf{T_i}, \mathbf{T_2}), \ldots, d(\mathbf{T_i}, \mathbf{T_N})] \tag{13}$$

It is important to note that each feature vector represents a single tensor $\mathbf{T_i}$ by means of the distances to all the tensors in the set $\mathbf{S}$. Now, as shown in [36], a kernel $\mathbf{K}$ can be defined as the inner-product between the feature vectors:

$$K_{ij} =< \mathbf{f_i}, \mathbf{f_j} >= \sum_k d(\mathbf{T_i}, \mathbf{T_k}) d(\mathbf{T_j}, \mathbf{T_k}) \tag{14}$$

where K_{ij} is the element in row i and column j of the kernel matrix $\mathbf{K}$. In order to perform a comparison between different tensor distances, a normalization step is performed in the feature space, as proposed in [17]:

$$\tilde{K}_{ij} = \frac{K_{ij}}{\sqrt{K_{ii} K_{jj}}} \tag{15}$$

A method to assess the quality of a binary clustering was proposed in [11], where Cristianini et al. defined the Kernel Target Alignment (KTA), a measure that describes how good a kernel is with respect of a set of labeled objects. The alignment between two arbitrary kernel matrices $\mathbf{K_1}$ and $\mathbf{K_2}$ is

$$KTA(\mathbf{K_1}, \mathbf{K_2}) = \frac{< \mathbf{K_1}, \mathbf{K_2} >_F}{\sqrt{< \mathbf{K_1}, \mathbf{K_1} >_F < \mathbf{K_2} \mathbf{K_2} >_F}} \tag{16}$$

where $< \mathbf{A}, \mathbf{B} >_F = \sum_{ij} a_{ij} b_{ij}$ is the Frobenius product between two matrices. In order to measure the quality of the clustering defined by the labeled set $\mathbf{S}$, we first define a target matrix $\mathbf{P}$. From the vector of labels $\mathbf{l}$, the target is obtained using the matrix product $\mathbf{P} = \mathbf{l}^T \mathbf{l}$. Now, the alignment between the kernel matrix $\tilde{\mathbf{K}}$ and the target matrix $\mathbf{P}$ describes the quality of the clustering:

$$KTA(\tilde{\mathbf{K}}, \mathbf{P}) = \frac{< \tilde{\mathbf{K}}, \mathbf{P} >_F}{\sqrt{< \tilde{\mathbf{K}}, \tilde{\mathbf{K}} >_F < \mathbf{P}, \mathbf{P} >_F}} \tag{17}$$

For our comparison purposes, a higher KTA value for a certain tensor distance with respect to others represents a better clustering of the tensors inside and outside the region of interest, indicating a better separability between both regions.

3.2.2 Intra/Inter Class Distances

The KTA described before describes the quality of the clustering of tensors into region of interest and background. In order to complement this measure, we next define an alternative descriptor of the goodness of such a clustering. To that end, let us consider again a set of tensors $\mathbf{S}$. From that, we can compute the distance matrix $\mathbf{D}$, where D_{ij} is the distance between two tensors, $D_{ij} = d(\mathbf{T_i}, \mathbf{T_j})$.

We can now describe the goodness of the binary clustering in terms of the relation between the mean intraclass distance and the interclass distance. From the distance matrix **D**, the mean distance between tensors that belong to the region of interest (intraclass distance) can be computed as

$$\bar{d}_{AA} = \frac{\sum_i \sum_j D_{ij} \cdot \Lambda_{ij} \cdot L_{ij}}{\sum_i \sum_j \Lambda_{ij} \cdot L_{ij}} \tag{18}$$

where $\Lambda_{ij} = 1$ if $i > j$ and $L_{ij} = 1$ if both $\mathbf{T_i}$ and $\mathbf{T_j}$ belong to the region of interest, and 0 otherwise.

Similarly, the mean interclass distance can be computed as

$$\bar{d}_{AB} = \frac{\sum_i \sum_j D_{ij} \cdot \Lambda_{ij} \cdot L'_{ij}}{\sum_i \sum_j \Lambda_{ij} \cdot L'_{ij}} \tag{19}$$

where $L'_{ij} = 1$ if $\mathbf{T_i}$ and $\mathbf{T_j}$ belong to different classes, and 0 otherwise.

From the mean intraclass and the interclass distances, a descriptor of the goodness of a clustering result depending of the distance employed can be defined, for instance, as the ratio between these two quantities:

$$\beta = \frac{\bar{d}_{AA}}{\bar{d}_{AB}} \tag{20}$$

A clustering showing a low mean intraclass distance when compared to its mean interclass distance is considered to represent better the uniqueness of a region of interest than a cluster showing a higher ratio. Therefore, for a given region of interest, the lower this ratio, the better performance of a certain tensor distance to extract that particular region of interest.

4 Results and Discussion

In this section we provide results on the analysis of the behaviour of the different tensor distances on the two atlases that were employed for this study. Results will be shown, for each analysis, for both atlases, so as to increase the consistency of the study.

First, we show in Fig. 1 the Kernel Target Aligment values for the different tensor distances, considering the corpus callosum, corona radiata and cingulum as regions of interest. As obtaining the KTA is computationally very expensive, 100 random samples were drawn for both inside the regions of interest and the background, which is the remaining of the white matter of the brain. Five repetitions were made, and results are given as boxplots. For the hybrid distances, the number inside the parenthesis indicates the parameter α in Eq. 12.

There is a number of important findings that can be extracted from the KTA values. First, results from the JHU and UCLA atlases are fairly coherent, which

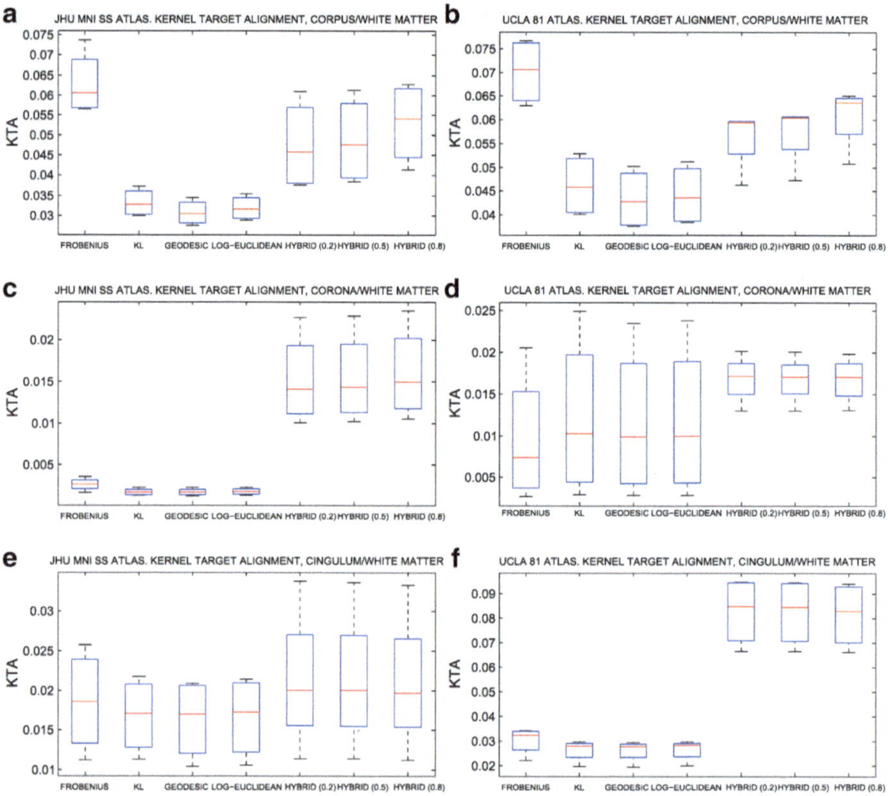

Fig. 1 KTA values for the different tensor distances for the corpus callosum (*top row*), corona radiata (*medium row*) and cingulum (*bottom row*), for the JHU MNI SS atlas (*left*) and UCLA 81 atlas (*right*). The background is defined as all the white matter outside the region of interest

supports the conclusions that can be derived from the study. Second, the hybrid distance shows a better performance than the other, more traditional, tensor distances. There is an important exception, which is the performance of the Frobenius distance for the corpus callosum. In this case, the Frobenius distance outperforms all the other dissimilarity measures. Finally, the behaviour of the KL, Information Geodesic and Log-Euclidean distances is very similar in this comparison.

The results shown in Fig. 1 consider as a background the whole white matter of the brain. However, many segmentation algorithms that have been used for anatomical structures in DT-MRI work, or could work, locally. This means that the propagation of an evolving level set, for instance, can depend on the comparison between the features inside the region of interest and the surrounding region, instead of considering the whole white matter or the whole brain. A region growing algorithm would also work this way. Then, it is interesting to find out what happens

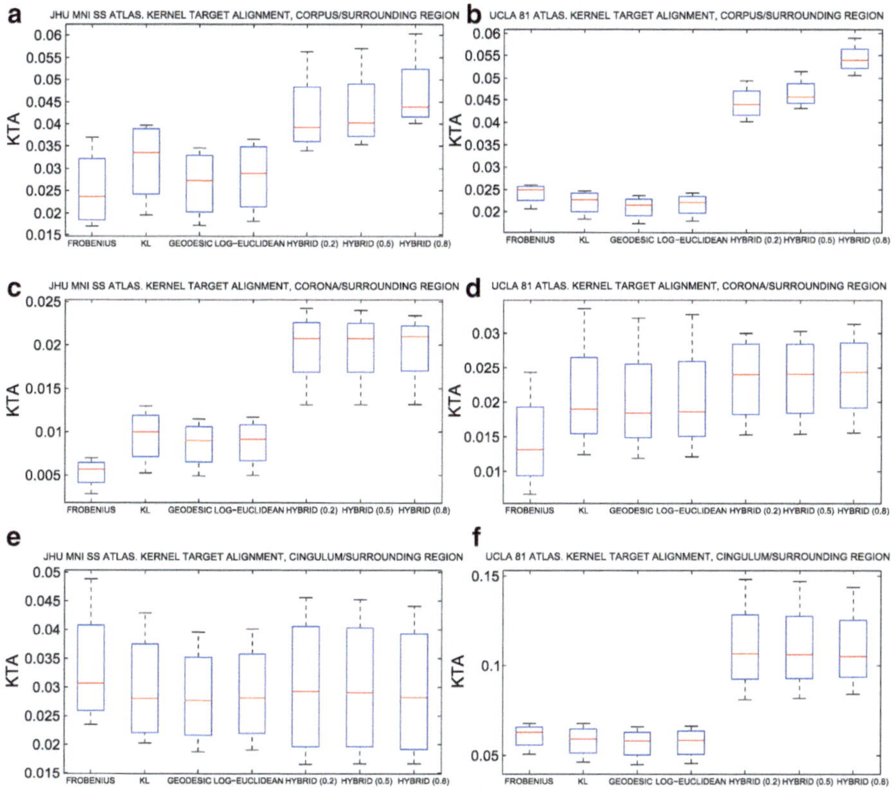

Fig. 2 KTA values for the different tensor distances for the corpus callosum (*top row*), corona radiata (*medium row*) and cingulum (*bottom row*), for the JHU MNI SS atlas (*left*) and UCLA 81 atlas (*right*). The background is defined as the area surrounding the region of interest, with a width 14 voxels

if the KTA is computed considering only a surrounding region as a background instead of the whole white matter. This comparison can be seen in Fig. 2.

Interestingly, results do not differ much, in general, with respect to those in Fig. 1. However, for the corpus callosum, the Frobenius distance decreases its performance in a significant manner. As will be seen later, this fact is related to the distribution of similar regions than the corpus callosum, in terms of the Frobenius distance, within the white matter. With respect to the hybrid distance, a higher weighting in the invariants component seems to work better for the corpus callosum, whereas almost no difference can be found for the corona radiata or the cingulum.

Next, we show in Figs. 3 and 4 the results for the intraclass/interclass ratio described in Sect. 3.2.2. In this case, the lower the ratio, the better the separation between clases. It should be expected that these results confirmed previous findings, and they partially do, while arising new questions.

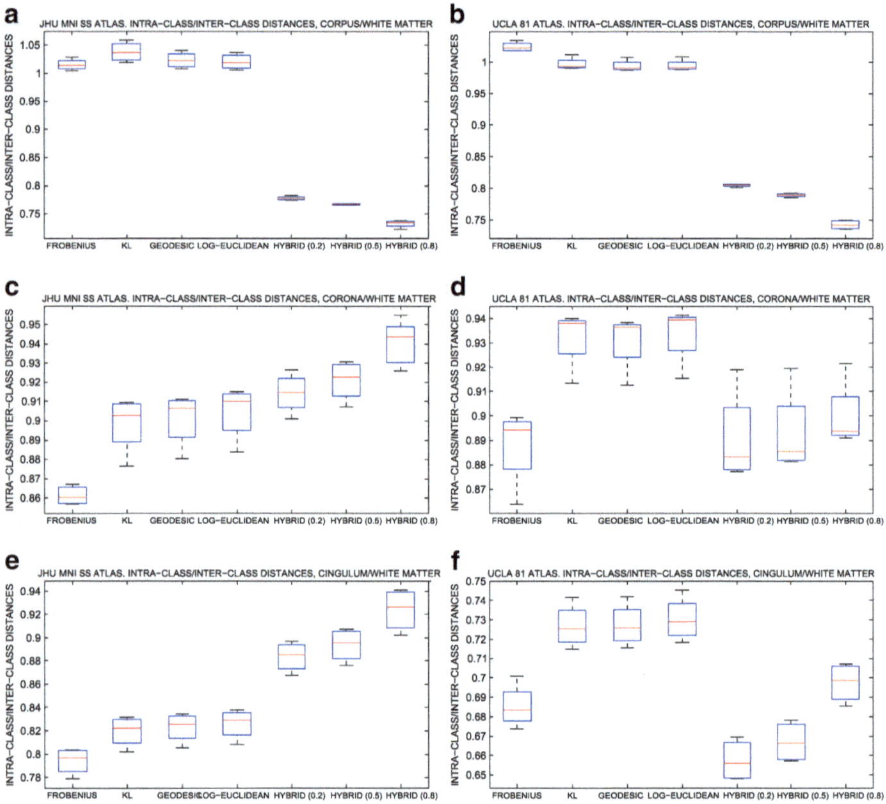

Fig. 3 Intraclass/interclass distance ratios for the different tensor distances for the corpus callosum (*top row*), corona radiata (*medium row*) and cingulum (*bottom row*), for the JHU MNI SS atlas (*left*) and UCLA 81 atlas (*right*). The background is defined as all the white matter outside the region of interest

For the corpus callosum, results show an important superiority of the hybrid distances with respect to the rest of dissimilarity measures, both for the consideration of the whole white matter or only a surrounding region as background. Furthermore, a higher relative weight of the invariant term improves separability, as also found in Figs. 1 and 2. Unlike the KTA values, the intraclass/interclass ratio does not show a good performance of the Frobenius distance for the corpus/white matter case. It is interesting to note that in Fig. 3a and b, the ratios have, for some distances, values above the unity, meaning that the mean intraclass is higher than the mean interclass distance. This implies that, with these distances, there is a really bad separation between the corpus and the background, which would make the segmentation very complicated. However, when only a surrounding region is considered as background (Fig. 4a, b), the values fall below 1. The reason is that in the white matter there are, for some tensor distances, regions where the tensors are very similar to those in the corpus, and these regions happen to be quite distant from the corpus. This is

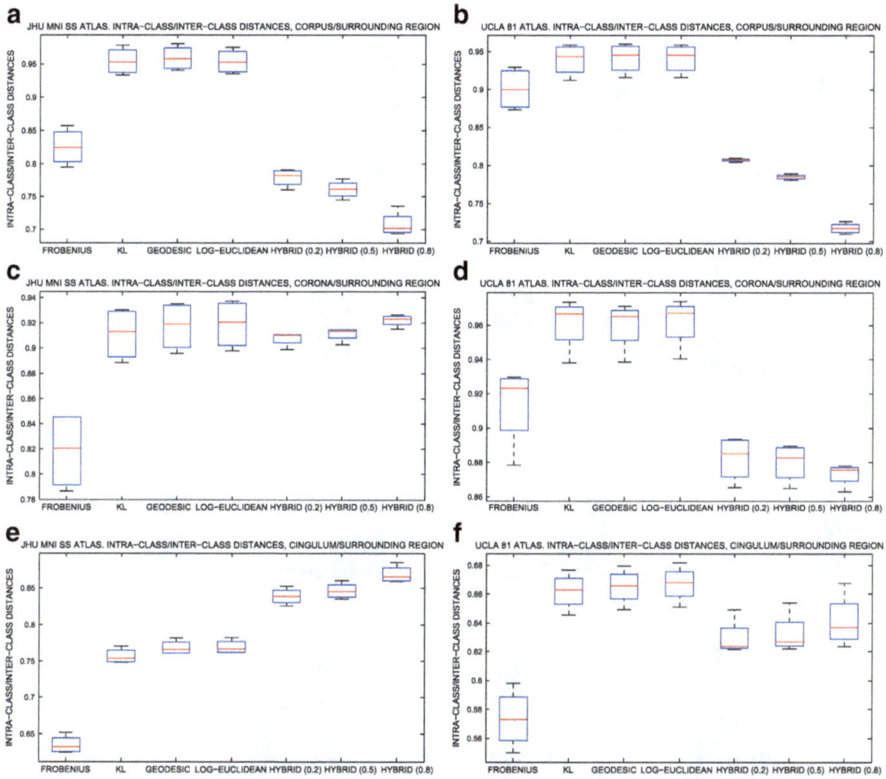

Fig. 4 Intraclass/interclass distance ratios for the different tensor distances for the corpus callosum (*top row*), corona radiata (*medium row*) and cingulum (*bottom row*), for the JHU MNI SS atlas (*left*) and UCLA 81 atlas (*right*). The background is defined as the area surrounding the region of interest, with a width 14 voxels

why the ratios decrease in Fig. 4 when compared to Fig. 3, not only for the corpus callosum but also for the other structures. Indeed, this constitutes a very important finding, meaning that segmentation algorithms will benefit from considering only surrounding areas instead of the whole white matter.

With regard to the relative behaviour of the different hybrid distances, the intraclass/interclass ratio also shows a better performance of an increased weighting for the invariants component in the corpus callosum. As for the corona radiata and the cingulum, there appears to be a slight improvement for a lower factor, meaning a higher relative weight of the orientation component.

A relevant point of disagreement between the KTA and the intraclass/interclass ratio arises for the corona radiata and, especially, for the cingulum. Although there is some discrepancy between the JHU and the UCLA atlases, there are clues that indicate that the Frobenius, KL, Information Geodesic and Log-Euclidean distances perform in some cases better than the hybrid distances. This effect is especially

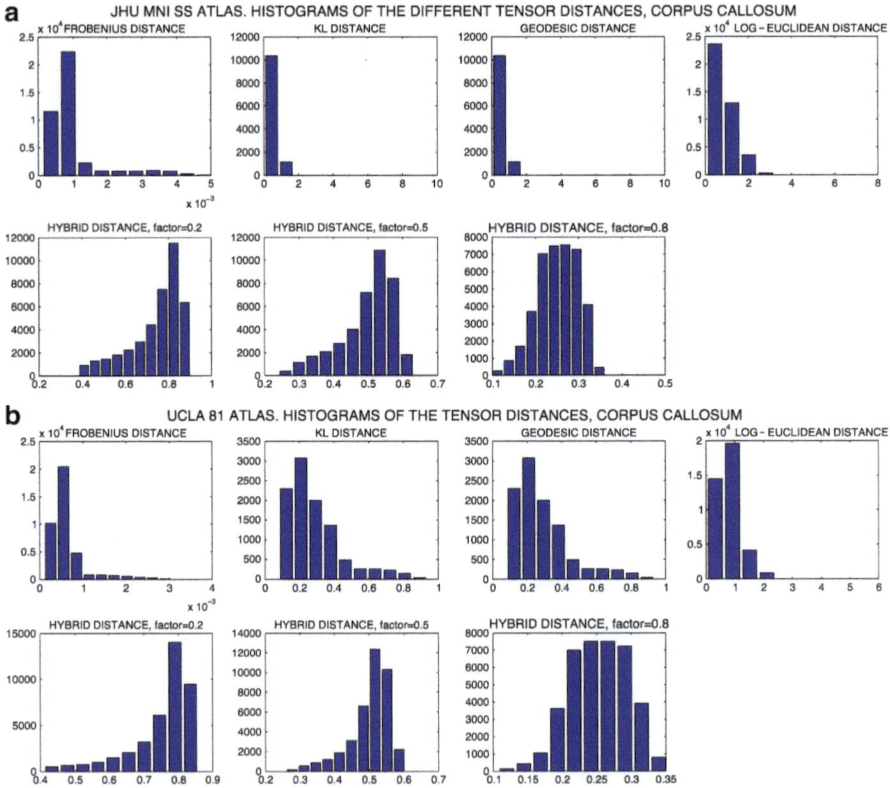

Fig. 5 Histograms of the tensor distances from the voxels inside an area surrounding the region of interest (with a width of 14 voxels) and the mean tensor inside the corpus callosum, for the different tensor distances. (**a**) JHU MNI SS atlas; (**b**) UCLA 81 atlas

important for the Frobenius distance in the cingulum where, particularly for the region of interest/surrounding region scenario, this metric performs better than any other distance. However, as there are important differences between both atlases and depending on the chosen background, no clear interpretation can be made at this point.

In order to gain a deeper insight on the behaviour of the different distances, we next show in Figs. 5–7 the histograms of the tensor distances between the mean tensor inside the region of interest[4] and all the tensors in the region surrounding it. This information can be useful in order to better understand the performance of the distances for the cases where no clear statements could be made based on the results shown before. For a certain region of interest, if most of the voxels surrounding this region have tensor values very similar to the values inside the region of interest, then

[4]The mean tensor is computed according to the employed distance.

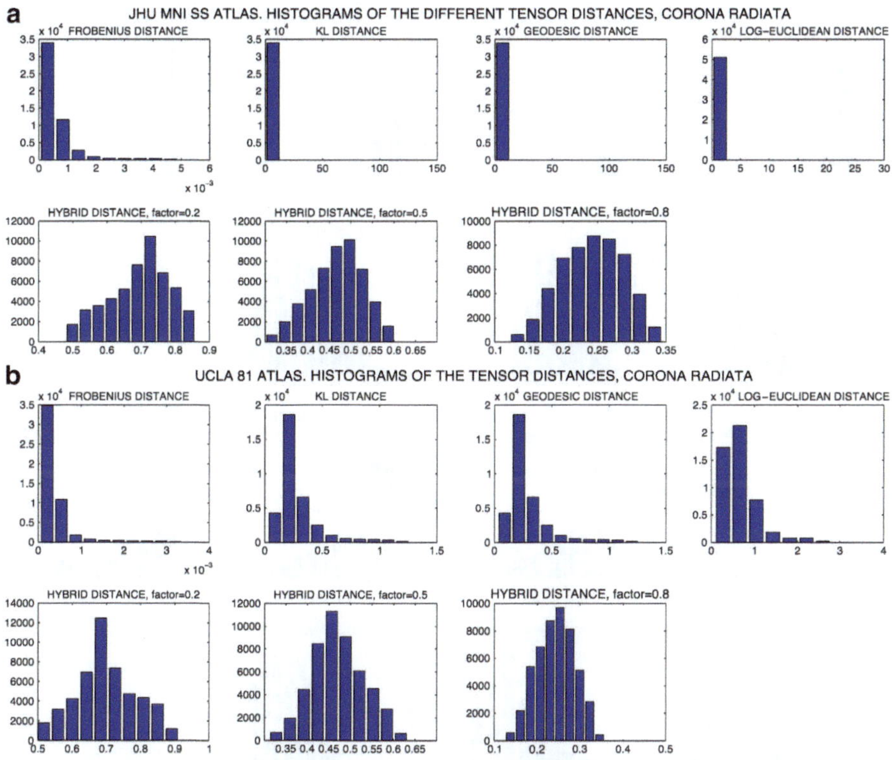

Fig. 6 Histograms of the tensor distances from the voxels inside an area surrounding the region of interest (with a width of 14 voxels) and the mean tensor inside the corona radiata, for the different tensor distances. (**a**) JHU MNI SS atlas; (**b**) UCLA 81 atlas

the separation between these two classes becomes a complicated problem. A narrow histogram, concentrated towards lower distances, will reflect this situation. On the other hand, if for the same region of interest a more suitable tensor distance is chosen and, consequently, the distances between the tensor values inside and outside the region of interest spread far away from the lower end, the segmentation problem will be easier to solve.

The histograms corresponding to the corpus callosum, in Fig. 5, reflect well the situation described before. Whereas the distributions for the Frobenius, KL, Information Geodesic and Log-Euclidean distances show a high number of tensors very similar to the mean tensor, much wider distributions can be observed for the hybrid distances, where the peaks are located at higher distances.

A similar situation is depicted in Figs. 6 and 7. Even though the results corresponding to the intraclass/interclass distances for the corona radiata and cingulum were hard to interpret and could arise doubts on the initial findings, the histograms clearly support the statement that the use of the hybrid distances yields wider and more favorable distributions of the distances, which should in turn improve the segmentation.

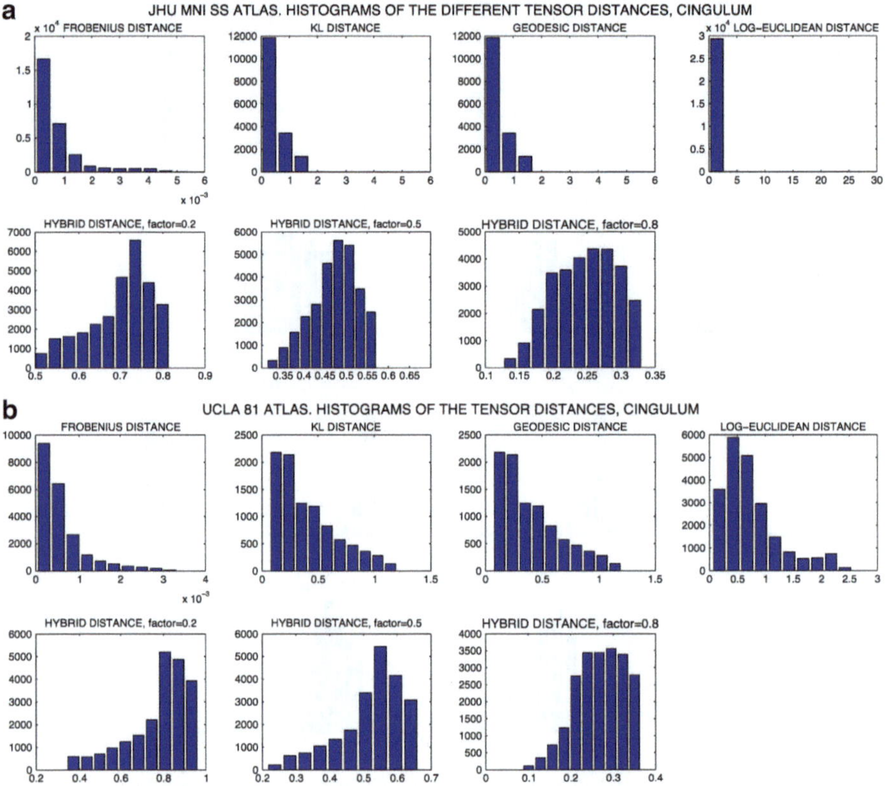

Fig. 7 Histograms of the tensor distances from the voxels inside an area surrounding the region of interest (with a width of 14 voxels) and the mean tensor inside the cingulum, for the different tensor distances. (**a**) JHU MNI SS atlas; (**b**) UCLA 81 atlas

As stated before, most of the segmentation methods designed for anatomical structures in DT-MRI make use of spatial information. Therefore, the distribution through the white matter of tensors that are similar or dissimilar to the tensors in the region of interest is a key issue for the segmentation to be successful. We investigate this aspect in Figs. 8 and 9, where axial, coronal and sagittal slices are shown depicting the distances from each voxel to the mean tensor inside the corpus callosum. As can be seen, when Frobenius distance is employed, the tensors closely surrounding the corpus are very similar to the tensors inside the corpus, except from a region just below the corpus where the red color indicates a higher distance. In Figs. 8b–d it can be seen that the KL, Information Geodesic and Log-Euclidean distances perform in an almost indistinguishable manner. The areas around the corpus show tensors slightly more distant from the mean tensor than using the Frobenius distance, but the tensors above the corpus are still very similar to those in the corpus.

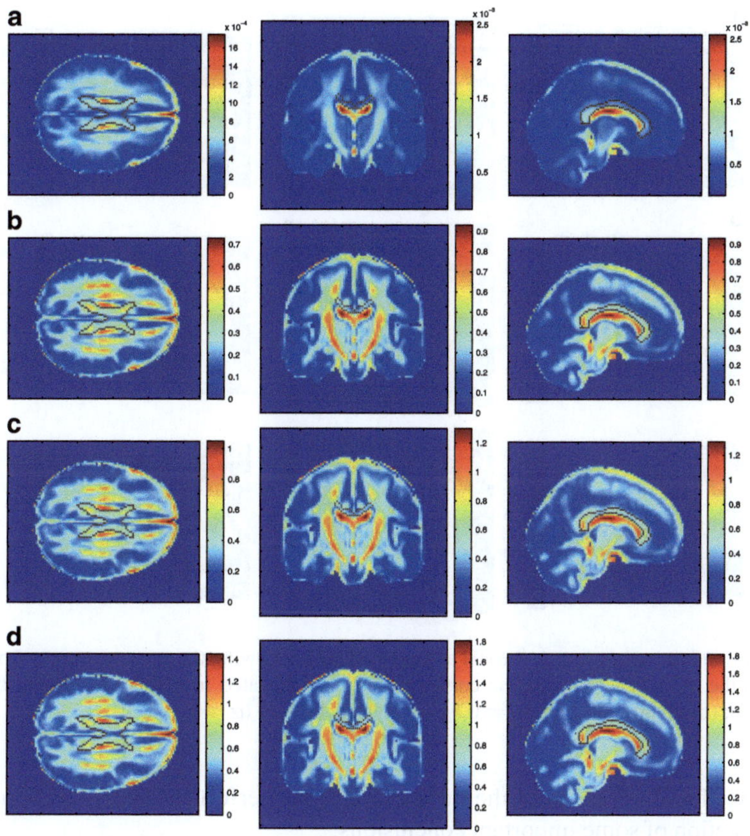

Fig. 8 From left to right, axial, coronal and saggital slices showing the tensor distances from each voxel to the mean tensor inside the corpus callosum (delineated in *black*). (**a**) Frobenius distance; (**b**) KL distance; (**c**) Geodesic distance; (**d**) Log-Euclidean distance

Finally, as shown in Fig. 9, this behaviour changes substantially when the hybrid distance is employed. When $\alpha = 0.2$ (Fig. 9a), all regions surrounding the corpus have very distant tensors from the mean tensor inside. However, even inside the corpus the tensors show a significant distance from the mean tensor, meaning that this hybrid distance is extremely sensitive to small changes in the orientation of the tensors that occur inside the corpus.

If $\alpha = 0.5$ (Fig. 9b), this behaviour is slightly alleviated, with tensors inside the corpus being more similar and tensors outside the corpus being marginally less distant from the mean tensor. Finally, when $\alpha = 0.8$ (Fig. 9c), the same tendency holds. Now, the tensors inside the corpus are extremely similar, while tensors outside are less dissimilar than before, although considerably more than when using Frobenius, KL, Information Geodesic and Log-Euclidean distances.

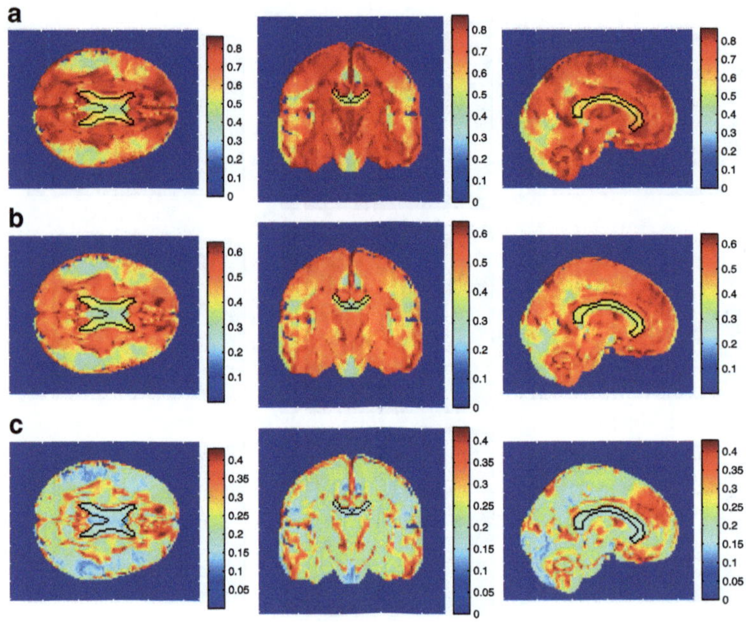

Fig. 9 From left to right, axial, coronal and saggital slices showing the tensor distances from each voxel to the mean tensor inside the corpus callosum (delineated in *black*). (**a**) Hybrid distance ($\alpha = 0.2$); (**b**) Hybrid distance ($\alpha = 0.5$); (**c**) Hybrid distance ($\alpha == 0.8$)

Altogether, the results of the different analysis performed in this section allow the extraction of some important conclusions:

- Results from both atlases are considerably coherent, which supports the validity of the conclusions that can be made from the analysis.
- The KL, Information Geodesic and Log-Euclidean distances show an extremely similar behaviour. Whereas some works in the literature suggest the superiority of some of them, our findings indicate that their performance with regard to the separability of the different classes is almost equivalent.
- The hybrid distances show a good performance, yielding the best results with the exception, for some cases, of the Frobenius distance. With respect of the relative importance of the invariants and orientation components, results seem to indicate that a higher weight of the invariants performs better for the corpus callosum,[5] while no clear interpretations can be made for the corona radiata and cingulum.

[5]While increasing the importance of the invariants produces better results in terms of KTA and intraclass/interclass distances, increasing the weight of the orientation increases the tensor distance of surrounding tensors with respect to the mean tensor inside. However, as the inside variability also increases due to the higher sensitivity of the distance to changes in orientation, this effect is probably atenuated.

- The interpretation of the Frobenius distance's behaviour is more complicated. On the one hand, it performs similarly than the KL, Information Geodesic and Log-Euclidean distances for most of the depicted results. On the other hand, it shows a better performance in terms of the KTA using the whole white matter as background, and results for the intraclass/interclass distances are not consistant. All in all, given the histograms of the distances and the views of the tensor distance to the mean tensor inside the corpus in different slices, it appears that, although possibly better than the KL, Information Geodesic and Log-Euclidean distances, the Frobenius distance does not separate the different classes as nicely as the hybrid distances do.

5 Summary

In this chapter, a comparative study between different tensor distances for DT-MRI segmentation has been carried out. Two tensor atlases of human white matter of the brain were employed for the study as they provide a valuable ground truth for the comparisons. Frobenius, KL, Information Geodesic, Log-Euclidean and Hybrid distances were compared in terms of the separability between the corpus callosum, corona radiata and cingulum and the background.

Results, which are supported by the strong coherence between the outcome of comparisons using both atlases, indicate that the Frobenius, KL, Information Geodesic and Log-Euclidean distances perform in a very similar manner. These findings are controversial to several works in the literature claiming the superiority of some tensor dissimilarity measures. With regard to the Hybrid distance, the different analyses performed suggest a superior behaviour compared to the traditional metrics.

Although the results in this study are consistent, more research needs to be done in order to undoubtedly assess the superiority of a certain tensor distance among the others. Future work will focus on the validation of the conclusions of this chapter by means of a thorough comparison between tensor distances in terms of DT MRI segmentation results.

Acknowledgements The authors acknowledge Junta de Castilla y León for grant VA0339A10-2 and grant belonging to order SAN/103/2011,[6] and Ministerio de Ciencia e Innovación for grant TEC2010-17982. The first author was funded, at the time part of this research was carried out, by the Spanish MEC/Fulbright Commission grant 2007–1238. This is also acknowledged.

[6]Grant from the Consejería de Sanidad, entitled "Resonancia Magnética de Difusión como Ayuda al Diagnóstico y Seguimiento en Neurología y Neurocirugía: Estudio en Alzheimer e Hidrocefalia Crónica".

References

1. Arsigny, V., Fillard, P., Pennec, X., Ayache, N.: Fast and simple calculus on tensors in the Log-Euclidean framework. In: Proceedings of the MICCAI'05, Volume 3749 of Lecture Notes in Computer Science, Palm Springs, CA, USA, pp. 115–122, Oct 2005
2. Arsigny, V., Fillard, P., Pennec, X., Ayache, N.: Log-Euclidean metrics for fast and simple calculus on diffusion tensors. Magn. Reson. Med. **56**(2), 411–421 (2006)
3. Atkinson, C., Mitchell, A.: Rao's distance measure. Sankhya **43**(A), 345–365 (1981)
4. Awate, S.P., Gee, J.C.: A fuzzy, nonparametric segmentation framework for DTI and MRI analysis. In: Proceedings of Information Processing in Medical Imaging (IPMI), Kerkrade, The Netherlands (2007)
5. Awate, S.P., Zhang, H., Gee, J.C.: A fuzzy, nonparametric segmentation framework for DTI and MRI analysis: with applications to DTI-tract extraction. IEEE Trans. Med. Imaging **16**(1 $\frac{1}{4}$), 1525–1536 (2007)
6. Basser, P., Mattiello, J., Bihan, D.L.: MR diffusion tensor spectroscopy and imaging. Biophysica **66**, 259–267 (1994)
7. Le Bihan, D., Breton, E., Lallemand, D., Grenier, P., Cabanis, E., Laval-Jeantet, M.: Mr imaging of intravoxel incoherent motions: application to diffusion and perfusion in neurologic disorders. Radiology **161**(2), 401–407 (1986)
8. Brox, T., Rousson, M., Deriche, R., Weickert, J.: Unsupervised segmentation incorporating colour, texture, and motion. Technical Report 4760, INRIA, Mar. 2003
9. Caselles, V., Kimmel, R., Sapiro, G.: Geodesic active contours. Int. J. Comput. Vis. **22**, 61–79 (1997)
10. Chefd'hotel, C., Tschumperlé, D., Deriche, R., Faugeras, O.: Regularizing flows for constrained matrix-valued images. J. Math. Imaging Vis. **10**(1–2), 147–162 (2004)
11. Cristianini, N., Shawe-Taylor, J., Kandola, J.: On kernel target alignment. Adv. Neural Inf. Process. Syst. **1**(14), 367–374 (2001)
12. de Luis-Garcia, R., Alberola-Lopez, C.: Mixtures of Gaussians on tensor fields for the segmentation of DT-MRI. In: Proceeding of the MICCAI'07, Volume 4791 of Lecture Notes in Computer Science, Brisbane, Australia, pp. 319–326 (2007)
13. de Luis-Garcia, R., Alberola-Lopez, C., Kindlmann, G., Westin, C.-F.: Automatic segmentation of white matter structures from DTI using tensor invariants and tensor orientation. In: Proceedings of the ISMRM Scientific Meeting, Honolulu, HI, USA, April 2009
14. Duan, Y., Li, X., Xi, Y.: Thalamus segmentation from diffusion tensor magnetic resonance imaging. Int. J. Biom. Imaging **2007**, 1–5 (2007)
15. Ennis, D.B., Kindlmann, G.: Orthogonal tensor invariants and the analysis of diffusion tensor magnetic resonance images. Magn. Reson. Med. **55**(1–2), 136–146 (2006)
16. Feddern, C., Weickert, J., Burgeth, B.: Level set methods for tensor-valued images. In: Proceedings of the 9th IEEE Workshop on Variational, Geometric and Level Set Methods in Computer Vision, Nice, France, pp. 65–72, Oct. 2003
17. Graf, A.B.A., Borer, S.: Normalization in support vector machines. In: Proceedigns of the 23rd DAGM Symposium on Pattern Recognition, London, UK, pp. 277–282 (2001)
18. Horsfield, M., Jones, D.: Applications of diffusion-weighted and diffusion tensor MRI to white matter diseases-a review. NMR Biom. **15**(7–8), 570–577 (2002)
19. Grandlund, G.H., Bigun, J.: Optimal orientation detection of linear symmetry. In: Proceedings of the 1st IEEE International Conference on Computer Vision, London, June 1987
20. Jonasson, L., Hagmann, P., Bresson, X., Meuli, R., Cuisenaire, O., Thiran, J.-P.: White matter mapping in DT-MRI using geometric flows. In: Proceedings. Eurocast 2003, Volume 2809 of Lecture Notes in Computer Science, pp. 585–595, Las Palmas de Gran Canaria, Spain, Feb. 2003
21. Jonasson, L.: Segmentation of diffusion weighted MRI using the level set framework. PhD dissertation, École Polytechnique Fédérale de Lausanne, Lausanne, Switzerland (2005).

22. Jonasson, L., Bresson, X., Hagmann, P., Cuisenaire, O., Meuli, R., Thiran, J.-P.: White matter fiber tract segmentation in DT-MRI using geometric flows. Med. Image Anal. **9**, 223–236 (2005)
23. Kichenassamy, S, Kumar, A., Olver, P., Tannenbaum, A., Yezzi, A.: Gradient flows and geometric active contour models. In: Proceedings of the 5th International Conference on Computer Vision, Orlando, FL, USA, pp. 810–815, June 1995
24. Knutsson, H.: A tensor representation of 3-d structures. In: Proceedings 5th IEEE-ASSP and EURASIP Workshop on Multidimensional Signal Processing, Noordwijkerhout, The Netherlands, Sept. 1987
25. Kubicki, M., McCarley, R., Westin, C., Park, H., Maier, S., Kikinis, R., Jolesz, F., Shenton, M. E.: A review of diffusion tensor imaging studies in schizophrenia. J. Psychiatr. Res. **41**(1–2), 15–30 (2007)
26. Lenglet, C., Rousson, M., Deriche, R.: Segmentation of 3d probability density fields by surface evolution: application to diffusion mri. In: Proceedings of the Conference on Medical Image Computing and Computer Assisted Intervention (MICCAI), Saint Malo, France, Sept. 2004
27. Lenglet, C., Rousson, M., Deriche, R., Faugeras, O.: Statistics on multivariate normal distributions: a geometric approach and its application to diffusion tensor MRI. Technical Report 5242, INRIA, June 2004
28. Lenglet, C., Rousson, M., Deriche, R., Faugeras, O.: Toward segmentation of 3D probability density fields by surface evolution: application to diffusion MRI. Technical Report 5243, INRIA, June 2004
29. Lenglet, C., Rousson, M., Deriche, R.: DTI segmentation by statistical surface evolution. IEEE Trans. Med. Imaging **25**(6), 685–700 (2006)
30. Lenglet, C., Rousson, M., Deriche, R.: DTI segmentation by statistical surface evolution. Technical Report 5843, INRIA, Feb. 2006
31. Lenglet, C., Rousson, M., Deriche, R., Faugeras, O.: Statistics on the manifold of multivariate normal distributions: theory and application to diffusion tensor MRI processing. J. Math. Imaging Vis. **25**, 423–444 (2006)
32. Mori, S., Wakana, S., van Zijl, P.C.M., Nagae-Poetscher, L.M.: MRI Atlas of Human White Matter, 1st edn. Elsevier, Amsterdam (2005)
33. Paragios, N., Deriche, R.: Geodesic active regions: a new framework to deal with frame partition problems in computer vision. J. Vis. Commun. Image Represent. **13**, 249–268 (2002)
34. Pasternak, O., Sochen, N., Basser, P.J.: The effect of metric selection on the analysis of diffusion tensor MRI data. Neuroimage **49**(3), 2190–2204 (2010)
35. Peeters, T.H.J.M., Rodrigues, P.R., Vilanova, A., ter Haar Romeny, B.M.: Analysis of Distance/Similarity Measures for Diffusion Tensor Imaging. Springer, Berlin (2008)
36. Pekalaska, E., Paclik, P., Duin, R.: A generalized kernel approach to dissimilarity based classification. J. Mach. Learn. Res. **2**, 175–211 (2001)
37. Peled, S., Gudbjartsson, H., Westin, C.-F., Kikinis, R., Jolesz, F.A.: Magnetic resonance imaging shows orientation and asymmetry of white matter fiber tracts. Brain Res. **780**, 27–33 (1998)
38. Rao, C.R.: Information and accuracy attainable in the estimation of statistical parameters. Bull. Calcutta Math. Soc. **37**, 81–91 (1945)
39. Rousson, M., Brox, T., Deriche, R.: Active unsupervised texture segmentation on a diffusion based feature space. Technical Report 4695, INRIA, Jan. 2003
40. Rousson, M., Brox, T., Deriche, R.: Active unsupervised texture segmentation on a diffusion based feature space. In: Proceedings of the IEEE Conference on Computer Vision and Pattern Recognition (CVPR), Madison, Wisconsin, USA, Jun. 2003
41. Rousson, M., Lenglet, C., Deriche, R.: Level set and region based surface propagation for diffusion tensor mri segmentation. In: Proceedings of the Computer Vision Approaches to Medical Image Analysis Workshop, Prague, Czech Republic, May 2004
42. Salmenpera, T.M., Simister, R.J., Bartlett, P., Symms, M.R., Boulby, P.A., Free, S.L., Barker, G.J., Duncan, J.S.: High-resolution diffusion tensor imaging of the hippocampus in temporal lobe epilepsy. Epilepsy Res. **71**, 102–106 (2006)

43. Scholkopf, B.: Learning with Kernels. MIT, Cambridge MA (2002)
44. Shawe-Taylor, J., Cristianini, N.: Kernel Methods for Pattern Analysis. Cambridge University Press, New York (2004)
45. Sotak, C.H.: The role of diffusion tensor imaging in the evaluation of ischemic brain injury – a review. NMR Biom. **15**(7–8), 561–569 (2002)
46. Sundgren, P.C., Dong, Q., Gómez-Hassan, D., Mukherji, S.K., Maly, P., Welsh, R.: Diffusion tensor imaging of the brain: review of clinical applications. Neuroradiology **46**(5), 339–350, Aug. 2004
47. Wakana, S., Jiang, H., Nagae-Poetscher, M., van Zijl, P.C.M., Mori, S.: A fiber-tract based atlas of human white matter anatomy. Radiology **230**, 77–87 (2004)
48. Wang, Z., Vemuri, B.: An affine invariant tensor dissimilarity measure and its applications to tensor-valued image segmentation. In: Proceedings of the IEEE Conference on Computer Vision and Pattern Recognition, Washington DC, USA, pp. 228–233 (2004)
49. Wang, Z., Vemuri, B.: Tensor field segmentation using region based active contour model. In: Proceedings of the European Conference on Computer Vision, Prague, Czech Republic, pp. 304–315 (2004)
50. Weldeselassie, Y.T., Hamarneh, G.: DT-MRI segmentation using graph cuts. In: Pluim, J.P.W., Reinhardt, J.M. (eds.) Proceedings of Medical Imaging, SPIE, San Diego, CA, USA, Feb. 2007
51. Wiegell, M.R., Tuch, D.S., Larsson, H.B.W., Wedeen, V.J.: Automatic segmentation of thalamic nuclei from diffusion tensor magnetic resonance imaging. NeuroImage **19**, 391–401 (2003)
52. Zhukov, L., Museth, K., Breen, D., Whitaker, R., Barr, A.H.: Level set segmentation and modeling of dt-mri human brain data. J. Electron. Imaging **12**(1), 125–133 (2003)
53. Ziyan, U., Tuch, D., Westin, C.-F.: Segmentation of thalamic nuclei from DTI using spectral clustering. In: Proceedings of the MICCAI'06, Volume 4191 of Lecture Notes in Computer Science, Copenhagen, Denmark, pp. 807–814 (2006)

Divergence Measures and Means of Symmetric Positive-Definite Matrices

Maher Moakher

Abstract The importance of symmetric positive-definite matrices can hardly be exaggerated as they play fundamental roles in many disciplines such as mathematics, numerical analysis, probability and statistics, engineering, and biological and social sciences. On the other hand, in the last few years there has been a renewed interest in developing the theory of means of symmetric positive-definite matrices. In this work we present several divergence functions for measuring closeness between symmetric positive-definite matrices. We then study the invariance properties of these divergence functions as well as the matrix means based on them. We show that the means based on the various divergence functions of a finite collection of symmetric positive-definite matrices are bounded below by their harmonic mean and above by their arithmetic mean. Furthermore, the means based on the studied divergence functions of two symmetric positive-definite matrices are given in closed forms. In particular, we show that the mean based on the Bhattacharyya divergence function of a pair of symmetric positive-definite matrices coincides with their geometric mean.

1 Introduction

Driven by many applications dealing with the processing of symmetric positive-definite matrices, in the last few years, there has been a renewed interest in developing the theory of means of symmetric positive-definite matrices, see e.g., [2, 4, 6, 15, 19]. We mention in particular the notion of metric-based means which has been successfully used in many applications that require the averaging,

M. Moakher (✉)
Laboratory for Mathematical and Numerical Modeling in Engineering Science, National Engineering School at Tunis, University of Tunis El Manar, ENIT-LAMSIN, B.P. 37, 1002 Tunis-Belvédère, Tunisia
e-mail: maher.moakher@enit.rnu.tn

D.H. Laidlaw and A. Vilanova (eds.), *New Developments in the Visualization and Processing of Tensor Fields*, Mathematics and Visualization,
DOI 10.1007/978-3-642-27343-8_16, © Springer-Verlag Berlin Heidelberg 2012

interpolation, regularization, segmentation, classification and clustering of datasets consisting of positive-definite matrices. Such is the case for the processing of diffusion-tensor magnetic resonance imaging (DT-MRI) [3, 11, 17]. In the mean time, some divergence measures of symmetric positive-definite matrices have been successfully used, and have become more and more popular, in measuring the similarities between such matrices [17, 18, 21, 22]. These divergence functions, as we shall see, can be used instead of distance functions as they share with them many nice properties and in some cases can be easily computed.

The notion of "divergence function" was introduced in 1930 by Mahalanobis to measure the discrimination between two correlated multivariate distributions [13]. Since then, closeness between two probability distributions p and q on an event space Ω is usually measured by a divergence function (also called dissimilarity measure). To cite but a few, we mention here the Kullback-Leibler divergence [10]

$$KL(p,q) = \int_{\Omega} p(x) \ln \frac{p(x)}{q(x)} dx, \tag{1}$$

and the Bhattacharyya divergence [5]

$$B(p,q) = -\ln \int_{\Omega} \sqrt{p(x)q(x)}dx. \tag{2}$$

Now, when the distributions p and q are n-dimensional multivariate normal distributions with means a and b, and (symmetric positive-definite) covariance matrices A and B, the Kullback-Leibler divergence (1) becomes [9]

$$KL(p,q) = \frac{1}{2}\left((a-b)^T B^{-1}(a-b) - n + \operatorname{tr}(B^{-1}A) - \ln\det(B^{-1}A)\right), \tag{3}$$

and the Bhattacharyya divergence (2) becomes [9]

$$B(p,q) = \frac{1}{4}(a-b)^T (A+B)^{-1}(a-b) + \frac{1}{2}\ln\frac{\det\frac{1}{2}(A+B)}{\sqrt{\det A \det B}}, \tag{4}$$

where the superscript ^{T} indicates the transpose and tr denotes the trace. Furthermore, if the normal distributions p and q have zero means, i.e., when $a = b = 0$, the Kullback-Leibler divergence reduces to

$$KL(p,q) = \frac{1}{2}\left(\operatorname{tr}(B^{-1}A) - n - \ln\det(B^{-1}A)\right), \tag{5}$$

and the Bhattacharyya divergence reduces to

$$B(p,q) = \frac{1}{2}\ln\frac{\det\frac{1}{2}(A+B)}{\sqrt{\det A \det B}}. \tag{6}$$

In what follows, we shall use the expressions given in the right hand sides of (5) and (6) as definitions of two divergence functions between A and B in the space of symmetric positive-definite matrices. We will introduce in Sect. 3 another divergence function on this space. We then study invariance properties of these divergence functions as well as means based on them. But before we proceed, we fix some notations and give some background materials on divergence functions and means.

Let $\mathscr{M}(n)$ be the linear space of $n \times n$ real matrices, and let $GL(n)$ denote the group of invertible matrices in $\mathscr{M}(n)$. The space $\mathscr{M}(n)$ is endowed with the Frobenius inner product defined by $\langle A, B\rangle = \text{tr}(A^T B)$. The associated norm is $\|A\| = (\text{tr}(A^T A))^{1/2}$.

A matrix $A \in \mathscr{M}(n)$ is symmetric if $A^T = A$. A symmetric matrix $A \in \mathscr{M}(n)$ is said to be positive semidefinite if $x^T Ax \geq 0$ for all $x \in \mathbb{R}^n$ and positive definite if in addition it is invertible. Let $\mathscr{S}(n)$ be the space of symmetric matrices in $\mathscr{M}(n)$ and $\mathscr{P}(n)$ denote the set of symmetric positive-definite matrices in $\mathscr{M}(n)$. It is well known that $\mathscr{P}(n)$ is an open convex cone. By a cone we mean a set K satisfying $K + K \subset K$ and $tK \subset K$ for all $t > 0$. Finally, on $\mathscr{S}(n)$ there is a natural partial order, called the Löwner partial order, which is defined by $A \leq B$ if $B - A$ is positive semidefinite and $A < B$ if $B - A$ is positive definite.

Definition 1. A divergence function on $\mathscr{P}(n)$ is a non-negative function $D(\cdot,\cdot)$ on $\mathscr{P}(n)\times\mathscr{P}(n)$ which vanishes only on the diagonal. That is for all A and B in $\mathscr{P}(n)$ we have $D(A, B) \geq 0$ and $D(A, B) = 0$ if and only if $A = B$.

Remark 1. The divergence is almost a distance function except that it needs not to be symmetric with respect to its arguments nor to satisfy the triangle inequality. In some sense, a divergence function is a generalization of squared distances.

Following Bregman [7], given a convex and differentiable function one can construct a divergence function called Bregman divergence.

Proposition 1. *Let $f : \mathscr{P}(n) \to \mathbb{R}$ be a differentiable and strictly convex function. Then*

$$B_f(A, B) = f(A) - f(B) - \langle \nabla f(B), (A - B)\rangle, \tag{7}$$

is a divergence function on $\mathscr{P}(n)$.

The non negativity of the $B_f(\cdot,\cdot)$ is a consequence of the convexity of f, see e.g. [20, Theorem 25.1]. Furthermore, by the strict convexity of f we conclude that $B_f(A, B)$ vanishes only when $A = B$.

We recall that the mean relative to a distance function $d(\cdot,\cdot)$ on $\mathscr{P}(n)$ is defined as the minimizer of the sum of the squared distances $\sum_{i=1}^{N} d(P_i, P)^2$, over all P in $\mathscr{P}(n)$. When instead of a distance function we use a divergence function for measuring closeness between two symmetric positive-definite matrices we can introduce the notion of divergence-based means. But because a divergence function is in some sense a generalization of a squared distance and the lack of symmetry, we propose the following definitions of divergence-based means.

Definition 2. We define the *right mean* of a finite set of symmetric positive-definite matrices $P_1, \cdots, P_N$, relative to a divergence function $D(\cdot,\cdot)$, to be the symmetric positive-definite matrix that minimizes $F^r(P) = \sum_{i=1}^{N} D(P_i, P)$, over all P in $\mathscr{P}(n)$.

Similarly, we define the *left mean* of $P_1, \cdots, P_N$, relative to $D(\cdot,\cdot)$, to be the symmetric positive-definite matrix that minimizes $F^l(P) = \sum_{i=1}^{N} D(P, P_i)$, over all P in $\mathscr{P}(n)$.

Of course, when the divergence function $D(\cdot,\cdot)$ is symmetric the right and left means coincide and in this case we simply use the term *mean*. We note that when the divergence function is not symmetric one can define a mean, relative to $D(\cdot,\cdot)$, by the symmetric positive-definite matrix that minimizes over all P in $\mathscr{P}(n)$ the function $F^l(P) + F^r(P)$. This is exactly the mean relative to the symmetrized version of the divergence function $D(A, B)$ defined by $(D(A, B) + D(A, B))/2$. Other (non trivial) symmetrization procedures, such as the one discussed in Sect. 4, can also be used.

In the sequel we present several divergence functions for measuring closeness between symmetric positive-definite matrices. We then study the invariance properties of these divergence functions as well as the matrix means based on them.

2 Bhattacharyya Divergence

The Bhattacharyya divergence of two symmetric positive-definite matrices is defined as

$$D_{\mathrm{B}}(A, B) = \ln \frac{\det(A + B)}{2^n \sqrt{\det A \det B}}. \tag{8}$$

It is convenient to write the above as

$$D_{\mathrm{B}}(A, B) = \ln \frac{\det(A \triangle B)}{\det(A \# B)}, \tag{9}$$

where $A \triangle B = \frac{1}{2}(A + B)$ is the arithmetic mean of A and B, and $A\#B$ is their geometric mean given by one of the following equivalent expressions (see e.g., [1, 4, 15])

$$A\#B = A^{\frac{1}{2}}(A^{-\frac{1}{2}}BA^{-\frac{1}{2}})^{\frac{1}{2}}A^{\frac{1}{2}} = A(A^{-1}B)^{\frac{1}{2}} = (BA^{-1})^{\frac{1}{2}}A \tag{10}$$

$$= B^{\frac{1}{2}}(B^{-\frac{1}{2}}AB^{-\frac{1}{2}})^{\frac{1}{2}}B^{\frac{1}{2}} = B(B^{-1}A)^{\frac{1}{2}} = (AB^{-1})^{\frac{1}{2}}B. \tag{11}$$

Proposition 2. *The function defined by (8) is a symmetric divergence function on the space of symmetric positive-definite matrices.*

Proof. The matrix geometric–arithmetic mean inequality states that for all matrices A and B in $\mathscr{P}(n)$ we have [1]

$$A\#B \leq A\triangle B, \tag{12}$$

with equality holds if and only if $A = B$. Therefore, we deduce that[1] $\det(A\#B) \leq \det(A\triangle B)$, and since the logarithm is a monotone increasing function we obtain

$$\ln \frac{\det(A\triangle B)}{\det(A\#B)} \geq 0.$$

Alternatively, using $(10)_2$ we rewrite (9) as

$$\begin{aligned} D_{\mathrm{B}}(A,B) &= \ln \frac{\det\left[\frac{1}{2}(I + A^{-1}B)\right]}{\det[(A^{-1}B)^{\frac{1}{2}}]} \\ &= \ln \prod_{i=1}^{n} \left(\frac{1+\lambda_i}{2\sqrt{\lambda_i}}\right) = \ln \prod_{i=1}^{n} \frac{1}{2}\left(\sqrt{\lambda_i} + \frac{1}{\sqrt{\lambda_i}}\right), \end{aligned}$$

where λ_i, $i = 1, \ldots, n$ are the (positive) eigenvalues of $A^{-1}B$. Since all factors in the product are bigger than 1, it follows that $D_{\mathrm{B}}(A,B) \geq 0$ and that equality to 0 holds if and only if all factors are equal to 1, or equivalently $A = B$. □

We note that the Bhattacharyya divergence satisfies the following desirable properties that one would expect from a divergence function on $\mathscr{P}(n)$.

Proposition 3. *The Bhattacharyya divergence function (8) satisfies:*

1. *Invariance under inversion*

$$D_{\mathrm{B}}(A^{-1}, B^{-1}) = D_{\mathrm{B}}(A,B), \quad \forall A, B \in \mathscr{P}(n).$$

2. *Invariance under congruence transformations*

$$D_{\mathrm{B}}(CAC^T, CBC^T) = D_{\mathrm{B}}(A,B), \quad \forall A, B \in \mathscr{P}(n), \forall C \in GL(n).$$

These properties follow from the fact that the arithmetic and geometric means of matrices in $\mathscr{P}(n)$ satisfy the stated invariances [15]. In some applications it is important to have the divergence (or the distance) between two symmetric positive-definite matrices equal to that between their inverses [16]. Also, invariant under

[1]We use the fact that if $P, Q \in \mathscr{P}(n)$ are such that $P \leq Q$, then $I \leq P^{-\frac{1}{2}}QP^{-\frac{1}{2}}$. Form which it follows that all eigenvalues of $P^{-\frac{1}{2}}QP^{-\frac{1}{2}}$ are bigger than one and hence $\det P \leq \det Q$.

congruent transformation is desirable as it expresses invariance under change of coordinate system.

Remark 2. The Bhattacharyya divergence function is not a distance as the triangle inequality may not be satisfied even in the scalar case. For two positive numbers a and b we have $D_{\mathrm{B}}(a,b) = \ln \frac{a+b}{2\sqrt{ab}}$ whereas $D_{\mathrm{B}}(a,1) + D_{\mathrm{B}}(1,b) = \ln \frac{(a+1)(b+1)}{4\sqrt{ab}}$. If we choose a and b such that $a < 1 < b$, then the triangle inequality does not hold.

The Bhattacharyya divergence is closely related to the Riemannian metric on $\mathscr{P}(n)$ given by $g_P(X,Y) = \operatorname{tr}(P^{-1}XP^{-1}Y)$, which is defined for all X and Y on the tangent space to $\mathscr{P}(n)$ at the base point P. In fact, the following Proposition shows that the Bhattacharyya divergence behaves as the Riemannian distance for sufficiently close matrices.

Proposition 4. *For all X, Y in $\mathscr{S}(n)$ and any P in $\mathscr{P}(n)$ we have*

$$\left.\frac{\partial^2}{\partial s \partial t} D_{\mathrm{B}}(P, P + tX + sY)\right|_{t=s=0} = \frac{1}{4}\operatorname{tr}(P^{-1}XP^{-1}Y).$$

Proof. We note that the derivative of $D_{\mathrm{B}}(P, P + tX + sY)$ with respect to t writes

$$\frac{\partial}{\partial t} D_{\mathrm{B}}(P, P+tX+sY) = \frac{\frac{\partial}{\partial t}\det\left(P + \frac{1}{2}tX + \frac{1}{2}sY\right)}{\det\left(P + \frac{1}{2}tX + \frac{1}{2}sY\right)} - \frac{1}{2}\frac{\frac{\partial}{\partial t}\det(P + tX + sY)}{\det(P + tX + sY)}.$$

By recalling that

$$\frac{\partial}{\partial t}\det(P + tC) = \det P \frac{\partial}{\partial t}\det(I_n + tCP^{-1}) = \det P \operatorname{tr}(CP^{-1}),$$

we obtain

$$\left.\frac{\frac{\partial}{\partial t}\det\left(P + \frac{1}{2}tX + \frac{1}{2}sY\right)}{\det\left(P + \frac{1}{2}tX + \frac{1}{2}sY\right)}\right|_{t=0} = \frac{1}{2}\operatorname{tr}\left(X\left(P + \frac{1}{2}sY\right)^{-1}\right),$$

and similarly

$$\left.\frac{\frac{\partial}{\partial t}\det(P + tX + sY)}{\det(P + tX + sY)}\right|_{t=0} = \operatorname{tr}(X(P + sY)^{-1}).$$

Finally, by noting that $\frac{\partial}{\partial s}(C(s))^{-1} = -(C(s))^{-1}\frac{\partial}{\partial s}C(s)(C(s))^{-1}$, it follows that

$$\frac{\partial^2}{\partial s \partial t} D_{\mathrm{B}}(P, P + tX + sY)\bigg|_{t=s=0} = \frac{1}{4}\operatorname{tr}(P^{-1}XP^{-1}Y). \qquad \square$$

2.1 Bhattacharyya Divergence-Based Mean

Proposition 5. *The mean relative to the Bhattacharyya divergence of a finite set of matrices $P_1, \ldots, P_N$ in $\mathscr{P}(n)$ is given by P in $\mathscr{P}(n)$ that solves the nonlinear matrix equation*

$$\frac{1}{N}\sum_{i=1}^{N}(P + P_i)^{-1} = \frac{1}{2}P^{-1}. \tag{13}$$

Proof. We start by recalling that the condition of optimality of a real-valued function g defined on the nonlinear space $\mathscr{P}(n)$ is given by the vanishing of its gradient. The gradient of $g(P)$ is the unique element $\nabla g(P)$ of $\mathscr{S}(n)$ (the tangent space to $\mathscr{P}(n)$ at P) such that

$$\frac{d}{dt}g(P + tX)\bigg|_{t=0} = \langle \nabla g(P), X\rangle, \quad \forall X \in \mathscr{S}(n).$$

The derivative of $D_{\mathrm{B}}(P_i, P)$ in the direction X in $\mathscr{S}(n)$ is

$$\begin{aligned}
&\frac{d}{dt} D_{\mathrm{B}}(P_i, P + tX)\bigg|_{t=0} \\
&\quad = \frac{d}{dt}\left\{\ln\left(\det\frac{1}{2}(P_i + P + tX)\right) - \frac{1}{2}\ln(\det(P_i)\det(P + tX))\right\}\bigg|_{t=0} \\
&\quad = \operatorname{tr}\left(X\left[(P_i + P)^{-1} - \frac{1}{2}P^{-1}\right]\right).
\end{aligned}$$

Therefore, the gradient of $\sum_{i=1}^{N} D_{\mathrm{B}}(P_i, P)$ is $\sum_{i=1}^{N}(P + P_i)^{-1} - \frac{1}{2}NP^{-1}$, and hence the mean relative to the Bhattacharyya divergence satisfies the nonlinear matrix equation (13). $\square$

Corollary 1. *The Bhattacharyya divergence-based mean of two matrices A and B in $\mathscr{P}(n)$ is given by their geometric mean $A\#B$.*

Proof. To prove this we need to show that $A\#B$ satisfies the nonlinear equation

$$(P + A)^{-1} + (P + B)^{-1} = P^{-1}.$$

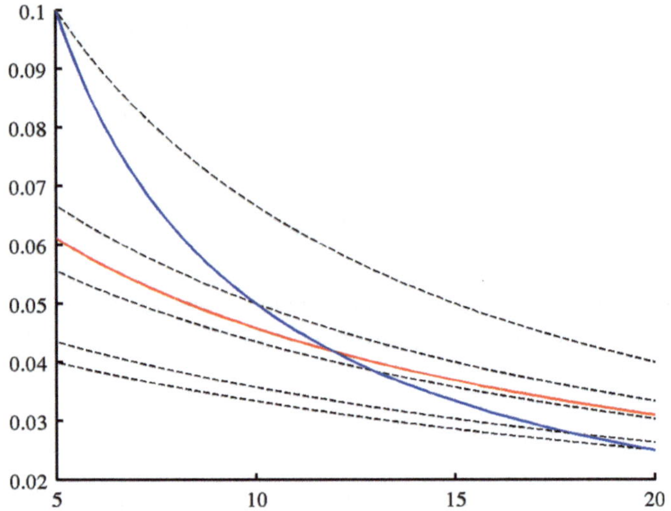

Fig. 1 Plots of the functions $1/(x + x_i)$, $i = 1, \ldots, 5$ (*dashed lines*), $\frac{1}{5}\sum_{i=1}^{5} 1/(x + x_i)$ (*blue line*) and $\frac{1}{2x}$ (*red line*), where $x_1 = 5$, $x_2 = 10$, $x_3 = 13$, $x_4 = 18$ and $x_5 = 20$. The Bhattacharyya mean of the x_i's is the intersection of the *blue* and *red lines*

Left multiplication of the above by $(P + A)$ followed by right multiplication by $(P + B)$ and after simplification yields the equation $P = AP^{-1}B$, whose unique solution in $\mathscr{P}(n)$ is of course $A\#B$. □

In general, the Bhattacharyya divergence-based mean of more than two symmetric positive-definite matrices cannot be obtained in closed form. In the scalar case, the Bhattacharyya divergence-based mean satisfies the equation

$$\frac{1}{N}\sum_{i=1}^{N}\frac{1}{x + x_i} = \frac{1}{2x}, \tag{14}$$

which can be written as a polynomial equation of degree N. The mean which is the positive root of this polynomial equation cannot be given in closed form for $N > 4$. For a graphical illustration of the solution of (14) see Fig. 1.

Proposition 6. *The mean $\mathscr{M}_B(P_1, \ldots, P_N)$ relative to the Bhattacharyya divergence of a finite set of matrices $P_1, \ldots, P_N$ in $\mathscr{P}(n)$ satisfies*

$$\mathscr{H}(P_1, \ldots, P_N) \leq \mathscr{M}_B(P_1, \ldots, P_N) \leq \mathscr{A}(P_1, \ldots, P_N), \tag{15}$$

where

$$\mathscr{H}(P_1, \ldots, P_N) = N\left[\sum_{i=1}^{N} P_i^{-1}\right]^{-1}, \qquad \mathscr{A}(P_1, \ldots, P_N) = \frac{1}{N}\sum_{i=1}^{N} P_i,$$

are the harmonic and arithmetic means of $P_1, \ldots, P_N$.

Proof. Form (13) and the harmonic–arithmetic mean inequality

$$\mathscr{H}(P_1,\ldots,P_N) \leq \mathscr{A}(P_1,\ldots,P_N),$$

we obtain

$$\frac{1}{2}P^{-1} = \frac{1}{N}\sum_{i=1}^{N}(P+P_i)^{-1} \geq N\left[\sum_{i=1}^{N}(P+P_i)\right]^{-1},$$

which after inversion and simplification yields the second inequality in (15). Now again from the harmonic–arithmetic mean inequality we have $4(P+P_i)^{-1} \leq P^{-1} + P_i^{-1}$. Then using (13) we obtain

$$2P^{-1} = \frac{4}{N}\sum_{i=1}^{N}(P+P_i)^{-1} \leq P^{-1} + \frac{1}{N}\sum_{i=1}^{N}P_i^{-1}.$$

After simplification and inversion of this inequality we get the first inequality in (15). □

To solve the nonlinear (13) we used the two fixed-point iterations

$$\begin{cases} P^{(0)} = \mathscr{H}(P_1,\ldots,P_N), \\ P^{(k+1)} = F(P^{(k)}), \quad k = 0,1,\ldots \end{cases} \qquad \begin{cases} Q^{(0)} = \mathscr{H}(P_1,\ldots,P_N)^{-1}, \\ Q^{(k+1)} = G(Q^{(k)}), \quad k = 0,1,\ldots \end{cases}$$

where

$$F(P) := \frac{N}{2}\left[\sum_{i=1}^{N}(P+P_i)^{-1}\right]^{-1}, \quad G(Q) := \frac{2}{N}\sum_{i=1}^{N}(Q^{-1}+P_i)^{-1}.$$

The first fixed point algorithm converges to the Bhattacharyya mean and the second one converges to the inverse of the Bhattacharyya mean. Extensive numerical experiments showed that both algorithms converge rapidly to their fixed point.

3 Modified Bhattacharyya Divergence

Motivated by the nice expression (9) that we gave for the Bhattacharyya divergence, we were interested to see whether the function

$$(A,B) \mapsto \operatorname{tr}(A \triangle B) - \operatorname{tr}(A \# B)$$

defines a divergence function on $\mathscr{P}(n)$. Indeed, by using the geometric–arithmetic mean inequality for matrices we can show that this function is positive. Furthermore, this function vanishes if and only if $A = B$. For simplicity, in the following, we introduce the slightly modified version

$$D_{\mathrm{M}}(A, B) = \mathrm{tr}(A \triangle B) - \mathrm{tr}(A^{\frac{1}{2}} B^{\frac{1}{2}}), \tag{16}$$

which coincides with the aforementioned divergence function when the matrices A and B commute.

Proposition 7. *The function defined by (16) is a symmetric divergence function on the space of symmetric positive-definite matrices.*

Proof. The divergence function (16) can be written as

$$D_{\mathrm{M}}(A, B) = \frac{1}{2}\,\mathrm{tr}((A^{\frac{1}{2}} - B^{\frac{1}{2}})^2) = \frac{1}{2}\|A^{\frac{1}{2}} - B^{\frac{1}{2}}\|^2, \tag{17}$$

which shows that it is always positive and equal to zero if and only if A and B are equal. Because of (17) we shall call this divergence function the modified Bhattacharyya divergence. □

Proposition 8. *The modified Bhattacharyya divergence satisfies the relaxed triangle inequality*

$$D_{\mathrm{M}}(A, B) \leq 2(D_{\mathrm{M}}(A, C) + D_{\mathrm{M}}(C, B)), \quad \forall A, B, C \in \mathscr{P}(n).$$

Proof. As $D_{\mathrm{M}}(A, B) = \frac{1}{2}\|A^{\frac{1}{2}} - B^{\frac{1}{2}}\|^2$, it follows from the triangle inequality for the Frobenius norm that

$$\sqrt{D_{\mathrm{M}}(A, B)} \leq \sqrt{D_{\mathrm{M}}(A, C)} + \sqrt{D_{\mathrm{M}}(C, B)}$$

which when squared yields

$$D_{\mathrm{M}}(A, B) \leq D_{\mathrm{M}}(A, C) + D_{\mathrm{M}}(C, B) + 2\sqrt{D_{\mathrm{M}}(A, C) D_{\mathrm{M}}(C, B)}.$$

The results follows by use of the geometric–arithmetic mean inequality

$$\sqrt{D_{\mathrm{M}}(A, C) D_{\mathrm{M}}(C, B)} \leq \frac{1}{2}(D_{\mathrm{M}}(A, C) + D_{\mathrm{M}}(C, B)).$$ □

We note that the modified Bhattacharyya divergence function is not invariant under inversion, i.e., $D_{\mathrm{M}}(A^{-1}, B^{-1})$ need not to be equal to $D_{\mathrm{B}}(A, B)$ in general. It is not difficult to show that the modified Bhattacharyya divergence function is indeed invariant under congruence transformations.

Proposition 9. *The modified Bhattacharyya divergence function (16) is invariant under congruence transformations:*

$$D_{\mathrm{M}}(CAC^T, CBC^T) = D_{\mathrm{M}}(A, B), \quad \forall A, B \in \mathscr{P}(n), \forall C \in GL(n).$$

3.1 Mean Based on the Modified Bhattacharyya Divergence

Proposition 10. *The mean relative to the modified Bhattacharyya divergence, of a finite set of matrices $P_1, \ldots, P_N$ in $\mathscr{P}(n)$ is given by*

$$\left(\frac{1}{N}\sum_{i=1}^{N} P_i^{\frac{1}{2}}\right)^2. \tag{18}$$

This proposition follows directly from the fact that the square root of the mean relative to the modified Bhattacharyya divergence is the arithmetic mean of the square root of the P_i's. Furthermore, it is straightforward to show that the mean relative to the modified Bhattacharyya divergence, of a finite set of matrices $P_1, \ldots, P_N$ in $\mathscr{P}(n)$ is bounded from below by their harmonic mean and from above by their arithmetic mean.

4 Kullback-Leibler Divergence

The Kullback-Leibler divergence of two symmetric positive-definite matrices is defined as

$$D_{\mathrm{KL}}(A, B) = \mathrm{tr}(B^{-1}A - I_n) - \ln\det(B^{-1}A). \tag{19}$$

Proposition 11. *The divergence function defined by (19) is the Bregman divergence associated with the log-barrier function $f(A) = -\ln\det A$.*

Proof. The Bregman divergence associated with the above function is

$$\begin{aligned} f(A) - f(B) - \langle \nabla f(B), A - B\rangle &= -\ln\det A + \ln\det B + \mathrm{tr}(B^{-1}(A - B)) \\ &= -\ln\det(B^{-1}A) + \mathrm{tr}(B^{-1}A - I_n). \end{aligned} \qquad \square$$

This divergence function (19) is not symmetric. It can be symmetrized in many different ways. For instance, one can symmetrize it by simply adding $D_{\mathrm{KL}}(A, B)$ and $D_{\mathrm{KL}}(B, A)$ to obtain the divergence function

$$D_{\mathrm{KL}}^s(A, B) = \mathrm{tr}(B^{-1}A + A^{-1}B - 2I_n) = \mathrm{tr}[(A - B)(B^{-1} - A^{-1})], \tag{20}$$

which, following the literature on divergence of probability distributions, we call it the Jeffreys divergence [8]. Cauchy's inequality implies that

$$D^s_{\mathrm{KL}}(A,B) \leq \|A-B\| \cdot \|B^{-1}-A^{-1}\| \leq \frac{1}{2}(\|A-B\|^2 + \|B^{-1}-A^{-1}\|^2).$$

Another way to symmetrize (19) is to consider the function

$$D_{\mathrm{JS}}(A,B) = \frac{1}{2}(D_{\mathrm{KL}}(A, A\triangle B) + D_{\mathrm{KL}}(B, A\triangle B)). \tag{21}$$

which, by analogy to the probability distribution case, we call the Jensen-Shannon divergence [12, 14]. A simple calculation shows that $D_{\mathrm{JS}}(A,B) = D_{\mathrm{B}}(A,B)$. It is worthy to note that this is not true for the probability distribution setting, i.e., $\frac{1}{2}(KL(p, \frac{1}{2}(p+q)) + KL(q, \frac{1}{2}(p+q)))$ is in general different from $B(p,q)$.

Proposition 12. *The function defined by (20) is a symmetric divergence function on the space of symmetric positive-definite matrices.*

Proof. As A and B are symmetric matrices there exists an invertible matrix C that simultaneously diagonalizes A and B

$$A = C\,\mathrm{diag}(\lambda_1, \ldots \lambda_n)C^T, \quad B = C\,\mathrm{diag}(\mu_1, \ldots, \mu_n)C^T,$$

where λ_i and μ_i $i = 1, \ldots, n$ are the (positive) eigenvalues of A and B, respectively. Then

$$(A-B)(B^{-1}-A^{-1}) = C\,\mathrm{diag}\left(\frac{(\lambda_1-\mu_1)^2}{\lambda_1\mu_1}, \ldots, \frac{(\lambda_n-\mu_n)^2}{\lambda_n\mu_n}\right)C^T,$$

and hence it follows that (20) is non negative and equal to zero if and only if A and B are equal. □

Proposition 13. *The symmetrized Kullback-Leibler divergence (20) satisfies the following invariance properties:*

1. *Invariance under inversion*

$$D^s_{\mathrm{KL}}(A^{-1}, B^{-1}) = D^s_{\mathrm{KL}}(A,B), \quad \forall A, B \in \mathscr{P}(n).$$

2. *Invariance under congruence transformations*

$$D^s_{\mathrm{KL}}(CAC^T, CBC^T) = D^s_{\mathrm{KL}}(A,B), \quad \forall A, B \in \mathscr{P}(n), \forall C \in GL(n).$$

4.1 Kullback-Leibler Divergence-Based Mean

Proposition 14. *The right mean relative to the Kullback-Leibler divergence (19), of a finite set of matrices $P_1, \ldots, P_N$ in $\mathscr{P}(n)$ is given by their arithmetic mean.*

Proof. The derivative of $D_{\mathrm{KL}}(P_i, P)$ in the direction X in $\mathscr{S}(n)$ is

$$\frac{d}{dt} D_{\mathrm{KL}}(P_i, P + tX)\Big|_{t=0} = -\operatorname{tr}(XP^{-1}(P_i P^{-1} - I_n)).$$

Therefore, the gradient of $\sum_{i=1}^{N} D_{\mathrm{KL}}(P_i, P)$ is $-P^{-1}\left(\sum_{i=1}^{N} P_i P^{-1} - N I_n\right)$, which vanishes when P is the arithmetic mean of $P_1, \ldots, P_N$. □

Proposition 15. *The left mean relative to the Kullback-Leibler divergence (19), of a finite set of matrices $P_1, \ldots, P_N$ in $\mathscr{P}(n)$ is given by their harmonic mean.*

Proof. The derivative of $D_{\mathrm{KL}}(P, P_i)$ in the direction X in $\mathscr{S}(n)$ is

$$\frac{d}{dt} D_{\mathrm{KL}}(P + tX, P_i)\Big|_{t=0} = \operatorname{tr}(X(P_i^{-1} - P^{-1})).$$

Therefore, the gradient of $\sum_{i=1}^{N} D_{\mathrm{KL}}(P, P_i)$ is $\sum_{i=1}^{N} P_i^{-1} - NP^{-1}$, which vanishes when P is the harmonic mean of $P_1, \ldots, P_N$. □

Proposition 16. *The mean relative to the symmetrized Kullback-Leibler divergence (20), of a finite set of matrices $P_1, \ldots, P_N$ in $\mathscr{P}(n)$ is given by the geometric mean of the arithmetic and harmonic means of $P_1, \ldots, P_N$.*

Proof. The derivative of $D^s_{\mathrm{KL}}(P, P_i)$ in the direction X in $\mathscr{S}(n)$ is

$$\frac{d}{dt} D^s_{\mathrm{KL}}(P + tX, P_i)\Big|_{t=0} = \operatorname{tr}(X(P_i^{-1} - P^{-1} P_i P^{-1})).$$

Therefore, the gradient of $\sum_{i=1}^{N} D^s_{\mathrm{KL}}(P, P_i)$ is $\sum_{i=1}^{N} (P_i^{-1} - P^{-1} P_i P^{-1})$, which vanishes when $\sum_{i=1}^{N} P_i^{-1} = P^{-1} \sum_{i=1}^{N} P_i P^{-1}$, i.e., when P is the geometric mean of the arithmetic and harmonic means of $P_1, \ldots, P_N$. □

This is just another proof that the geometric mean of two symmetric positive-definite matrices is also the geometric mean of their arithmetic and harmonic means [17]. Therefore, the mean $\mathscr{M}^s_{\mathrm{KL}}(P_1, \ldots, P_N)$ relative to the symmetrized Kullback-Leibler divergence (20) of $P_1, \ldots, P_N$ is bounded from below by $2(\mathscr{A}(P_1, \ldots, P_N)^{-1} + \mathscr{H}(P_1, \ldots, P_N)^{-1})^{-1}$ and from above by $\frac{1}{2}(\mathscr{A}(P_1, \ldots, P_N) + \mathscr{H}(P_1, \ldots, P_N))$.

5 Conclusion

In this chapter we used some divergence functions for measuring the similarities between probability distributions to define divergence functions on the space of symmetric positive-definite matrices. We showed that some of these divergence functions are invariant under inversion and under congruent transformations which is an important property to have in certain applications. We then used these divergence functions to define means of symmetric positive-definite matrices. We presented numerical algorithms for computing a divergence-based mean when a closed-form expression cannot be obtained. We showed that all these divergence-based means of a finite number of symmetric positive-definite matrices is always bounded below and above by their harmonic and arithmetic means, respectively.

References

1. Ando, T.: Concavity of certain maps of positive definite matrices and applications to Hadamard products. Linear Algebra Appl. **26**, 203–241 (1979)
2. Ando, T., Li, C.K., Mathias, R.: Geometric means. Linear Algebra Appl. **385**, 305–334 (2004)
3. Arsigny, V., Fillard, P., Pennec, X., Ayache, N.: Log-Euclidean metrics for fast and simple calculus on diffusion tensors. Magn. Reson. Med. **56**, 411–421 (2006)
4. Bhatia, R., Holbrook, J.: Riemannian geometry and matrix geometric means. Linear Algebra Appl. **413**(2–3), 594–618 (2006)
5. Bhattacharyya, A.: On a measure of divergence between two statistical populations defined by their probability distributions. Bull. Calcutta Math. Soc. **35**, 99–109 (1943)
6. Bini, D., Iannazzo, B.: A note on computing matrix geometric means. Adv. Comput. Math. **35**(2–4), pp. 175–192 (2011)
7. Bregman, L.M.: The relaxation method of finding the common point of convex sets and its application to the solutions of problems in convex programming. U.S.S.R. Comput. Math. Math. Phys. **7**(3), 200–217 (1967)
8. Jeffreys, H.: An invariant form for the prior probability in estimation problems. Proc. R. Soc. Lond. A **186**, 453–461 (1946)
9. Kailath, T.: The divergence and Bhattacharyya distance measures in signal selection. IEEE Trans. Commun. Technol. **15**(1), 52–68 (1967)
10. Kullback, S., Leibler, R.A.: On information and sufficiency. Ann. Math. Stat. **22**(1), 79–86 (1951)
11. Lenglet, C., Rousson, M., Deriche, R., Faugeras, O.: Statistics on the manifold of multivariate normal distributions: theory and application to diffusion tensor MRI processing. J. Math. Imaging Vis. **25**, 423–444 (2006)
12. Lin, J.: Divergence measures based on the Shanon entropy. IEEE Trans. Inf. Theory **37**(1), 145–151 (1991)
13. Mahalanobis, P.C.: On tests and measures of group divergence. J. Asiatic Soc. Bengal **26**(4), 541–588 (1930)
14. Menéndez, M.L., Pardo, J.A., Pardo, L., Pardo, M.C.: The Jensen-Shannon divergence. J. Frankl. Inst. **334**(2), 307–318 (1997)
15. Moakher, M.: A differential-geometric approach to the geometric mean of symmetric positive-definite matrices. SIAM J. Matrix Anal. Appl. **26**(3), 735–747 (2005)
16. Moakher, M.: On the averaging of symmetric positive-definite tensors. J. Elast. **82**, 273–296 (2006)

17. Moakher, M., Batchelor, P.G.: The symmetric space of positive definite tensors: from geometry to applications and visualization. In: Weickert, J., Hagen, H. (eds.) Visualization and Processing of Tensor Fields, Chapter 17, pp. 285–298. Springer, Berlin (2006)
18. Peeters, T.H.J.M., Rodrigues, P.R., Vilanova, A., ter Haar Romeny, B.M.: Analysis of distance/similarity measures for diffusion tensor imaging. In: Laidlaw, D., Weickert, J. (eds.) Visualization and Processing of Tensor Fields: Advances and Perspectives, pp. 113–136. Springer, Berlin (2009)
19. Petz, D., Temesi, R.: Means of positive numbers and matrices. SIAM J. Matrix Anal. Appl. **27**, 712–720 (2005)
20. Rockafellar, R.T.: Convex Analysis. Princeton Landmarks in Mathematics and Physics. Princeton University Press, Princeton (1997)
21. Wang, Z., Vemuri, B.C.: DTI segmentation using an information theoretic tensor dissimilarity measure. IEEE Trans. Med. Imaging **24**(10), 1267–1277 (2005)
22. Ziyan, U., Sabuncu, M.R., Grimson, W.E.L., Westin, C.-F.: Consistency clustering: A robust algorithm for group-wise registration, segmentation and automatic atlas construction in diffusion MRI. Int. J. Comput. Vis. **85**, 279–290 (2009)

Metric Selection and Diffusion Tensor Swelling

Ofer Pasternak, Nir Sochen, and Peter J. Basser

Abstract The measurement of the distance between diffusion tensors is the foundation on which any subsequent analysis or processing of these quantities, such as registration, regularization, interpolation, or statistical inference is based. Euclidean metrics were first used in the context of diffusion tensors; then geometric metrics, having the practical advantage of reducing the "swelling effect," were proposed instead. In this chapter we explore the physical roots of the swelling effect and relate it to acquisition noise. We find that Johnson noise causes shrinking of tensors, and suggest that in order to account for this shrinking, a metric should support swelling of tensors while averaging or interpolating. This interpretation of the swelling effect leads us to favor the Euclidean metric for diffusion tensor analysis. This is a surprising result considering the recent increase of interest in the geometric metrics.

1 Introduction

Metric selection defines how we compare entities and is the basic step for most, if not all, data processing methods. In diffusion tensor imaging (DTI), the entity of interest is the diffusion tensor [7], and a metric for diffusion tensors has to be explicitly provided or implicitly assumed for all related image processing, including

O. Pasternak (✉)
Brigham and Women's Hospital, Harvard Medical School, Boston, MA, USA
e-mail: ofer@bwh.harvard.edu

N. Sochen
Department of Applied Mathematics, Tel Aviv University, Tel Aviv, Israel

P.J. Basser
The Eunice Kennedy Shriver National Institute of Child Health and Human Development (NICHD), National Institutes of Health, Bethesda, MD, USA

D.H. Laidlaw and A. Vilanova (eds.), *New Developments in the Visualization and Processing of Tensor Fields*, Mathematics and Visualization,
DOI 10.1007/978-3-642-27343-8_17,

averaging, interpolation, registration, clustering and any statistical inference performed on the tensor quantities. DTI was introduced almost 20 years ago, and in that time the method has successfully delineated white matter structures in the brain, recognized many types of brain disorders, and established connectivity measures to help us better understand brain function [3]. As the uses of DTI increase, the accuracy of the measurements becomes more important; researchers would now like to distinguish subtle differences reflected in DTI-derived quantities, or that require the grouping of a large cohort of subjects to increase the significance of the observed results [18]. Making an appropriate metric selection therefore becomes more critical in order to address a wide variety of clinical, biological and neuroscience questions.

A rotation-invariant Euclidean metric for diffusion tensors was initially proposed as part of the tensor-variate statistical analysis framework [6]. It was followed by the introduction of affine-invariant geometric metrics for diffusion tensors that were designed to account for tensors having only positive eigenvalues, and which had the practical advantage of reducing the "swelling effect" [8, 13, 19, 22, 25]. Swelling is described as an increase in the determinant of a tensor, obtained by interpolating or averaging tensors with a lower determinant.

In our previous work we were able to link a Euclidean metric with diffusion tensors through the expected distribution caused by thermal acquisition noise [24]. At the same time our theoretical framework predicted that geometric metrics, such as the Affine-invariant metric and the Log-Euclidean metric, would cause bias in diffusion tensor estimations and processing. Nevertheless, there is one aspect—swelling—in which the geometric metric has consistently been shown to provide superior results compared with the Euclidean metric [2]. In this chapter we extend our previous work to find why swelling occurs in the Euclidean analysis; how is it circumvented using the geometric metrics; and which of these two approaches is more appropriate for the analysis of diffusion tensors. The chapter briefly introduces diffusion tensor metrics, the problem of metric selection, and tensor swelling. In Sect. 3 we discuss the determinant and trace as invariants of the different metrics, and provide physical considerations for their preservation. In Sect. 4 we relate the invariants to the swelling problem through the effect of acquisition noise, and demonstrate that trace preservation through the Euclidean metric reduces statistical biases which are encountered when preserving the determinant using the geometric metrics. We conclude in Sect. 5, and provide insights into possible future improvements to diffusion metrics.

2 Riemannian Metrics for Diffusion Tensors

A Riemannian metric, $\mathbf{G}(\mathbf{x}) = \{g_{ij}(\mathbf{x})\}$, defines the infinitesimal distance over a Riemannian manifold [11] as:

$$ds^2 = d\mathbf{x}^T \mathbf{G}(\mathbf{x}) d\mathbf{x},$$

where $\mathbf{x}$ is the coordinate of a point on the manifold for a chosen coordinate system or the units in which the tensor is measured. Any positive-definite and symmetric metric is admissible. The distance function is defined as the geodesic, i.e., the shortest path on the manifold. To define the geometric distance between tensors, a metric and a local coordinate system for tensor representation are chosen.

2.1 The Euclidean and Geometric Metrics

Two main families of Riemannian metrics are commonly used with diffusion tensors: the Euclidean metric family and the geometric metric family. The Euclidean family places the diffusion tensors on a Euclidean manifold and its most common member defines the metric to be a constant $\mathbf{G(x)} = \mathbf{I}$, resulting in

$$ds^2 = tr((d\mathbf{D})^T d\mathbf{D}),$$

where $\mathbf{D}$ is the tensor coordinates in the canonical tensor coordinate system, and tr denotes the matrix trace, which is equivalent to summing the eigenvalues:

$$tr(\mathbf{D}) = \sum_i \lambda_i.$$

The geodesic between any two tensors, $\mathbf{D}_1$ and $\mathbf{D}_2$, with this metric, is simply a straight line, or the Euclidean distance

$$Dist_{Euc}(\mathbf{D}_1, \mathbf{D}_2) = ||\mathbf{D}_1 - \mathbf{D}_2|| \,, \tag{1}$$

where $||\cdot||$ denotes the Frobenius norm. The Euclidean metric is defined over the entire space of symmetric matrices and is rotation-invariant, which makes it invariant for the selection of orthogonal coordinates, but not for the selection of non-orthogonal tensor coordinate systems.

The geometric metric restricts the distance function to be affine-invariant (which includes rotation, scale, shear, and inversion invariance), and operates only on tensors belonging to the space of positive definite symmetric matrices, S^+ [8, 13, 14, 19, 22, 25]. The Affine-invariant metric [25], a Riemannian metric that satisfies these requirements, has an infinitesimal distance [21]

$$ds^2 = tr((\mathbf{D}^{-1} d\mathbf{D})^2.$$

Since this distance is affine-invariant it does not depend on the choice of tensor coordinate system. The corresponding geodesic is found by integration [21]:

$$Dist_{Aff}(\mathbf{D}_1, \mathbf{D}_2) = \sqrt{tr\left(log^2(\mathbf{D}_1^{-1}\mathbf{D}_2)\right)} \,. \tag{2}$$

The notation $log(\mathbf{D})$ stands for the matrix logarithm. The Log-Euclidean metric with its corresponding geodesic [2],

$$Dist_{LogEuc} = ||log(\mathbf{D}_1) - log(\mathbf{D}_2)|| , \tag{3}$$

was proposed as an efficient approximation for the computationally demanding Affine-invariant metric.

2.2 *Metric Selection*

In our previous work we showed that the Euclidean metrics are related to a normal distribution and that the geometric metrics are related to a log-normal distribution [24]. We further showed that many of the relevant sources of variability in the acquisition lead to a normal distribution of the diffusion coefficients or any linear combination of them. We concluded that an unbiased estimator for diffusion quantities should be based on the Euclidean metric, and demonstrated this point with synthetic and real datasets where diffusivity and variability estimations were shown to be biased when using a geometric metric, and less so when using the Euclidean metric. This lead to the conclusion that the Euclidean metric is a more appropriate choice for diffusion tensor analysis.

But in practice, previous studies pointed out that the main effect of selecting a geometric metric rather than a Euclidean metric is encountered when interpolating or averaging between two anisotropic tensors [8]. The Euclidean metric does not preserve the determinant (which is proportional to the volume of the diffusion ellipsoid) and, as a result, the interpolated tensor may have a determinant larger than the initial tensors, i.e., it may be "swollen." With the introduction of the Affine-invariant and Log-Euclidean metrics it was shown that the swelling effect is reduced [2]. In practice, the swelling effect is usually obviated by applying piece-wise smoothed operators, or pre-segmentation that will avoid interpolating initially distant tensors [14]. In theory, however, it is still interesting to understand why the swelling effect occurs.

3 Determinant Versus Trace

The determinant of the diffusion tensor is proportional to the volume of the diffusion ellipsoid. It is a rotation-invariant measure, and is equivalent to the product of the eigenvalues:

$$det(\mathbf{D}) = \prod_i \lambda_i .$$

Fig. 1 Determinant Versus Trace. All *blue* tensors have the same trace while all *red* tensors have the same determinant as the tensor on the *left*. Tensors with a preserved trace have a higher volume (swollen) than those with a preserved determinant

It was noticed early on that the geometric metrics preserve the determinant of interpolated tensors, while a Euclidean metric preserves the trace of the tensors [8]. Figure 1 demonstrates the difference between preserving the trace and preserving the determinant. All blue ellipsoids have the same trace ($\text{tr} = 0.8 \times 10^{-3}$ mm^2/s), while all red ellipsoids have the same determinant ($\det = 0.24 \times 10^{-3}$ (mm^2/s)^3). Clearly both trace and determinant do not preserve the fractional anisotropy (FA), as the ellipsoids vary from elongated to isotropic. When preserving the trace, the tensors are indeed more "swollen," i.e., their volume, or determinant grows. Deciding between the geometric metric and the Euclidean metric therefore first requires preference for trace over determinant preservation.

3.1 Physical Considerations

The Einstein equation for free diffusion caused by Brownian motion establishes the fundamental defining relationship between the diffusion coefficient and the mean-squared displacement along an axis [10]:

$$\sigma^2(t) = \mathrm{E}[(\mathbf{x}_t - \mathbf{x}_0)^2] = 2dt \ , \tag{4}$$

where E is the expectation operator. The position along the axis at time t is $\mathbf{x}_t$; $\mathbf{x}_0$ is the position at the origin. This relationship defines the diffusion coefficient, d, as proportional to the variance of particle displacements, $\sigma^2(t)$, at time t, and arises from the normal distribution of particles expected for Brownian motion, $\mathbf{x}_t \sim N(\mathbf{x}_0, \sigma^2(t))$. The diffusion tensor is a 3D generalization of the diffusion coefficient [9]: it compactly quantifies the variance and covariance of particle displacements along any axis, with eigenvalues that are the diffusion coefficients along a set of three orthogonal principal axes, described by the eigenvectors and aligned with the maximal variance (or maximal diffusivity) orientations [5]. Therefore the diffusion coefficient, any element of the diffusion tensor, the eigenvalue of the diffusion tensor, and the trace of the tensor are all measures of diffusion with units of (*distance*^{2}/*time*). The determinant of a matrix is a coefficient that describes a scale factor. In the case where the matrix describes distances, the determinant describes a volume, but in the diffusion tensor case, the determinant—with units of (*distance*^{2}/*time*)^3—describes the volume of the ellipsoid that represents the

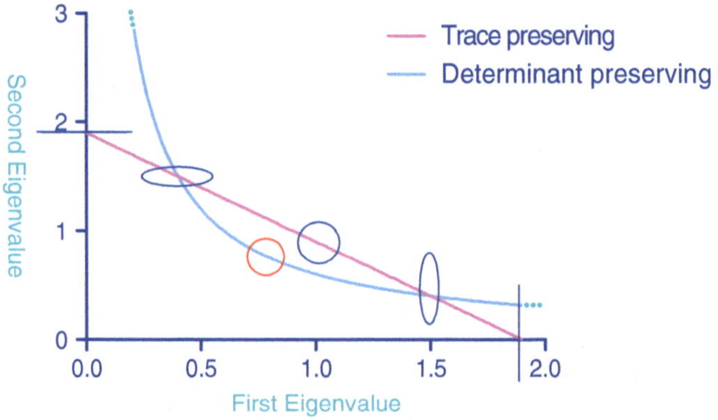

Fig. 2 Interpolations. The graph shows the first (λ_1) over the second (λ_2) eigenvalues of a 2D tensor interpolated while preserving trace, and while preserving determinant. The two original tensors are diag[1.5 0.4] and diag[0.4 1.5] and are shown along with the isotropic tensors that the interpolation goes through. When further extrapolating while preserving the trace, a degenerate tensor (*line*) with a single finite eigenvalue is found. When extrapolating while preserving the determinant, a tensor with an infinitely large principal-eigenvalue and infinitely small minor-eigenvalue is predicted; this kind of tensor is experimentally not feasible

diffusion tensor, not the volume of diffusion itself. Thus there is a clear physical distinction between the trace and the determinant of a diffusion tensor, which should be accounted for when deciding which of these two to preserve.

When introducing the Log-Euclidean metric for diffusion tensors, Arsigny et al. [2] identified the determinant of the diffusion tensor as a *"direct measure of the dispersion of the local diffusion process,"* the preservation of which is critical since *"introducing more dispersion in computations amounts to introducing more diffusion, which is physically unrealistic."* We agree that a computation should preserve the amount of dispersion of the local diffusion process, however, we believe that in the context of diffusion tensors, the determinant is not directly measuring dispersion. Moreover, the Einstein equation above (Eq. 4) shows that the direct measure of dispersion (variance) is the diffusivity itself captured by the diffusion tensor, its eigenvalues, or their sum, i.e., the trace. Indeed the trace—not the determinant—quantifies particle dispersion, and introducing more diffusion by computations is physically unrealistic, encouraging the preservation of trace over the preservation of determinant.

Figure 2 further illustrates that preserving the determinant may lead to physically unrealistic results, unlike preserving the trace. The figure shows trace-preserving and determinant-preserving interpolations between two 2D tensors, diag[1.5 0.4] and diag[0.4 1.5]. Extrapolating beyond these tensors, the trace-preserving extrapolation predicts a degenerate tensor (line) restricted to a single orientation having a single, finite eigenvalue; physically it predicts that when restricting diffusion to a single orientation, the average displacement will be larger than any other non-restricted orientation, yet finite. On the other hand, the determinant-preserving extrapolation predicts that one of the eigenvalues diminishes to an infinitesimally small value while the other eigenvalue grows to an infinitesimally large value—physically this means that it is predicted that when restricting diffusion to a single orientation the diffusion coefficient grows to infinity, meaning that particles will on average diffuse infinitely fast. This is clearly physically infeasible. In water for instance at body temperature, the diffusivity cannot be more than about 3×10^{-3} mm^2/s.

4 Tensor Swelling

In the previous section we concluded that preserving the trace of a diffusion tensor has a better physical justification than preserving the determinant of a diffusion tensor. In this section we further explore the practical implications of using the trace preserving Euclidean metric versus the determinant-preserving Log-Euclidean metric, concentrating on the main effect of the Log-Euclidean metric [2], namely reducing tensor swelling.

4.1 Variability Caused by Johnson Noise

In order to test how acquisition noise affects the determinant and trace we ran Monte Carlo simulations that create multiple noisy realizations of a given tensor. The methods follow the synthetic experiments reported in [23, 24]: noisy replications of tensor images were generated by selecting a reference tensor and introducing Johnson noise (Rician distributed) using Monte Carlo simulations to reproduce noisy diffusion weighted images. Tensors were fitted to each replicate using a conventional tensor estimation [4] and using a fitting procedure in [12] that ensures positive definite tensors. We note that when the conventional fitting procedure yielded a positive tensor, this tensor was identical to the one obtained by the positive restricting fitting, hence the difference between the two fitting procedures is restricted only to tensors having one or more negative eigenvalues. Trace and determinant were then calculated for each noisy replicate and the collection of all values was plotted as a probability distribution function.

Figure 3 shows the distribution of determinants and traces for 100 K noisy replicates of an anisotropic tensor representing white matter that has eigenvalues $1.5, 0.4$, and 0.4×10^{-3} mm^2/s, i.e., with a determinant of 0.24×10^{-3} (mm^2/s)^3 and a trace of 2.3×10^{-3} mm^2/s and of an isotropic tensor, representing gray matter, that has eigenvalues of 0.8×10^{-3} mm^2/s, i.e., a determinant of 0.512×10^{-3} (mm^2/s)^3 and a trace of 2.4×10^{-3} (mm^2/s). As expected noise can increase or decrease both trace and determinant. However, while the trace shows equal probability to be either higher or lower than the initial trace value, the determinant shows a tendency to shrink following the introduction of Johnson noise. For the anisotropic tensor,

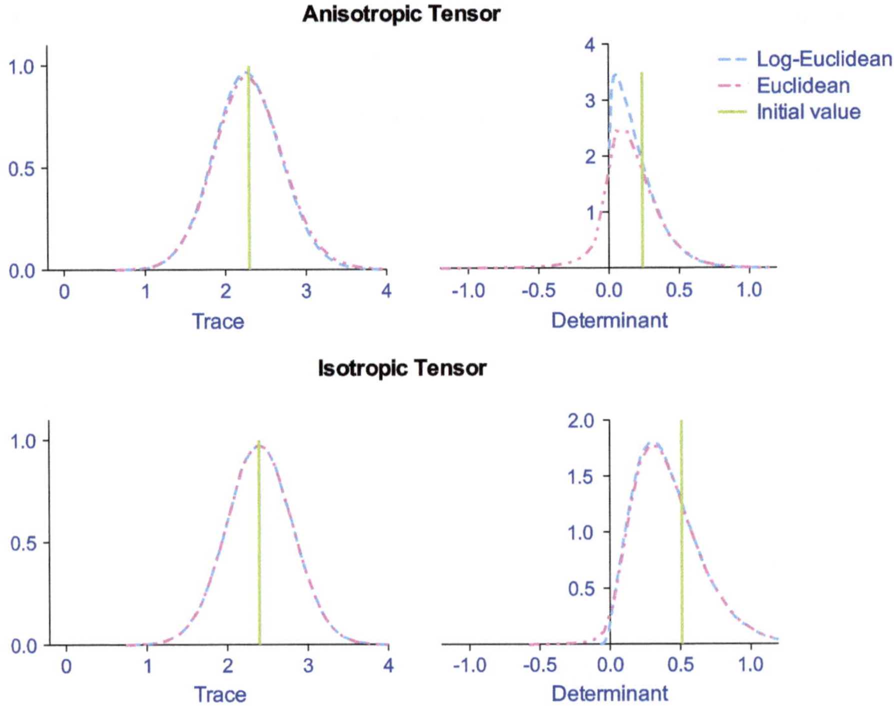

Fig. 3 Distributions. Monte Carlo simulations were used to generate empirical trace and determinant probability density functions in the presence of noise for a given anisotropic tensor, and a given isotropic tensor. Trace tends to have equal probability to have higher or lower values than the initial value. The determinant is more likely to have a lower value compared to the initial value, i.e., the tensors are expected to shrink

51.22% of the trace values are below the initial trace value, while 70.62% of the tensors shrink with determinant values that are below the initial determinant value; for the isotropic tensor, 49.999% of the trace values are below the initial trace value and 70.5% of the tensors shrink. When restricting the fitting of the noisy tensors to be positive, the determinant distribution for the anisotropic tensor changes considerably, compared with the non-restricting fitting, yet the tendency to have a lower determinant than the initial value remains, as 69.23% of the tensors shrink. For the isotropic tensor, representing gray matter, the number of negative eigenvalue occurrences decreases, and the two fits are very much aligned with each other both for the trace distribution and for the determinant distribution. Similar to the anisotropic case, the determinant in the isotropic case is more likely to have a lower value than the initial value following the introduction of Johnson noise.

The conclusion of this finding is that for these two types of tensors—representing white and gray matter—the introduction of noise causes shrinking, i.e., a decrease in the determinant. This finding is not dependent on the type of tensor fitting used.

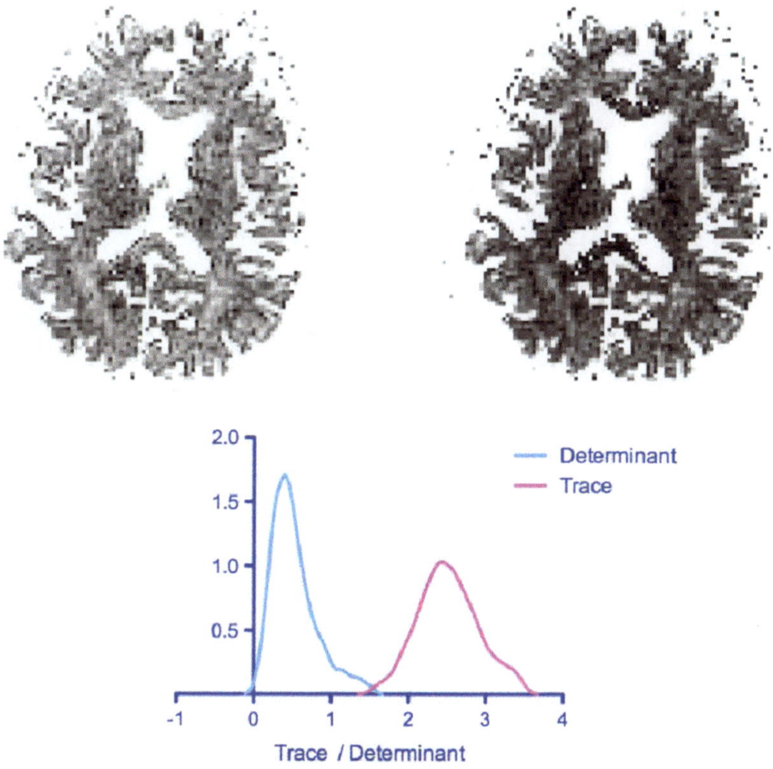

Fig. 4 **Ground truth.** A slice of diffusion data was taken and fitted to positive tensors, then masked to obtain only gray and white matter. The determinant (*left*) and trace (*right*) maps have similar contrast that does not distinguish gray matter from white matter. The slice contains tensors ranging around a trace value of 2.4×10^{-3} mm^2/s and a determinant value of 0.4×10^{-3} (mm^2/s)^3

4.2 The Extent of Tensor Shrinking in the Brain

The finding in the previous section suggests that tensors have the tendency to shrink as a result of thermal acquisition noise, yet in order to get a better notion of the extent of the shrinking in a more realistic setting we will next look at the effect of noise over an entire brain slice. We took a diffusion imaging dataset, and fitted it with the positive restricting method to yield a positive tensor field. We treated this field as a given "ground truth" for a Monte Carlo simulation, where 100 noisy replicates for each voxel were generated.

Figure 4 shows the determinant and trace maps for the ground truth tensors (maps are masked to include only gray and white matter). The contrast of both maps is very similar, not distinguishing between gray (isotropic) and white (anisotropic) matter. The trace is distributed around the value of 2.4×10^{-3} mm^2/s, and the determinant around the value 0.4×10^{-3} (mm^2/s)^3. The original values were compared with the

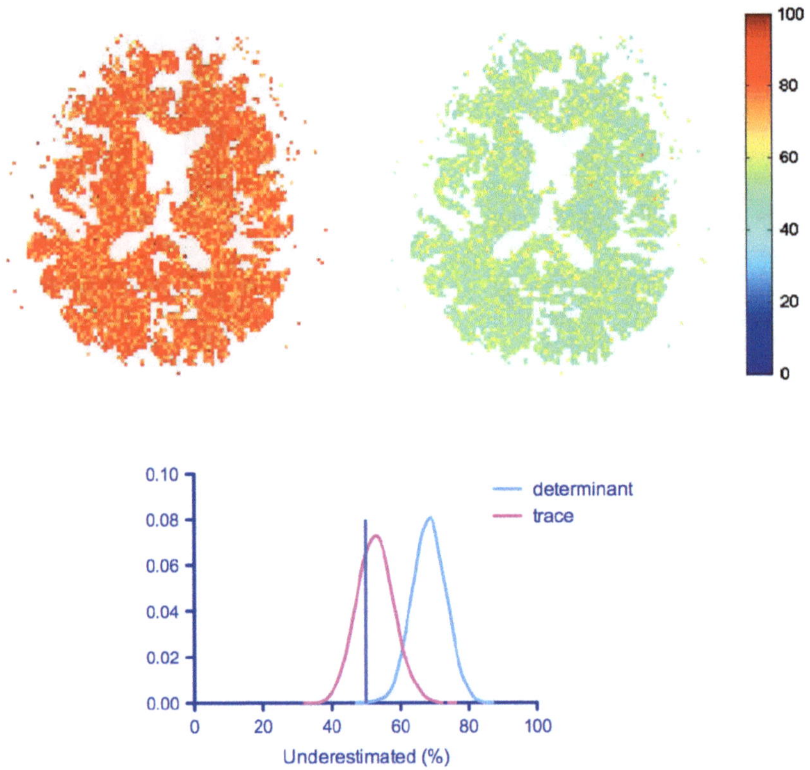

Fig. 5 Extent of Shrinking. The maps show the percent of noisy replicates that had a lower determinant (*left*) and trace value (*right*) than the ground truth values. The value for the determinant map for most voxels is larger than 50% indicating that the noisy replicates there tend to shrink. The values for the trace map are closer to 50%, indicating almost equal probability to find an increased or decreased trace in the noisy replicates

trace and determinant values of all noisy replicates to yield the maps in Fig. 5. In these maps the value of each voxel is weighted by the percentage of noisy replicates that had a value lower than the ground truth value for the same voxel. For example, for the determinant map (left), a voxel with a value higher than 50% (yellow to red) means that most of the noisy replicates in this voxel were shrinking since they had a lower determinant value than the ground truth. For the determinant map it is clear that almost all tensors in this slice tend to shrink. The trace map shows a slight tendency to a decrease in value, yet most tensors are around the 50% mark (green), meaning that the noisy replicates had equal probability to either increase or decrease trace.

The conclusion of these findings is that following the type of noise expected in diffusion MRI acquisition, tensors are expected on average to decrease their determinants while preserving their trace. This conclusion provides an additional support for favoring a trace-preserving metric over a determinant-preserving metric.

4.3 Why Do Tensors Swell?

Having seen that tensors are expected to shrink when noise is introduced we can now better understand why tensors swell during averaging and interpolations. Averaging is usually performed when a single tensor is not considered informative enough to provide an accurate estimate of local properties. When noise is considerable, averaging is used to increase SNR, which ideally would provide a better and an unbiased estimate of the true value. In the estimation of diffusion tensors from noisy replicates, our simulations predict that the noisy replicates will have a lower determinant than the original determinant. Therefore, in order to obtain an unbiased estimate of the tensor and its trace, the estimation method should cause the determinant to increase, i.e., an unbiased estimate for a diffusion tensor is one that causes noisy replicates of the tensor to swell. Indeed we have previously shown that the main ill-effect accompanied with using the determinant-preserving geometric metric is a consistent bias to the estimation [24]. We can now understand that the bias is there since the method preserves the determinant, while the noise properties dictate preservation of the trace. One way of interpreting our results would be saying that both the Euclidean and the geometric metrics introduce biases, but as it happens the Euclidean metric bias compensates for noise-related-bias. This is a valid claim that could be further explored.

Another way of interpreting the results would consider the acquisition step itself as an averaging mechanism. When measuring diffusion images, the signal is affected by many different tissue compartments and by various noise sources, each of these components may be modeled as a tensor, and the final signal averages all of the components into a single tensor [15]. We hypothesize that the Euclidean metric might be the native metric of this kind of averaging mechanism, yet testing this hypothesis requires further experimental results. At this time we can only point at the way diffusion itself is defined via Fick's Law

$$J = -D\nabla\phi(x),$$

where J is the diffusion flux and ϕ is the concentration, or the displacement probability function. In Fick's law the gradient is measuring variations of the probability function using a Euclidean norm. Analogous equations are believed to govern many other physical phenomena, such as electrical flux (Ohm's law) and heat flux (the heat equation). Taking the geometric approach further step ahead would claim that these laws themselves should be defined using geometric metrics to become

$$J = -D\nabla\phi(log(x)).$$

This claim was made, for example, in [29]. It is our belief that this form of Fick's Law is highly problematic, and not generally supported by experimental data, yet it is consistent with the use of the geometric metric.

5 Beyond Riemannian Metrics

We have shown that Johnson noise that is expected in the acquisition of diffusion images is more likely to cause a decrease in the determinant of the noisy tensors. Johnson noise is expected to be encountered in all voxels and under all scanning conditions [1, 17], but there are many other sources of variability besides thermal noise. To name a few, there are eddy currents, which depend on the gradient magnitude and direction and specific acquisition sequence used [27]; reconstruction artifacts originating from the use of multiple surface coils [17]; and motion artifacts, due to rigid head motion or stationary cardiac pulsation effects [26, 28]. Other types of smaller scale motion are ubiquitous in MRI acquisitions. In addition, there is biological variability, for example, tensors that are part of the same white matter tract are expected to have similar diffusivity values and similar shapes, but to express higher variability in the orientation of the tensors as they follow the trajectory of the tract. In this example, preserving the determinant of the tensors when interpolating or averaging among them is required. But not only the determinant must be preserved: Since the variability source dictates rotation of the reference frame, all rotation-invariant tensor properties must be preserved [8]. For this all eigenvalues, and their related quantities (such as trace, determinant, and FA) must be maintained. As we show above in Fig. 1, preserving trace alone or determinant alone does not guarantee preservation of FA.

The Riemannian metrics such as those we have analyzed here have the property of being global, meaning the distance between two given tensors remains the same no matter where these tensors are located within the image volume or within the tissue. With the absence of any prior geometric information (such as the expected anisotropy or orientation), the swelling effect is predicted by the MR measurement, and there is no physical reason to preserve the determinant. But when additional geometric information is available, local metrics, especially designed for the type of variability expected within that tissue are needed. These tissue-specific metrics have yet to be developed, but we note that the family of metrics proposed in [16, 20] might be of interest, as they are able to separate the information encoded in the tensor to orthogonal features embedded in a Euclidean space, allowing weighting of the influence of each feature on the final distance measure. Some more relevant information can be found in chapter "On the choice of a tensor distance for DTI white matter segmentation".

The diffusion tensor is proportional to the variance of normally distributed particle displacements (4). But as experiments dictate, when a voxel contains multiple components, each normally distributed on its own, the composition may deviate from the gaussian distribution and the tensor model. This brings up an interesting question of whether the result of averaging diffusion tensors, which actually means mixing two separate diffusivity components, should yield a tensor, or a different statistical representation. Some interesting ideas as to what this type of representation can be are found in chapter "Representation and Estimation of Tensor-Pairs".

The choice of metrics remains an interesting, important and controversial area of research. Although Johnson or Rician Noise is a dominant contributor to the measured MRI signal, other sources of noise and variability have illuded a similarly compact description. For the time being, we are left with having to use metrics that are the simplest and does the least harm to the data. Using these criteria as well, also leads us to the use of the Euclidean metric in analyzing DTI data.

6 Summary

We have tested how the determinant and trace of a diffusion tensor change when Johnson noise is introduced. We have found that the determinant is likely to be reduced in all types of tensors, and that the trace is equally likely to either reduce or increase following the introduction of Johnson noise. This lead us to the conclusion that in order to provide an unbiased estimate of the trace, an average or interpolation tensor operator should result in an increased determinant compared with the determinant of the initial noisy replicates. This provides a physical explanation for the swelling effect, and implies that in the most general case, where no additional geometric information is provided, there is no physical justification for a determinant-preserving metric, but there is a justification for a trace-preserving metric. This explanation provides a practical reason to prefer the Euclidean metric over the geometric metrics, in addition to previous theoretic considerations [24] that are also in favor of the Euclidean choice.

Acknowledgements We thank Liz Salak for reviewing the manuscript. OP wishes to acknowledge with thanks that part of the research on which this publication is based was supported by a Fulbright Post-doctoral Scholar Fellowship, awarded by the Fulbright commission for Israel, the United States-Israel Educational Foundation. PJB was supported by the Intramural Research Program of the Eunice Kennedy Shriver National Institute of Child Health and Human Development.

References

1. Andersson, J.L.: Maximum a posteriori estimation of diffusion tensor parameters using a Rician noise model: why, how and but. NeuroImage **42**(4), 1340–1356 (2008)
2. Arsigny, V., Fillard, P., Pennec, X., Ayache, N.: Log-Euclidean metrics for fast and simple calculus on diffusion tensors. Magn. Reson. Med. **56**(2), 411–421 (2006). doi:10.1002/mrm.20965
3. Assaf, Y., Pasternak, O.: Diffusion tensor imaging (DTI)-based white matter mapping in brain research: a review. J. Mol. Neurosci. **34**(1), 51–61 (2008)
4. Basser, P.J., Pierpaoli, C.: A simplified method to measure the diffusion tensor from seven MR images. Magn. Reson. Med. **39**, 928–934 (1998)
5. Basser, P.J., Jones, D.K.: Diffusion-tensor MRI: theory, experimental design and data analysis—a technical review. NMR Biom. **15**, 456–467 (2002)

6. Basser, P.J., Pajevic, S.: A normal distribution for tensor-valued random variables to analyze diffusion tensor MRI data. IEEE TMI **22**, 785–794 (2003)
7. Basser, P.J., Mattiello, J., LeBihan, D.: MR diffusion tensor spectroscopy and imaging. Biophys. J. **66**, 259–267 (1994)
8. Batchelor, P.G., Moakher, M., Atkinson, D., Calamante, F., Connelly, A.: A rigorous framework for diffusion tensor calculus. Magn. Reson. Med. **53**, 221–225 (2005)
9. Crank, J.: The Mathematics of Diffusion. Oxford University Press, New York, USA (1975)
10. Einstein, A.: Investigations on the Theory of the Brownian Movement. Dover, New York (1926)
11. Eisenhart, L.: Differential Geometry. Princeton University Press, Princeton (1940)
12. Fillard, P., Arsigny, V., Pennec, X., Ayache, N.: Clinical DT-MRI estimation, smoothing and fiber tracking with log-Euclidean metrics. IEEE Trans. Med. Imaging **26**(11), 1472–1482 (2007). doi:10.1109/TMI.2007.899173. PMID: 18041263
13. Fletcher, P.T., Joshi, S.: Riemannian geometry for the statistical analysis of diffusion tensor data. Signal Process. **87**(2), 250–262 (2007)
14. Gur, Y., Pasternak, O., Sochen, N.: Fast gl(n)-invariant framework for tensors regularization. Int. J. Comput. Vis. **85**(3) (2009)
15. Jian, B., Vemuri, B.C., Ozarslan, E., Carney, P.R., Mareci, T.H.: A novel tensor distribution model for the diffusion-weighted MR signal. NeuroImage **37**(1), 164–176 (2007)
16. Kindlmann, G., Ennis, D.B., Whitaker, R.T., Westin, C.F.: Diffusion tensor analysis with invariant gradients and rotation tangents. IEEE Trans. Med. Imaging **26**(11), 1483–1499 (2007). doi:10.1109/TMI.2007.907277
17. Koay, C.G., Basser, P.J.: Analytically exact correction scheme for signal extraction from noisy magnitude MR signals. J. Magn. Reson. **179**(2), 317–322 (2006)
18. Kubicki, M., McCarley, R., Westin, C.F., Park, H.J., Maier, S., Kikinis, R., Jolesz, F.A., Shenton, M.E.: A review of diffusion tensor imaging studies in schizophrenia. J. Psychiatr. Res. **41**, 15–30 (2007)
19. Lenglet, C., Rousson, M., Deriche, R., Faugeras, O.: Statistics on the manifold of multivariate normal distributions: theory and application to diffusion tensor MRI processing. J. Math. Imaging Vis. **25**(3), 423–444 (2006)
20. de Luis-Garcia, R., Alberola-Lopez, C., Kindlmann, G., Westin, C.F.: Automatic segmentation of white matter structures from DTI using tensor invariants and tensor orientation. In: Proc 17th Annual Meeting ISMRM. International Society for Magnetic Resonance in Medicine, Honolulu, USA (2009)
21. Maaß, H.: Siegel's Modular Forms and Dirichlet Series. Springer, Berlin (1971)
22. Moakher, M.: On the averaging of symmetric positive-definite tensors. J. Elast. **82**, 273–296 (2006)
23. Pajevic, S., Basser, P.J.: Parametric and non-parametric statistical analysis of DT-MRI data. J. Magn. Reson. **161**(1), 1–14 (2003)
24. Pasternak, O., Sochen, N., Basser, P.: The effect of metric selection on the analysis of diffusion tensor MRI data. Neuroimage **49**, 2190–2204 (2010)
25. Pennec, X.: Intrinsic statistics on Riemannian manifolds: basic tools for geometric measurements. J. Math. Imaging Vis. **25**(1), 127–154 (2006). doi:10.1007/s10851-006-6228-4
26. Pierpaoli, C., Marenco, S., Rohde, G., Jones, D., Barnett, A.: Analyzing the contribution of cardiac pulsation to the variability of quantities derived from the diffusion tensor. In: Proc 11th Annual Meeting ISMRM, p. 70. International Society for Magnetic Resonance in Medicine, Toronto, Canada (2003)
27. Rohde, G.K., Barnett, A.S., Basser, P.J., Marenco, S., Pierpaoli, C.: Comprehensive approach for correction of motion and distortion in diffusion-weighted MRI. Magn. Reson. Med. **51**(1), 103–114 (2004)
28. Skare, S., Anderson, J.: On the effects of gating in diffusion imaging of the brain using single shot EPI. Magn. Reson. Imaging **19**, 1125–1128 (2001)
29. Tarantola, A.: Elements for Physics: Quantities, Qualities, and Intrinsic Theories. Springer, Berlin (2006)

Part VII
Tensor Analysis

$\mathscr{H}^2$-Matrix Compression

Steffen Börm

Abstract Representing a matrix in a hierarchical data structure instead of the standard two-dimensional array can offer significant advantages: submatrices can be compressed efficiently, different resolutions of a matrix can be handled easily, and even matrix operations like multiplication, factorization or inversion can be performed in the compressed representation, thus saving computation time and storage. $\mathscr{H}^2$-matrices use a subdivision of the matrix into a hierarchy of submatrices in combination with a hierarchical basis, similar to a wavelet basis, to handle $n \times n$ matrices in $\mathscr{O}(nk)$ units of storage, where k is a parameter controlling the compression error. This chapters gives a short introduction into the basic concepts of the $\mathscr{H}^2$-matrix method, particularly concerning the compression of arbitrary matrices.

1 Overview

The $\mathscr{H}^2$-matrix technique can be considered a refinement of well-known techniques for dealing with non-local interactions relevant, e.g., when dealing with gravitational or electromagnetic forces connecting a large number of points in space.

One of the basic ideas is already present in the Ewald summation technique [18] for crystallographic research: since long-range electrostatic interactions are "smooth", it is possible to approximate them by polynomials in order to separate variables and therefore reduce the computational complexity. Short-range interactions have to be treated separately.

The panel-clustering technique [31, 35] and the well-known multipole approach [27, 28, 34] not only distinguish between long-range and short-range interactions,

S. Börm (✉)
Institute of Computer Science, Christian-Albrechts-Universität zu Kiel, D-24118, Kiel, Germany
e-mail: sb@informatik.uni-kiel.de

D.H. Laidlaw and A. Vilanova (eds.), *New Developments in the Visualization and Processing of Tensor Fields*, Mathematics and Visualization,
DOI 10.1007/978-3-642-27343-8_18, © Springer-Verlag Berlin Heidelberg 2012

but introduce a hierarchy of pairs of subdomains based on their relative distance. Using this approach, it is possible to approximate the interactions of n points in space using $\mathcal{O}(nk \log n)$ or even only $\mathcal{O}(nk)$ operations, where k is a parameter controlling the accuracy of the approximation.

Panel-clustering and multipole techniques are typically constructed for a particular problem, e.g., by expanding the underlying forces into a rapidly converging series or applying interpolation. The hierarchical matrix technique [16,19,22,29,30, 32] uses an algebraic approach: instead of approximating the interaction of two subdomains by a degenerate function, a block of a matrix is approximated by a low-rank matrix. Since low-rank matrices can be constructed efficiently by the singular value decomposition or rank-revealing factorizations, hierarchical matrices can be applied to problems ranging from elliptic partial differential equations [3, 11], boundary integral equations [1,4,14,15], preconditioning techniques [2,19,22,23,25,26] to fast solvers for control problems [21,24].

Hierarchical matrices treat each block of a matrix as independent, i.e., not connected to the other blocks. In many practical applications, e.g., when treating boundary integral formulations of electrostatic or acoustic problems, the blocks are related, and the same basis of the expansion can be used for multiple blocks. $\mathcal{H}^2$-matrices [12, 13, 33] can use these connections between blocks in order to reduce storage requirements even further and improve the efficiency of matrix operations. While hierarchical matrices typically require $\mathcal{O}(nk \log n)$ units of storage, $\mathcal{O}(nk)$ units are sufficient for $\mathcal{H}^2$-matrix techniques, leading to significant improvements for large matrices.

The improved efficiency of $\mathcal{H}^2$-matrices compared to hierarchical matrices comes at a price: a compression algorithm has to be aware of the connections between blocks, and handling these connections properly can become a challenging task. Fortunately, it is possible to use a recursive approach [8, 9, 12, 13] to detect connections between blocks during the compression process, and the resulting family of algorithms is very efficient. Using a refined approach [10, 12], even the block structure can be chosen automatically by the compression algorithm.

$\mathcal{H}^2$-matrices can be compared to wavelet techniques [6, 17, 36]: the wavelet basis is also organized in a hierarchy, and a wavelet transformation combined with a truncation also leads to a data-sparse representation of blocks of a matrix, but while the wavelet basis has to be given before the compression algorithm can start, the $\mathcal{H}^2$-matrix technique can find a nearly optimal basis adaptively. This property offers advantages when approximating solution operators of equations with non-smooth coefficients, e.g., of electrostatic or groundwater flow problems in inhomogeneous media [11].

The following section introduces the data structures like cluster and block trees, cluster bases and $\mathcal{H}^2$-matrices using a simple model problem. The third section describes a compression algorithm that can approximate arbitrary matrices by nearly optimal $\mathcal{H}^2$-matrices and includes a simple error control strategy and a new refined variant of the algorithm that can reduce the computation time for the second phase of the algorithm significantly. The fourth section outlines improvements of the general algorithm: if the input matrix is already partially compressed, the complexity can be reduced substantially. Weighting strategies can be used to control the blockwise approximation error, e.g., to ensure that block-wise relative error bounds are satisfied. The algorithm can be extended to vector- and matrix-valued matrices and three-dimensional tensors. The fifth section gives a brief summary and considers possible future refinements of the technique.

2 $\mathscr{H}^2$-Matrix Representation

We consider a matrix $G \in \mathbb{R}^{\mathscr{I}\times\mathscr{J}}$. If we store the matrix in the usual way, i.e., as a two-dimensional array containing the coefficients $(g_{ij})_{i\in\mathscr{I},j\in\mathscr{J}}$, at least $(\#\mathscr{I})(\#\mathscr{J})$ units of storage are required. For large matrices, these storage requirements are too high, therefore we need a more efficient representation.

$\mathscr{H}^2$-matrices are motivated by numerical methods for treating integral operators: if the matrix entries have the form

$$g_{ij} = f(x_i, y_j)$$

for a function $f : \mathbb{R}^3 \times \mathbb{R}^3 \to \mathbb{R}$ and families $(x_i)_{i\in\mathscr{I}}$ and $(y_j)_{j\in\mathscr{J}}$ of points in three-dimensional space, an efficient representation of f will obviously lead to an efficient representation of G. In the case of integral operators, it is usually possible to find subsets $\tau, \sigma \subseteq \mathbb{R}^3$ such that $f|_{\tau\times\sigma}$ is smooth and can therefore be approximated by an m-th order Taylor expansion

$$f(x, y) \approx \sum_{|\nu+\mu|\le m} \frac{\partial^{\nu+\mu} f}{\partial^\nu x \partial^\mu y}(x_\tau, y_\sigma) \frac{(x - x_\tau)^\nu}{\nu!} \frac{(y - y_\sigma)^\mu}{\mu!} \qquad \text{for all } x \in \tau,\ y \in \sigma.$$

For the matrix G, this property means that for

$$\hat{\tau} := \{i \in \mathscr{I} : x_i \in \tau\}, \qquad \hat{\sigma} := \{j \in \mathscr{J} : y_j \in \sigma\},$$

we have

$$g_{ij} \approx \sum_{|\nu+\mu|\le m} \frac{\partial^{\nu+\mu} f}{\partial^\nu x \partial^\mu y}(x_\tau, y_\sigma) \frac{(x_i - x_\tau)^\nu}{\nu!} \frac{(y_j - y_\sigma)^\mu}{\mu!} \qquad \text{for all } i \in \hat{\tau},\ j \in \hat{\sigma}.$$

Introducing the multiindex set $M := \{\nu \in \mathbb{N}_0^3 : |\nu| \le m\}$ and the auxiliary matrices $V_\tau \in \mathbb{R}^{\hat{\tau}\times M}$, $W_\sigma \in \mathbb{R}^{\hat{\sigma}\times M}$ and $S_{\tau\sigma} \in \mathbb{R}^{M\times M}$ by

$$(V_\tau)_{i\nu} := \frac{(x_i - x_\tau)^\nu}{\nu!}, \qquad (W_\sigma)_{j\mu} := \frac{(y_j - y_\sigma)^\mu}{\mu!},$$

$$(S_{\tau\sigma})_{\nu\mu} := \begin{cases} \dfrac{\partial^{\nu+\mu} f}{\partial x^\nu \partial y^\mu}(x_\tau, y_\sigma) & \text{if } |\nu+\mu| \leq m, \\ 0 & \text{otherwise} \end{cases}$$

we can write this equation in the short form

$$G|_{\hat{\tau}\times\hat{\sigma}} \approx V_\tau S_{\tau\sigma} W_\sigma^*. \tag{1}$$

The matrix $S_{\tau\sigma}$ has only

$$k := \#M = \binom{m+3}{3} = \frac{(m+3)(m+2)(m+1)}{6}$$

entries, and the approximation requires $k(\#\hat{\tau} + \#\hat{\sigma} + k)$ units of storage. If $k \ll \#\hat{\tau}, \#\hat{\sigma}$, the approximation of the submatrix $G|_{\hat{\tau}\times\hat{\sigma}}$ is far more efficient than the standard representation by a two-dimensional array using $(\#\hat{\tau})(\#\hat{\sigma})$ units of storage.

The matrices $S_{\tau\sigma}$ can be considered "sufficiently small", since there size depends only on k, i.e., the order m. The size of the matrices V_τ and W_σ, on the other hand, depend on the sets $\hat{\tau}$ and $\hat{\sigma}$ and can therefore still require a large amount of storage. Fortunately, we can take advantage of their structure to improve the situation: if τ' is a subdomain of τ, we can choose another center $x_{\tau'}$ of expansion and get

$$\frac{(x-x_\tau)^\nu}{\nu!} = \frac{(x - x_{\tau'} + x_{\tau'} - x_\tau)^\nu}{\nu!} = \sum_{\nu' \leq \nu} \frac{(x-x_{\tau'})^{\nu'}}{\nu'!} \frac{(x_{\tau'} - x_\tau)^{\nu-\nu'}}{(\nu-\nu')!}$$

by the binomial equation. We introduce the auxiliary matrix $E_{\tau'}$ by

$$(E_{\tau'})_{\nu'\nu} := \begin{cases} \dfrac{(x_{\tau'} - x_\tau)^{\nu-\nu'}}{(\nu-\nu')!} & \text{if } \nu' \leq \nu, \\ 0 & \text{otherwise} \end{cases}$$

and have proven

$$V_\tau|_{\hat{\tau}'\times M} = V_{\tau'} E_{\tau'}, \tag{2}$$

i.e., if we already have the matrix $V_{\tau'}$, we only have to store the *transfer matrix* $E_{\tau'} \in \mathbb{R}^{M\times M}$ instead of the entire matrix V_τ.

In the $\mathcal{H}^2$-matrix method, a hierarchy of subdomains is used, and the Eq. (2) is applied recursively to all but the smallest subdomains. The hierarchy is represented by a tree:

Definition 1 (Cluster tree). Let $\mathcal{T}$ be a labeled tree with root r, and denote the label of $t \in \mathcal{T}$ by $\hat{t}$. $\mathcal{T}$ is called a *cluster tree* for the index set $\mathcal{I}$, if the following conditions hold:

- $\hat{r} = \mathscr{I}$, i.e., the root is labeled by $\mathscr{I}$,
- If $\text{sons}(t) \neq \emptyset$ for a $t \in \mathscr{T}$, we have

$$\hat{t} = \bigcup_{t' \in \text{sons}(t)} \hat{t}', \qquad \text{for all } t \in \mathscr{T}$$

i.e., the union of the labels of sons equals the label of the father, and
- We have

$$t_1 \neq t_2 \Rightarrow \hat{t}_1 \cap \hat{t}_2 = \emptyset \qquad \text{for all } t \in \mathscr{T},\ t_1, t_2 \in \text{sons}(t),$$

i.e., the labels of sons are pairwise disjoint.

A cluster tree for an index set $\mathscr{I}$ is denoted by $\mathscr{T}_{\mathscr{I}}$. The set of *leaves* of $\mathscr{T}_{\mathscr{I}}$ is denoted by

$$\mathscr{L}_{\mathscr{I}} := \{t \in \mathscr{T}_{\mathscr{I}} \ : \ \text{sons}(t) = \emptyset\}.$$

The elements $t \in \mathscr{T}_{\mathscr{I}}$ of a cluster tree are called *clusters*.

The change in notation from $\hat{\tau}$ to $\hat{t}$ corresponds to the fact that clusters are not necessarily connected to subdomains, and this generalization allows us to use sophisticated methods [22, 25, 26] for constructing efficient cluster trees in quite general situations. An example is the simple algorithm described in [22] that assumes that each degree of freedom $i \in \mathscr{I}$ corresponds to a location in d-dimensional space and tries to form clusters of small diameter.

We introduce the *set of descendants*

$$\text{sons}^*(t) := \begin{cases} \{t\} & \text{if } \text{sons}(t) = \emptyset, \\ \{t\} \cup \bigcup_{t' \in \text{sons}(t)} \text{sons}^*(t') & \text{otherwise} \end{cases} \qquad \text{for all } t \in \mathscr{T}_{\mathscr{I}}$$

and note that a simple induction can be used to prove that for arbitrary $t, s \in \mathscr{T}_{\mathscr{I}}$ we have that $\hat{t} \cap \hat{s} \neq \emptyset$ implies $t \in \text{sons}^*(s)$ or $s \in \text{sons}^*(t)$, i.e., if the labels of two clusters are not disjoint, one of them is a descendant of the other. Among other useful consequences, this result implies that the labels of all leaf clusters have to be disjoint and therefore form a partition of the index set $\mathscr{I}$.

In order to approximate a matrix $G \in \mathbb{R}^{\mathscr{I} \times \mathscr{J}}$, we require cluster trees $\mathscr{T}_{\mathscr{I}}$ and $\mathscr{T}_{\mathscr{J}}$ for the row indices $\mathscr{I}$ and the column indices $\mathscr{J}$, respectively, and we have to split the index set $\mathscr{I} \times \mathscr{J}$ into suitable subblocks $\hat{t} \times \hat{s}$ with $t \in \mathscr{T}_{\mathscr{I}}$ and $s \in \mathscr{T}_{\mathscr{J}}$. Since $\mathscr{T}_{\mathscr{I}}$ and $\mathscr{T}_{\mathscr{J}}$ are trees, the blocks are usually also organized in a tree:

Definition 2 (Block tree). Let $\mathscr{T}_{\mathscr{I}}$ be a cluster tree for $\mathscr{I}$ with root $r_{\mathscr{I}}$, and let $\mathscr{T}_{\mathscr{J}}$ be a cluster tree for $\mathscr{J}$ with root $r_{\mathscr{J}}$. Let $\mathscr{T}$ be a labeled tree with root r. $\mathscr{T}$ is called a *block tree* for $\mathscr{T}_{\mathscr{I}}$ and $\mathscr{T}_{\mathscr{J}}$, if the following conditions hold:

- For each $b \in \mathscr{T}$, there are $t \in \mathscr{T}_{\mathscr{I}}$ and $s \in \mathscr{T}_{\mathscr{J}}$ with $b = (t, s)$, i.e., each element of the tree is a pair of two clusters,
- $r = (r_{\mathscr{I}}, r_{\mathscr{J}})$, i.e., the root is the pair of the roots of $\mathscr{T}_{\mathscr{I}}$ and $\mathscr{T}_{\mathscr{J}}$,

- For each $b=(t,s) \in \mathcal{T}$, we have $\hat{b}=\hat{t} \times \hat{s}$, i.e., the label is the Cartesian product of the label of t and the label of s,
- If $\operatorname{sons}(b) \neq \emptyset$ for a $b = (t,s) \in \mathcal{T}$, we have

$$\operatorname{sons}(b) = \begin{cases} \operatorname{sons}(t) \times \operatorname{sons}(s) & \text{if } \operatorname{sons}(t) \neq \emptyset, \operatorname{sons}(s) \neq \emptyset, \\ \{t\} \times \operatorname{sons}(s) & \text{if } \operatorname{sons}(t) = \emptyset, \operatorname{sons}(s) \neq \emptyset, \\ \operatorname{sons}(t) \times \{s\} & \text{if } \operatorname{sons}(t) \neq \emptyset, \operatorname{sons}(s) = \emptyset, \end{cases}$$

  i.e., the sons of b are pairs of sons of t and s, and if one of the clusters has no sons, the father takes their place.

A block tree for $\mathcal{T}_{\mathcal{I}}$ and $\mathcal{T}_{\mathcal{J}}$ is denoted by $\mathcal{T}_{\mathcal{I}\times\mathcal{J}}$, and its elements $b = (t,s) \in \mathcal{T}_{\mathcal{I}\times\mathcal{J}}$ are called *blocks*. The set of leaves of $\mathcal{T}_{\mathcal{I}\times\mathcal{J}}$ is denoted by

$$\mathcal{L}_{\mathcal{I}\times\mathcal{J}} := \{b \in \mathcal{T}_{\mathcal{I}\times\mathcal{J}} \ : \ \operatorname{sons}(b) = \emptyset\}.$$

Given a block $b = (t,s) \in \mathcal{T}_{\mathcal{I}\times\mathcal{J}}$, t is called the *row cluster* of b, while s is called the *column cluster*.

A closer look at the definition reveals that the block tree $\mathcal{T}_{\mathcal{I}\times\mathcal{J}}$ is in fact a special cluster tree for the index set $\mathcal{I} \times \mathcal{J}$, and we have already mentioned that the labels of the leaves of a cluster tree are a disjoint partition of the corresponding index set. Therefore we have

$$\mathcal{I} \times \mathcal{J} = \bigcup_{b=(t,s)\in\mathcal{L}_{\mathcal{I}\times\mathcal{J}}} \hat{t} \times \hat{s},$$

i.e., the leaves of the block tree describe a partition of the matrix G into non-overlapping submatrices $G|_{\hat{t}\times\hat{s}}$. Our goal is to approximate these submatrices efficiently.

Since we cannot hope to be able to represent all submatrices in the form (1), we introduce the concept of *admissible* and *inadmissible* blocks: the set $\mathcal{L}_{\mathcal{I}\times\mathcal{J}}$ of leaves of $\mathcal{T}_{\mathcal{I}\times\mathcal{J}}$ is split into two disjoint subsets $\mathcal{L}^+_{\mathcal{I}\times\mathcal{J}}$ and $\mathcal{L}^-_{\mathcal{I}\times\mathcal{J}}$ such that

$$\mathcal{L}^+_{\mathcal{I}\times\mathcal{J}} \cup \mathcal{L}^-_{\mathcal{I}\times\mathcal{J}} = \mathcal{L}_{\mathcal{I}\times\mathcal{J}}, \qquad \mathcal{L}^+_{\mathcal{I}\times\mathcal{J}} \cap \mathcal{L}^-_{\mathcal{I}\times\mathcal{J}} = \emptyset.$$

The blocks in $\mathcal{L}^+_{\mathcal{I}\times\mathcal{J}}$ correspond to "smooth" parts of the function f, i.e., we can apply (1). These blocks are called *admissible*. The blocks in $\mathcal{L}^-_{\mathcal{I}\times\mathcal{J}}$ cannot be approximated, so we represent them by storing the corresponding matrix coefficients directly. In order to reach an acceptable complexity, these *inadmissible* blocks should not be too large. We ensure this by requiring

$$b = (t,s) \in \mathcal{L}^-_{\mathcal{I}\times\mathcal{J}} \quad \Rightarrow \quad t \in \mathcal{L}_{\mathcal{I}},\ s \in \mathcal{L}_{\mathcal{J}}, \tag{3}$$

i.e., a block can only be inadmissible if both row and column cluster are leaves. For typical applications like elliptic partial differential equations or integral equations, good block trees can be constructed by fairly simple and yet very efficient algorithms [22]. In general cases, more sophisticated methods [20] can be employed to find a block tree that is perfectly adapted to the structure of the matrix coefficients.

The cluster trees $\mathscr{T}_{\mathscr{I}}$, $\mathscr{T}_{\mathscr{J}}$ and the block tree $\mathscr{T}_{\mathscr{I}\times\mathscr{J}}$ with admissible and inadmissible leaves $\mathscr{L}^+_{\mathscr{I}\times\mathscr{J}}$, $\mathscr{L}^-_{\mathscr{I}\times\mathscr{J}}$ describe the structure of the matrix. Now we consider the contents of the matrix.

Due to (2), we consider the V- and W-matrices of (1) not individually, but as hierarchical families.

Definition 3 (Cluster basis). Let $(K_t)_{t\in\mathscr{T}_{\mathscr{I}}}$ be a family of finite index sets, and let $(V_t)_{t\in\mathscr{T}_{\mathscr{I}}}$ be a family of matrices $V_t \in \mathbb{R}^{\hat{t}\times K_t}$. If there is a family $(E_t)_{t\in\mathscr{T}_{\mathscr{I}}}$ of matrices satisfying

$$E_{t'} \in \mathbb{R}^{K_{t'}\times K_t}, \qquad V_t|_{\hat{t}'\times K_t} = V_{t'}E_{t'} \qquad \text{for all } t \in \mathscr{T}_{\mathscr{I}},\ t' \in \operatorname{sons}(t),$$

we call $(V_t)_{t\in\mathscr{T}_{\mathscr{I}}}$ a *cluster basis* with *rank distribution* $(K_t)_{t\in\mathscr{T}_{\mathscr{I}}}$ and *transfer matrices* $(E_t)_{t\in\mathscr{T}_{\mathscr{I}}}$.

We represent a cluster basis by storing V_t only for leaf clusters $t \in \mathscr{L}_{\mathscr{I}}$, while E_t is stored for all $t \in \mathscr{T}_{\mathscr{I}}$. Using (2), we can reconstruct V_t for non-leaf clusters $t \in \mathscr{T}_{\mathscr{I}} \setminus \mathscr{L}_{\mathscr{I}}$ by using the transfer matrices. In practice, all important operations can be arranged in such a way that this is not necessary.

The important advantage of representing the cluster basis using transfer matrices is the reduction of storage requirements: the size of the transfer matrices is governed by the index sets K_t, and these index sets K_t depend only on the accuracy of the approximation, but not on the matrix dimension. In our example, $K_t = M$ depends only on the order of the Taylor expansion. In practice, $\#K_t$ can usually be chosen to be less than 100, even if matrices with millions of rows and columns are treated, so the transfer matrices are very small compared to the matrices V_t.

Definition 4 ($\mathscr{H}^2$-matrix). Let $(V_t)_{t\in\mathscr{T}_{\mathscr{I}}}$ and $(W_s)_{s\in\mathscr{T}_{\mathscr{J}}}$ be cluster bases with rank distributions $(K_t)_{t\in\mathscr{T}_{\mathscr{I}}}$ and $(L_s)_{s\in\mathscr{T}_{\mathscr{J}}}$, respectively. Let $\mathscr{T}_{\mathscr{I}\times\mathscr{J}}$ be a block tree with admissible leaves $\mathscr{L}^+_{\mathscr{I}\times\mathscr{J}}$ and inadmissible leaves $\mathscr{L}^-_{\mathscr{I}\times\mathscr{J}}$. A matrix $G \in \mathbb{R}^{\mathscr{I}\times\mathscr{J}}$ is an *$\mathscr{H}^2$-matrix*, if there is a family $(S_b)_{b\in\mathscr{L}^+_{\mathscr{I}\times\mathscr{J}}}$ of matrices satisfying

$$S_b \in \mathbb{R}^{K_t\times L_s}, \qquad G|_{\hat{t}\times\hat{s}} = V_t S_b W_s^* \qquad \text{for all } b = (t,s) \in \mathscr{L}^+_{\mathscr{I}\times\mathscr{J}}. \tag{4}$$

In this case, the matrices S_b are called *coupling matrices* for the $\mathscr{H}^2$-matrix representation of G.

If we let

$$\begin{aligned} n &:= \max\{\#\mathscr{I},\ \#\mathscr{J}\}, \\ k &:= \max\{\#K_t,\ \#L_s\ :\ t \in \mathscr{T}_{\mathscr{I}},\ s \in \mathscr{T}_{\mathscr{J}}\}, \\ r &:= \max\{\#\hat{t},\ \#\hat{s}\ :\ t \in \mathscr{L}_{\mathscr{I}},\ s \in \mathscr{L}_{\mathscr{J}}\}, \end{aligned}$$

it is possible to prove [13, 22] that the storage requirements for an $\mathcal{H}^2$-matrix representation are in $\mathcal{O}(n(k+r))$. In particular, the storage grows *linearly* with n, while it would grow *quadratically* if the standard representation by a two-dimensional array had been used. The parameter k controls the accuracy: in the model case, it is asymptotically proportional to m^3, where m is the order of the Taylor expansion. The parameter r depends on the choice of the cluster tree, and it is easy to ensure that $r \leq k$ holds by using a suitable stopping criterion during the construction of the cluster tree.

3 Compression

We have seen that an $\mathcal{H}^2$-matrix representation of a matrix requires only $\mathcal{O}(nk)$ units of storage with $k \ll n$, while the standard representation by a two-dimensional array typically requires $\mathcal{O}(n^2)$ units. Our goal is now to approximate a given matrix $G \in \mathbb{R}^{\mathcal{I}\times\mathcal{J}}$ by an $\mathcal{H}^2$-matrix. We assume that cluster trees $\mathcal{T}_{\mathcal{I}}$ and $\mathcal{T}_{\mathcal{J}}$ and the block tree $\mathcal{T}_{\mathcal{I}\times\mathcal{J}}$ with admissible leaves $\mathcal{L}^+_{\mathcal{I}\times\mathcal{J}}$ satisfying (3) are given. In practice, the cluster trees can be constructed based on simple geometric algorithms [22], while we can find a good block tree adaptively by starting with a very fine initial guess and coarsening blocks where the error is too low. More sophisticated techniques [10,20] are also available for this latter step. For now, we focus on finding good cluster bases $(V_t)_{t\in\mathcal{T}_{\mathcal{I}}}$ and $(W_s)_{s\in\mathcal{T}_{\mathcal{J}}}$ and coupling matrices $(S_b)_{b\in\mathcal{L}^+_{\mathcal{I}\times\mathcal{J}}}$.

In order to keep the presentation simple and to allow fine-grained control of the approximation error, we make use of *orthogonal* cluster bases:

Definition 5 (Orthogonal cluster basis). A cluster basis $(V_t)_{t\in\mathcal{T}_{\mathcal{I}}}$ is called *orthogonal* if

$$V_t^* V_t = I \qquad \text{holds for all } t \in \mathcal{T}_{\mathcal{I}}.$$

For an orthogonal cluster basis, we have

$$(V_t V_t^*)^2 = V_t V_t^* V_t V_t^* = V_t V_t^*,$$

so $V_t V_t^*$ is an orthogonal projection. If we are measuring the error in the *Frobenius norm*

$$\|X\|_F := \left(\sum_i \sum_j |x_{ij}|^2 \right)^{1/2},$$

the orthogonality implies

$$V_t^*(X - V_t V_t^* X) = 0,$$

i.e., the projection error $X - V_t V_t^* X$ is perpendicular to the range of V_t, and therefore we have

$$\begin{aligned} &\|X - V_t V_t^* X + V_t Y\|_F^2 \\ &\quad = \|X - V_t V_t^* X\|_F^2 + \|V_t Y\|_F^2 \qquad \text{for all } X \in \mathbb{R}^{K_t \times L},\ Y \in \mathbb{R}^{K_t \times M}, \end{aligned} \tag{5}$$

where L and M are arbitrary index sets. This equation is very useful when computing the approximation error.

We are interested in finding orthogonal matrices V_t and W_s such that the approximation error is sufficiently small. Since the orthogonal projection maps any matrix onto its best approximation in the range of the projection,

$$V_t V_t^* G|_{\hat{t}\times\hat{s}} W_s W_s^* = V_t S_b W_s^*, \qquad S_b := V_t^* G|_{\hat{t}\times\hat{s}} W_s$$

is the best approximation of $G|_{\hat{t}\times\hat{s}}$ in the required form (4). By using (5) and the fact that the Frobenius norm of a matrix is equal to that of its adjoint, we obtain

$$\begin{aligned} \|G|_{\hat{t}\times\hat{s}} - V_t S_b W_s^*\|_F^2 &= \|G|_{\hat{t}\times\hat{s}} - V_t V_t^* G|_{\hat{t}\times\hat{s}} W_s W_s^*\|_F^2 \\ &= \|G|_{\hat{t}\times\hat{s}} - V_t V_t^* G|_{\hat{t}\times\hat{s}} + V_t V_t^* (G|_{\hat{t}\times\hat{s}} - G|_{\hat{t}\times\hat{s}} W_s W_s^*)\|_F^2 \\ &= \|G|_{\hat{t}\times\hat{s}} - V_t V_t^* G|_{\hat{t}\times\hat{s}}\|_F^2 + \|V_t V_t^* (G|_{\hat{t}\times\hat{s}} - G|_{\hat{t}\times\hat{s}} W_s W_s^*)\|_F^2 \qquad (6) \\ &= \|G|_{\hat{t}\times\hat{s}} - V_t V_t^* G|_{\hat{t}\times\hat{s}}\|_F^2 + \|(G|_{\hat{t}\times\hat{s}}^* - W_s W_s^* G|_{\hat{t}\times\hat{s}}^*) V_t V_t^*\|_F^2 \\ &= \|G|_{\hat{t}\times\hat{s}} - V_t V_t^* G|_{\hat{t}\times\hat{s}}\|_F^2 + \|G|_{\hat{t}\times\hat{s}}^* V_t - W_s W_s^* G|_{\hat{t}\times\hat{s}}^* V_t\|_F^2, \qquad (7) \end{aligned}$$

i.e., we can split the error into the error introduced by applying the projection $V_t V_t^*$ to $G|_{\hat{t}\times\hat{s}}$ and the error introduced by applying $W_s W_s^*$ to $G|_{\hat{t}\times\hat{s}}^* V_t$.

This equation can be interpreted in two ways: since $V_t V_t^*$ is an orthogonal projection, (6) leads to the upper bound

$$\|G|_{\hat{t}\times\hat{s}} - V_t S_b W_s^*\|_F^2 \le \|G|_{\hat{t}\times\hat{s}} - V_t V_t^* G|_{\hat{t}\times\hat{s}}\|_F^2 + \|G|_{\hat{t}\times\hat{s}}^* - W_s W_s^* G|_{\hat{t}\times\hat{s}}^*\|_F^2,$$

i.e., if we have an algorithm for computing a suitable cluster basis $(V_t)_{t\in\mathscr{T}_\mathscr{I}}$, we can apply it to the adjoint matrix G^* to compute $(W_s)_{s\in\mathscr{T}_\mathscr{J}}$. This is currently the standard approach [13], since it is "nicely symmetric" and offers the advantages that essentially only one algorithm has to be considered and that rows and columns can be treated in parallel, but it has the disadvantages that computing both cluster bases takes roughly the same amount of time and has to be followed by the computation of $S_b = V_t^* G|_{\hat{t}\times\hat{s}} W_s$ for all admissible blocks.

In this paper, we are interested in a second approach: we work directly with (7). First the cluster basis $(V_t)_{t\in\mathscr{T}_\mathscr{I}}$ is computed. It turns out that we can arrange the algorithm in such a way that the matrices $G|_{\hat{t}\times\hat{s}}^* V_t$ appearing in the second term of (7) are prepared without additional cost. Using these matrices, we can then compute $(W_s)_{s\in\mathcal{T}_{\mathcal{J}}}$ and the coupling matrices $(S_b)_{b\in\mathcal{L}^+_{\mathcal{J}\times\mathcal{J}}}$ simultaneously. Since $G|_{\hat{t}\times\hat{s}}^* V_t$ has only $\#K_t \le \#\hat{t}$ columns, the second phase of the algorithm can be expected to require only a fraction of the time of the first phase, and there is no need for a third phase, since the coupling matrices are already available.

Let us consider the first phase: we have to find orthogonal matrices $V_t \in \mathbb{R}^{\hat{t}\times K_t}$ such that the rank $\#K_t$ is as small as possible and

$$\|G|_{\hat{t}\times\hat{s}} - V_t V_t^* G|_{\hat{t}\times\hat{s}}\|_F \le \boldsymbol{\varepsilon}_t. \tag{8}$$

holds for a suitable accuracy $\varepsilon_t \in \mathbb{R}_{>0}$. We call ε_t a *local accuracy*, since it depends only on the cluster t and the matrix. Since V_t depends only on t, but not on s, we have to find *one* matrix V_t that works well for *all* $s \in \mathcal{T}_{\mathcal{J}}$ with $b = (t,s) \in \mathcal{L}^+_{\mathcal{J}\times\mathcal{J}}$ simultaneously.

Due to the cluster basis property (2), anything that cannot be represented by V_t also cannot be represented by the matrix V_{t^+} corresponding to the father t^+ of t, and by induction by the matrices corresponding to all predecessors of t in the tree. This means that we also have to take all blocks $b = (t^+, s) \in \mathcal{L}^+_{\mathcal{J}\times\mathcal{J}}$ into account that are connected to a predecessor t^+ of t. Fortunately we can construct the set of relevant clusters $s \in \mathcal{T}_{\mathcal{J}}$ by a simple top-down procedure using the equation

$$r_t := \begin{cases} \{s \in \mathcal{T}_{\mathcal{J}} \ : \ (t,s) \in \mathcal{L}^+_{\mathcal{J}\times\mathcal{J}}\} \cup r_{t^+} & \text{if } t \text{ has a father } t^+, \\ \{s \in \mathcal{T}_{\mathcal{J}} \ : \ (t,s) \in \mathcal{L}^+_{\mathcal{J}\times\mathcal{J}}\} & \text{otherwise, i.e., if } t \text{ is the root.} \end{cases}$$

The task of ensuring (8) for *all* $s \in r_t$ can be simplified significantly if we use

$$N_t := \bigcup_{s\in r_t} \hat{s}$$

and look for an orthogonal matrix V_t satisfying

$$\|G|_{\hat{t}\times N_t} - V_t V_t^* G|_{\hat{t}\times N_t}\|_F \le \varepsilon_t. \tag{9}$$

Due to the definition of the Frobenius norm and since $\hat{s} \subseteq N_t$ holds for all $s \in r_t$, this inequality implies (9) for all $s \in r_t$.

Our goal is now to find an orthogonal matrix V_t satisfying (9) with the lowest possible number of columns, since the number of columns determines the efficiency of the compression. This problem is equivalent to finding a low-rank approximation of $G|_{\hat{t}\times N_t}$, and several algorithms have been proposed to perform this task, e.g., the singular value decomposition, rank-revealing LR- or QR-decompositions and cross approximation techniques. Among these techniques, the singular value decomposition yields the lowest possible rank and is also numerically stable. The approach is simple: the singular value decomposition

$$G|_{\hat{t}\times N_t} = P\Sigma Q^*, \qquad \Sigma = \operatorname{diag}(\sigma_1, \ldots, \sigma_\ell)$$

with orthogonal matrices P and Q and singular values $\sigma_1 \geq \sigma_2 \geq \ldots \geq \sigma_\ell$ is computed. Since the singular values are in decreasing order, we can easily find the smallest $k_t \in \{0, \ldots, \ell\}$ such that

$$\sum_{m=k_t+1}^{\ell} \sigma_m^2 \leq \varepsilon_t^2$$

holds. We drop the small singular values by replacing Σ with

$$\widetilde{\Sigma} := \operatorname{diag}(\sigma_1, \ldots, \sigma_{k_t}, 0, \ldots, 0)$$

in order to get

$$\|G|_{\hat{t}\times N_t} - P\widetilde{\Sigma}Q^*\|_F = \|P(\Sigma - \widetilde{\Sigma})Q^*\|_F = \|\Sigma - \widetilde{\Sigma}\|_F = \left(\sum_{m=k_t+1}^{\ell} \sigma_m^2\right)^{1/2} \leq \varepsilon_t. \tag{10}$$

In order to translate this estimate into the required form (9), we let V_t be the matrix consisting of the first k_t columns of P and obtain

$$V_t V_t^* G|_{\hat{t}\times N_t} = V_t V_t^* P\Sigma Q^* = P\widetilde{\Sigma}Q^*,$$

so we have indeed found an orthogonal matrix V_t satisfying (9). A closer look at the singular value decomposition reveals that no other orthogonal matrix satisfying this bound can have a lower number of columns, so V_t can be considered an optimal solution.

The technique outlined so far can be used if t is a leaf cluster. If t is not a leaf cluster, we have to take the cluster basis property (2) into account. To keep the presentation simple, we assume that a non-leaf cluster t has exactly two sons t_1 and t_2. We arrange our algorithm as a bottom-up recursion so that V_{t_1} and V_{t_2} have already been computed when we start looking for V_t, so according to (2) only the transfer matrices E_{t_1} and E_{t_2} have to be constructed. We introduce

$$\widehat{V}_t := \begin{pmatrix} E_{t_1} \\ E_{t_2} \end{pmatrix} \in \mathbb{R}^{M_t \times K_t}, \quad U_t := \begin{pmatrix} V_{t_1} & \\ & V_{t_2} \end{pmatrix} \in \mathbb{R}^{\hat{t}\times M_t}, \quad M_t := K_{t_1} \dot{\cup} K_{t_2} \tag{11}$$

and observe that (2) takes the form $V_t = U_t \widehat{V}_t$ and

$$U_t^* U_t = \begin{pmatrix} V_{t_1}^* V_{t_1} & \\ & V_{t_2}^* V_{t_2} \end{pmatrix} = \begin{pmatrix} I & \\ & I \end{pmatrix} = I, \quad \widehat{V}_t^* \widehat{V}_t = \widehat{V}_t^* U_t^* U_t \widehat{V}_t = V_t^* V_t = I$$

hold, so both U_t and $\widehat{V}_t$ are orthogonal matrices. In particular, $U_t U_t^*$ is an orthogonal projection, and we can use (5) to derive the error equation

$$\begin{aligned}
\|G|_{\hat{t}\times N_t} - V_t V_t^* G|_{\hat{t}\times N_t}\|_F^2 &= \|G|_{\hat{t}\times N_t} - U_t \widehat{V}_t \widehat{V}_t^* U_t^* G|_{\hat{t}\times N_t}\|_F^2 \\
&= \|G|_{\hat{t}\times N_t} - U_t U_t^* G|_{\hat{t}\times N_t} + U_t(U_t^* G|_{\hat{t}\times N_t} - \widehat{V}_t \widehat{V}_t^* U_t^* G|_{\hat{t}\times N_t})\|_F^2 \\
&= \|G|_{\hat{t}\times N_t} - U_t U_t^* G|_{\hat{t}\times N_t}\|_F^2 + \|U_t^* G|_{\hat{t}\times N_t} - \widehat{V}_t \widehat{V}_t^* U_t^* G|_{\hat{t}\times N_t}\|_F^2 \\
&= \left\| \begin{pmatrix} G|_{\hat{t}_1\times N_t} \\ G|_{\hat{t}_2\times N_t} \end{pmatrix} - \begin{pmatrix} V_{t_1} V_{t_1}^* G|_{\hat{t}_1\times N_t} \\ V_{t_2} V_{t_2}^* G|_{\hat{t}_2\times N_t} \end{pmatrix} \right\|_F^2 \\
&\quad + \|U_t^* G|_{\hat{t}\times N_t} - \widehat{V}_t \widehat{V}_t^* U_t^* G|_{\hat{t}\times N_t}\|_F^2 \\
&= \|G|_{\hat{t}_1\times N_t} - V_{t_1} V_{t_1}^* G|_{\hat{t}_1\times N_t}\|_F^2 + \|G|_{\hat{t}_2\times N_t} - V_{t_2} V_{t_2}^* G|_{\hat{t}_2\times N_t}\|_F^2 \\
&\quad + \|U_t^* G|_{\hat{t}\times N_t} - \widehat{V}_t \widehat{V}_t^* U_t^* G|_{\hat{t}\times N_t}\|_F^2.
\end{aligned}$$

The first two terms correspond to the errors introduced by the compression for the son clusters t_1 and t_2, and these terms we cannot change at this point of the algorithm. The third term, on the other hand, looks familiar: if we let

$$\widehat{G}_t := U_t^* G|_{\hat{t}\times N_t} = \begin{pmatrix} V_{t_1}^* G|_{\hat{t}_1\times N_t} \\ V_{t_2}^* G|_{\hat{t}_2\times N_t} \end{pmatrix},$$

we can see that it takes the form $\|\widehat{G}_t - \widehat{V}_t \widehat{V}_t^* \widehat{G}_t\|_F^2$, and we can construct the transfer matrices by looking for an orthogonal matrix $\widehat{V}_t$ satisfying

$$\|\widehat{G}_t - \widehat{V}_t \widehat{V}_t^* \widehat{G}_t\|_F \leq \varepsilon_t. \tag{12}$$

This problem is very similar to (9), and we already know how to solve it: by computing the singular value decomposition of $\widehat{G}_t$ and using the most important left singular vectors to construct $\widehat{V}_t$, and thereby the transfer matrices.

Since computing $\widehat{G}_t = U_t^* G|_{\hat{t}\times N_t}$ directly would be inefficient, we are looking for a more elegant way of constructing this matrix. The basic idea is to take advantage of the relationships between the individual blocks of G and $\widehat{G}_t$: we introduce

$$\bar{G}_{t,s} := V_t^* G|_{\hat{t}\times\hat{s}}, \quad \widehat{G}_{t,s} := \begin{cases} G|_{\hat{t}\times s} & \text{if } \operatorname{sons}(t) = \emptyset, \\ \begin{pmatrix} \bar{G}_{t_1,s} \\ \bar{G}_{t_2,s} \end{pmatrix} & \text{otherwise} \end{cases} \qquad \text{for all } t \in \mathcal{T}_{\mathcal{I}},\ s \in r_t.$$

Obviously, we have

$$\widehat{G}_t|_{M_t\times\hat{s}} = \widehat{G}_{t,s} \qquad \text{for all } t \in \mathcal{T}_{\mathcal{I}},\ s \in r_t,$$

procedure adaptive_row($t, r_{t^+}, G,$ **var** $\bar{G}, V$);
$r_t \leftarrow r_{t^+} \cup \{s \in \mathscr{T}_{\mathscr{J}} : (t,s) \in \mathscr{L}^+_{\mathscr{I}\times\mathscr{J}}\}$;
if $\mathrm{sons}(t) = \emptyset$ **then**
$M_t \leftarrow \hat{t}$;
for $s \in r_t$ **do** $\widehat{G}_{t,s} \leftarrow G|_{\hat{t}\times\hat{s}}$
else begin
$M_t \leftarrow \emptyset$;
for $t' \in \mathrm{sons}(t)$ **do begin**
adaptive_row($t', r_t, G, \bar{G}, V$);
$M_t \leftarrow M_t \cup K_{t'}$
end;
for $t' \in \mathrm{sons}(t), s \in r_t$ **do** $\widehat{G}_{t,s}|_{K_{t'}\times\hat{s}} \leftarrow \bar{G}_{t',s}$
end;
for $s \in r_t$ **do** $\widehat{G}|_{M_t\times\hat{s}} \leftarrow \widehat{G}_{t,s}$;
Find $\widehat{V}_t$ and $K_t \subseteq M_t$ satisfying (12);
for $s \in r_t$ **do** $\bar{G}_{t,s} \leftarrow \widehat{V}_t^* \widehat{G}_{t,s}$;
if $\mathrm{sons}(t) = \emptyset$ **then** $V_t \leftarrow \widehat{V}_t$
else
for $t' \in \mathrm{sons}(t)$ **do** $E_{t'} \leftarrow \widehat{V}_t|_{K_{t'}\times K_t}$

Fig. 1 Adaptive compression algorithm for the row basis $(V_t)_{t\in\mathscr{T}_{\mathscr{I}}}$

so we can construct the required matrix $\widehat{G}_t$ from the blocks $\widehat{G}_{t,s}$. On the other hand, we have

$$\bar{G}_{t,s} = V_t^* G|_{\hat{t}\times\hat{s}} = \widehat{V}_t^* U_t^* G|_{\hat{t}\times\hat{s}} = \widehat{V}_t^* \begin{pmatrix} V_{t_1}^* G|_{\hat{t}_1\times\hat{s}} \\ V_{t_2}^* G|_{\hat{t}_2\times\hat{s}} \end{pmatrix}$$
$$= \widehat{V}_t^* \begin{pmatrix} \bar{G}_{t_1,s} \\ \bar{G}_{t_2,s} \end{pmatrix} = \widehat{V}_t^* \widehat{G}_{t,s} \qquad \text{for all } t \in \mathscr{T}_{\mathscr{I}} \setminus \mathscr{L}_{\mathscr{I}},\ s \in r_t,$$

so we can compute $\bar{G}_{t,s}$ for non-leaf blocks using only $2(\#K_t)(\#M_t)(\#\hat{s})$ operations instead of the $2(\#K_t)(\#\hat{t})(\#\hat{s})$ operations needed for the direct approach.

In order to simplify the notation, we extend $\widehat{V}_t$ and M_t also to the case of leaf clusters by setting

$$\widehat{V}_t := V_t, \qquad \widehat{G}_t := G|_{\hat{t}\times N_t}, \qquad M_t := \hat{t} \qquad \text{for all } t \in \mathscr{L}_{\mathscr{I}} \tag{13}$$

in addition to (11), since then the inequality (9) is equivalent to (12) and we can treat both cases alike. The resulting algorithm is given in Fig. 1, the pattern of data reduction is illustrated in Fig. 2.

Since $N_t \subseteq \mathscr{J}$ holds for all $t \in \mathscr{T}_{\mathscr{I}}$, it is possible to prove that the number of operations performed for one cluster $t \in \mathscr{T}_{\mathscr{I}}$, without taking the recursions into account, is in $\mathscr{O}((\#M_t)^2\#\mathscr{J}) \subseteq \mathscr{O}(k^2 n)$ [12, Chap. 6]. Under mild assumptions, this implies that $\mathscr{O}(n^2 k)$ operations are sufficient to compute the entire cluster basis $(V_t)_{t\in\mathscr{T}_{\mathscr{I}}}$ and all matrices $\bar{G}_{t,s}$.

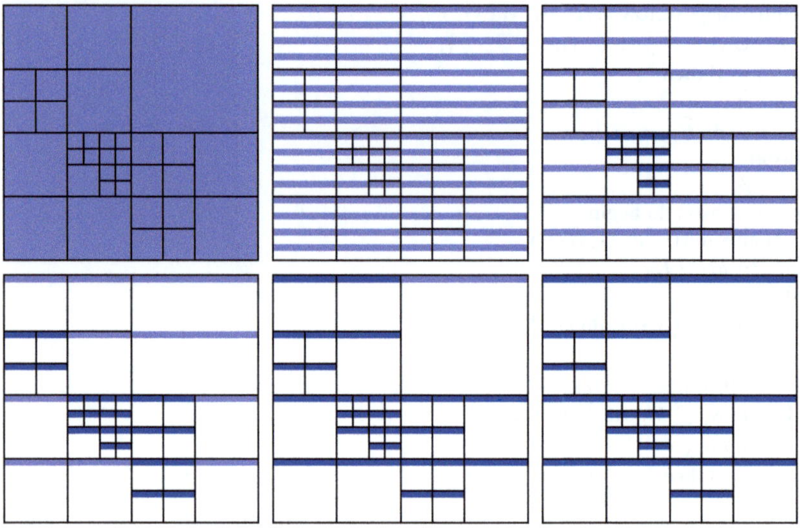

Fig. 2 Data reduction during the first phase of the compression: first the leaf clusters are compressed, followed by the higher-level clusters

We have already seen a glimpse of the structure of error estimates for the $\mathcal{H}^2$-matrix compression algorithm: the error for a cluster t is determined by the projection error introduced by $\widehat{V}_t \widehat{V}_t^*$ and the errors incurred in the sons of t. More precisely, we have

$$\begin{aligned}
\|G|_{\hat{t}\times\hat{s}} - V_t V_t^* G|_{\hat{t}\times\hat{s}}\|_F^2 &= \|G|_{\hat{t}\times\hat{s}} - U_t U_t^* G|_{\hat{t}\times\hat{s}} + U_t(U_t^* G|_{\hat{t}\times\hat{s}} - \widehat{V}_t V_t^* G|_{\hat{t}\times\hat{s}}\|_F^2 \\
&= \left\| \begin{pmatrix} G|_{\hat{t}_1\times\hat{s}} - V_{t_1} V_{t_1}^* G|_{\hat{t}_1\times\hat{s}} \\ G|_{\hat{t}_2\times\hat{s}} - V_{t_2} V_{t_2}^* G|_{\hat{t}_2\times\hat{s}} \end{pmatrix} \right\|_F^2 \\
&\quad + \|\widehat{G}_{t,s} - \widehat{V}_t \widehat{V}_t \widehat{G}_{t,s}\|_F^2 \\
&= \|G|_{\hat{t}_1\times\hat{s}} - V_{t_1} V_{t_1}^* G|_{\hat{t}_1\times\hat{s}}\|_F^2 + \|G|_{\hat{t}_2\times\hat{s}} - V_{t_2} V_{t_2}^* G|_{\hat{t}_2\times\hat{s}}\|_F^2 \\
&\quad + \|\widehat{G}_{t,s} - \widehat{V}_t \widehat{V}_t \widehat{G}_{t,s}\|_F^2,
\end{aligned}$$

and since the first two terms share the structure of the left-hand side, we can apply induction to prove

$$\|G|_{\hat{t}\times\hat{s}} - V_t V_t^* G|_{\hat{t}\times\hat{s}}\|_F^2 = \sum_{u\in\operatorname{sons}^*(t)} \|\widehat{G}_{u,s} - \widehat{V}_u \widehat{V}_u^* \widehat{G}_{u,s}\|_F^2. \tag{14}$$

Applying this equation to all $t \in \mathcal{T}_{\mathcal{I}}$ and $s \in r_t$ yields the following explicit representation of the error:

Theorem 1 (Approximation error). *We have*

$$\sum_{(t,s)\in\mathcal{L}^+_{\mathcal{I}\times\mathcal{J}}} \|G|_{\hat{t}\times\hat{s}} - V_t V_t^* G|_{\hat{t}\times\hat{s}}\|_F^2 = \sum_{t\in\mathcal{T}_{\mathcal{I}}} \|\widehat{G}_t - \widehat{V}_t \widehat{V}_t^* \widehat{G}_t\|_F^2.$$

Proof. Using (14) gives us

$$\begin{aligned}
\sum_{(t,s)\in\mathcal{L}^+_{\mathcal{I}\times\mathcal{J}}} \|G|_{\hat{t}\times\hat{s}} - V_t V_t^* G|_{\hat{t}\times\hat{s}}\|_F^2 &= \sum_{(t,s)\in\mathcal{L}^+_{\mathcal{I}\times\mathcal{J}}} \sum_{u\in\operatorname{sons}^*(t)} \|\widehat{G}_{u,s} - \widehat{V}_u \widehat{V}_u^* \widehat{G}_{u,s}\|_F^2 \\
&= \sum_{t\in\mathcal{T}_{\mathcal{I}}} \sum_{u\in\operatorname{sons}^*(t)} \sum_{\substack{s\in\mathcal{T}_{\mathcal{J}} \\ (t,s)\in\mathcal{L}^+_{\mathcal{I}\times\mathcal{J}}}} \|\widehat{G}_{u,s} - \widehat{V}_u \widehat{V}_u^* \widehat{G}_{u,s}\|_F^2 \\
&= \sum_{u\in\mathcal{T}_{\mathcal{I}}} \sum_{\substack{t\in\mathcal{T}_{\mathcal{I}} \\ u\in\operatorname{sons}^*(t)}} \sum_{\substack{s\in\mathcal{T}_{\mathcal{J}} \\ (t,s)\in\mathcal{L}^+_{\mathcal{I}\times\mathcal{J}}}} \|\widehat{G}_{u,s} - \widehat{V}_u \widehat{V}_u^* \widehat{G}_{u,s}\|_F^2 \\
&= \sum_{u\in\mathcal{T}_{\mathcal{I}}} \sum_{s\in r_u} \|\widehat{G}_{u,s} - \widehat{V}_u \widehat{V}_u^* \widehat{G}_{u,s}\|_F^2 \\
&= \sum_{u\in\mathcal{T}_{\mathcal{I}}} \|\widehat{G}_u - \widehat{V}_u \widehat{V}_u^* \widehat{G}_u\|_F^2,
\end{aligned}$$

where we have used the definition of N_t and the simple structure of the Frobenius norm in the last step. □

This means that we can not only bound the error, but we can compute it explicitly using quantities available during the course of the compression algorithm. This is particularly simple if we solve (12) by the singular value decomposition, since (10) allows us to compute the error explicitly using the singular values.

Computing $(V_t)_{t\in\mathcal{T}_{\mathcal{I}}}$ is only the first step towards finding an $\mathcal{H}^2$-matrix approximation of the given matrix G. We also have to find $(W_s)_{s\in\mathcal{T}_{\mathcal{J}}}$ and the coupling matrices $(S_b)_{b\in\mathcal{L}^+_{\mathcal{I}\times\mathcal{J}}}$. We denote the transfer matrices for the second cluster basis by $(F_s)_{s\in\mathcal{T}_{\mathcal{J}}}$ and the rank distribution by $(L_s)_{s\in\mathcal{T}_{\mathcal{J}}}$.

Fortunately, the first step provides us with all required information for the second step: we have to control the total error (7), and we already have found a way to control the first term of the error equation. For the second term we observe that

$$G|_{\hat{t}\times\hat{s}}^* V_t = (V_t^* G|_{\hat{t}\times\hat{s}})^* = \bar{G}_{t,s}^*$$

holds, so we only have to find $(W_s)_{s\in\mathcal{T}_{\mathcal{J}}}$ with

$$\|\bar{G}_{t,s}^* - W_s W_s^* \bar{G}_{t,s}^*\|_F \le \varepsilon_s \tag{15}$$

for a suitable accuracy $\varepsilon_s \in \mathbb{R}_{>0}$. This inequality is very similar to (8), so we can treat it by a very similar algorithm. The set $r_t \subseteq \mathcal{T}_{\mathcal{J}}$ of clusters connected to a cluster $t \in \mathcal{T}_{\mathcal{I}}$ is replaced by the set

$$c_s := \begin{cases} \{t \in \mathcal{T}_{\mathcal{I}} \, : \, (t,s) \in \mathcal{L}^+_{\mathcal{I}\times\mathcal{J}}\} \cup c_{s^+} & \text{if } s \text{ has a father } s^+ \in \mathcal{T}_{\mathcal{J}}, \\ \{t \in \mathcal{T}_{\mathcal{I}} \, : \, (t,s) \in \mathcal{L}^+_{\mathcal{I}\times\mathcal{J}}\} & \text{otherwise, i.e., if } s \text{ is the root.} \end{cases}$$

We have to ensure (15) for *all* $t \in c_s$, so we again collect all submatrices in a large matrix. Since the matrices $\bar{G}^*_{t,s}$ have already been partially compressed, the correct index set is

$$N_s := \bigcup_{t \in c_s} K_t,$$

and we can proceed as in the first phase: we let

$$\widehat{W}_s := \begin{cases} W_s & \text{if } \operatorname{sons}(s) = \emptyset, \\ \begin{pmatrix} W_{s_1} \\ W_{s_2} \end{pmatrix} & \text{otherwise,} \end{cases}$$

$$\bar{H}_{s,t} := \begin{cases} \bar{G}^*_{t,s} & \text{if } \operatorname{sons}(s) = \emptyset, \\ W_s^* \bar{G}^*_{t,s} & \text{otherwise,} \end{cases}$$

$$\widehat{H}_{s,t} := \begin{cases} \bar{H}_{s,t} & \text{if } \operatorname{sons}(s) = \emptyset, \\ \begin{pmatrix} \bar{H}_{s_1,t} \\ \bar{H}_{s_2,t} \end{pmatrix} & \text{otherwise,} \end{cases}$$

$$M_s := \begin{cases} \hat{s} & \text{if } \operatorname{sons}(s) = \emptyset, \\ L_{s_1} \dot{\cup} L_{s_2} & \text{otherwise} \end{cases} \qquad \text{for all } s \in \mathcal{T}_{\mathcal{J}},\ t \in c_s$$

and introduce the matrix $\widehat{H}_s \in \mathbb{R}^{M_s \times N_s}$ by

$$\widehat{H}_s|_{M_s \times K_t} = \widehat{H}_{s,t} \qquad \text{for all } s \in \mathcal{T}_{\mathcal{J}},\ t \in c_s.$$

If $s \in \mathcal{T}_{\mathcal{J}}$ is a leaf, $\widehat{H}_s$ is the row block matrix formed by combining all matrices $\bar{G}^*_{t,s}$ for $t \in c_s$. Otherwise, the submatrices are multiplied by $W^*_{s_1}$ and $W^*_{s_2}$, respectively, to take the projections corresponding to the son clusters into account. The condition (12) is replaced by

$$\|\widehat{H}_s - \widehat{W}_s \widehat{W}_s^* \widehat{H}_s\|_F \le \varepsilon_s, \tag{16}$$

and we arrive at the algorithm given in Fig. 3 (cf. Fig. 4 for the pattern of data reduction). It is very similar to the one used for $(V_t)_{t \in \mathcal{T}_{\mathcal{I}}}$ and differs only in one important point: it uses the "half-compressed" matrices $\bar{G}_{t,s}$ provided by the first phase instead of $G|_{\hat{t}\times\hat{s}}$.

procedure adaptive_col(s, c_{s^+}, $\bar{G}$, **var** $\bar{H}$, W);
$c_s \leftarrow c_{s^+} \cup \{t \in \mathscr{T}_{\mathscr{I}} \, : \, (t,s) \in \mathscr{L}^+_{\mathscr{I}\times\mathscr{J}}\}$;
if sons(s) $= \emptyset$ **then**
  $M_s \leftarrow \hat{s}$;
  for $t \in c_s$ **do** $\widehat{H}_{s,t} \leftarrow \bar{G}^*_{t,s}$
else begin
  $M_s \leftarrow \emptyset$;
  for $s' \in$ sons(s) **do begin**
    adaptive_col(s', c_s, $\bar{G}$, $\bar{H}$, W);
    $M_s \leftarrow M_s \cup L_{s'}$
  end;
  for $s' \in$ sons(s), $t \in c_s$ **do** $\widehat{H}_{s,t}|_{L_{s'}\times K_t} \leftarrow \bar{H}_{s',t}$
end;
for $t \in c_s$ **do** $\widehat{H}_s|_{M_s \times K_t} \leftarrow \widehat{H}_{s,t}$;
Find $\widehat{W}_s$ and $L_s \subseteq M_s$ satisfying (16);
for $t \in c_s$ **do** $\bar{H}_{s,t} \leftarrow \widehat{W}^*_s \widehat{H}_{s,t}$;
if sons(s) $= \emptyset$ **then** $W_s \leftarrow \widehat{W}_s$
else
  for $s' \in$ sons(s) **do** $F_{s'} \leftarrow \widehat{W}_s|_{L_{s'}\times L_s}$

Fig. 3 Adaptive compression algorithm for the column basis $(W_s)_{s\in\mathscr{T}_{\mathscr{J}}}$

Since V_t is an orthogonal matrix with $\#\hat{t}$ rows, it cannot have more than $\#\hat{t}$ columns, and we conclude that $\#K_t \leq \#\hat{t}$ holds. This implies $\#N_s \leq \#\mathscr{I}$, and we can proceed as for the first phase to show that $\mathscr{O}((\#M_s)^2\#\mathscr{I})$ operations are sufficient for each $s \in \mathscr{T}_{\mathscr{J}}$ and therefore $\mathscr{O}(n^2k)$ operations are sufficient for the entire construction of $(W_s)_{s\in\mathscr{T}_{\mathscr{J}}}$.

A closer look reveals that the second phase will in practice require far less operations than the first: under standard assumptions, c_s only contains $\mathscr{O}(\log n)$ clusters, and therefore $\#N_s$ is in $\mathscr{O}(k \log n)$. This means that $\mathscr{O}((\#M_s)^2 k \log n)$ operations are sufficient for each $s \in \mathscr{T}_{\mathscr{J}}$, leading to an improved estimate of $\mathscr{O}(nk^2 \log n)$ for the complete algorithm.

By definition, we have

$$H_{s,t} = W_s^* \bar{G}^*_{t,s} = W_s^* G|^*_{\hat{t}\times\hat{s}} V_t = S^*_b,$$

so the second phase provides us with the coupling matrices and the $\mathscr{H}^2$-matrix representation is complete.

Proceeding as in the proof of Theorem 1, we obtain

$$\sum_{(t,s)\in\mathscr{L}^+_{\mathscr{I}\times\mathscr{J}}} \|\bar{G}^*_{t,s} - W_s W_s^* \bar{G}^*_{t,s}\|_F^2 = \sum_{s\in\mathscr{T}_{\mathscr{J}}} \|\widehat{H}_s - \widehat{W}_s \widehat{W}_s^* \widehat{H}_s\|_F^2,$$

and combining (7) with (12) and (16) yields

$$\|G - \widetilde{G}\|_F^2 = \sum_{t\in\mathscr{T}_{\mathscr{I}}} \varepsilon_t^2 + \sum_{s\in\mathscr{T}_{\mathscr{J}}} \varepsilon_s^2 \tag{17}$$

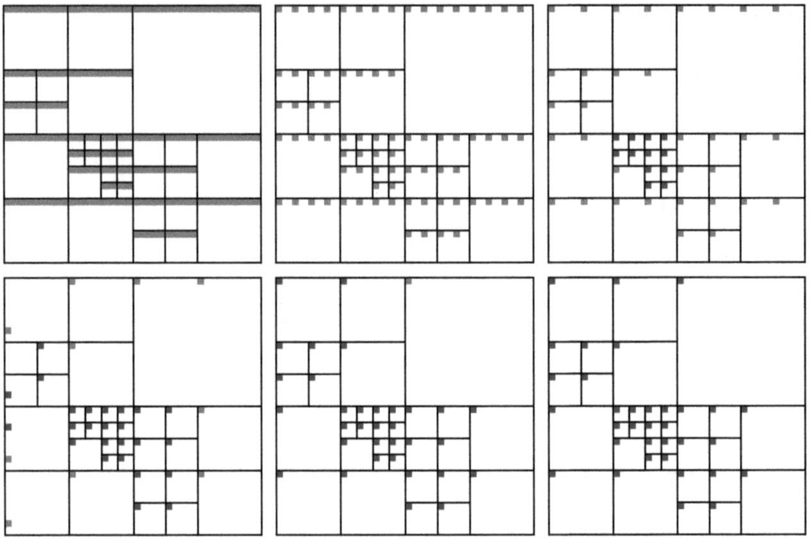

Fig. 4 Data reduction during the second phase of the compression: the rows have already been compressed, now the columns are treated starting with the finest level and working towards the root of the cluster tree

for the $\mathcal{H}^2$-matrix approximation of G defined by

$$\widetilde{G}|_{\hat{t}\times\hat{s}} = \begin{cases} V_t S_b W_s^* & \text{if } b \in \mathcal{L}^+_{\mathcal{I}\times\mathcal{J}}, \\ G|_{\hat{t}\times\hat{s}} & \text{otherwise} \end{cases} \qquad \text{for all } b = (t,s) \in \mathcal{L}_{\mathcal{I}\times\mathcal{J}}.$$

To summarize: combining the algorithms given in Fig. 1 and 3, we can compute an $\mathcal{H}^2$-matrix approximation $\widetilde{G}$ for any given matrix $G \in \mathbb{R}^{\mathcal{I}\times\mathcal{J}}$. The computation requires $\mathcal{O}(n^2k)$ operations, and the error can be controlled using (17) to indicate which singular values can be dropped during the compression process. High accuracy requirements lead to large ranks and therefore higher runtime and higher storage requirements for the compressed representation. Compared to the previous algorithm based on [13], we can expect the new algorithm to be approximately twice as fast, since the second phase of the new algorithm requires only $\mathcal{O}(nk^2 \log n)$ operations compared to $\mathcal{O}(n^2k)$ in the previous algorithm.

4 Improvements

The compression algorithm can be improved in several ways: the computational complexity can be reduced if the original matrix G is given in a data-sparse representation, the compression rate can be improved by using refined error control strategies, and the algorithm can be extended to cover vector- and tensor-valued coefficients.

4.1 Data-Sparse Input Matrices

We first consider the efficient treatment of data-sparse input matrices. A relatively simple example is a hierarchical matrix [22, 29, 30, 32]: each admissible block $b = (t, s) \in \mathscr{L}^{+}_{\mathscr{I}\times\mathscr{J}}$ is of low rank and represented in the form

$$G|_{\hat{t}\times\hat{s}} = A_b B_b^*, \qquad A_b \in \mathbb{R}^{\hat{t}\times K_b}, \qquad B_b \in \mathbb{R}^{\hat{s}\times K_b},$$

where K_b is an index set of low cardinality. Since the first phase of the compression algorithm uses only the left singular vectors of the matrix $\widehat{G}_t$, we can apply arbitrary orthogonal transformations to the columns without changing the result.

Let $b = (t, s) \in \mathscr{L}^{+}_{\mathscr{I}\times\mathscr{J}}$. Using Householder transformations again, we find an orthogonal matrix $P_b \in \mathbb{R}^{\hat{s}\times K_b}$ and a matrix $Z_b \in \mathbb{R}^{K_b\times K_b}$ satisfying

$$B_b = P_b Z_b$$

and obtain the more efficient *condensed* representation

$$G|_{\hat{t}\times\hat{s}} = A_b Z_b^* P_b^*.$$

Since P_b is orthogonal, the third factor can be ignored when computing the singular value decomposition, and an improved algorithm can work with the matrices

$$\widehat{A}_b := A_b Z_b^* \in \mathbb{R}^{\hat{t}\times K_b}$$

instead of $G|_{\hat{t}\times\hat{s}}$. In typical situations we have $\#K_b \ll \#\hat{s}$, so the new algorithm requires a significantly lower number of operations. There is one disadvantage: using the condensed representation means that the matrices $\bar{G}_{t,s} = V_t^* G|_{\hat{t}\times\hat{s}}$ are no longer computed during the course of the new algorithm, so we have to prepare them using

$$\bar{G}_{t,s} = V_t^* G|_{\hat{t}\times\hat{s}} = V_t^* A_b B_b^*.$$

Computing the left product first requires $\mathscr{O}((\#K_t)(\#\hat{t})(\#K_b))$ operations, and the second product takes $\mathscr{O}((\#K_t)(\#K_b)(\#\hat{s}))$. We conclude that setting up the matrices $\bar{G}_{t,s}$ for the second phase of the algorithm can be expected to be efficient if $\#K_b \leq \min\{\#\hat{t}, \#\hat{s}\}$ holds. Using standard assumptions, the condensed representation allows us to reduce the complexity from $\mathscr{O}(n^2k)$ to $\mathscr{O}(nk^2 \log n)$, and this "almost linear" complexity makes it possible to handle matrices with several million degrees of freedom within a few hours [12, Sects. 4.9 and 6.9].

A special case of this approach is the *unification* approach that takes a block matrix consisting of several $\mathscr{H}^2$-submatrices with different cluster bases and constructs cluster bases that can be used to turn the block matrix into a uniform $\mathscr{H}^2$-matrix. If we apply this technique recursively, the $\mathscr{H}^2$-matrix can be constructed by merging small submatrices, and it even becomes possible to determine the structure of the block tree $\mathcal{T}_{\mathcal{I}\times\mathcal{J}}$ during the course of the algorithm [10]. This method could be of interest for image compression applications where the optimal block tree depends on the structure of the image under consideration, although the improved flexibility comes at the price of a higher algorithmic complexity.

4.2 Refined Error Control

The efficiency of the compressed representation can be improved by taking into account that different parts of the matrix may require different accuracies. E.g., it may be attractive to use a relative error tolerance

$$\|G|_{\hat{t}\times\hat{s}} - V_t S_b W_s^*\|_F \leq \varepsilon_{\mathrm{rel}} \|G|_{\hat{t}\times\hat{s}}\|_F. \tag{18}$$

In the most general case, we would like to be able to prescribe tolerances for each submatrix $G|_{\hat{t}\times\hat{s}}$ appearing in the algorithm: for the first phase, we would like to ensure

$$\|\widehat{G}_{t,s} - \widehat{V}_t \widehat{V}_t^* \widehat{G}_{t,s}\|_F \leq \varepsilon_{t,s} \qquad \text{for all } t \in \mathcal{T}_{\mathcal{I}},\ s \in r_t. \tag{19}$$

In order to turn this requirement into a formulation that can be handled by the singular value decomposition, we rewrite it in the form

$$\|\varepsilon_{t,s}^{-1}\widehat{G}_{t,s} - \widehat{V}_t \widehat{V}_t^* \varepsilon_{t,s}^{-1}\widehat{G}_{t,s}\|_F \leq 1 \qquad \text{for all } t \in \mathcal{T}_{\mathcal{I}},\ s \in r_t,$$

i.e., we include the accuracy as a scaling factor applied to the matrices $\widehat{G}_{t,s}$. Replacing the matrix $\widehat{G}_t$ by $\widehat{G}_{t,\varepsilon} \in \mathbb{R}^{M_t\times N_t}$ given by

$$\widehat{G}_{t,\varepsilon}|_{M_t\times\hat{s}} = \varepsilon_{t,s}^{-1}\widehat{G}_{t,s} \qquad \text{for all } s \in r_t,$$

we can use the singular value decomposition to find an orthogonal matrix $\widehat{V}_t$ satisfying

$$\|\widehat{G}_{t,\varepsilon} - \widehat{V}_t \widehat{V}_t^* \widehat{G}_{t,\varepsilon}\|_F \leq 1,$$

and this implies (19) for all $s \in r_t$. Incorporating the scaling factors $\varepsilon_{t,s}$ into the first and second phase of the algorithm is straightforward, and the error estimate (14) takes the form

$$\|G|_{\hat{t}\times\hat{s}} - V_t V_t^* G|_{\hat{t}\times\hat{s}}\|_F^2 \leq \sum_{u\in\mathrm{sons}^*(t)} \varepsilon_{u,s}^2.$$

In order to ensure a blockwise relative error estimate like (18), we can let

$$\varepsilon_{u,s} := \frac{\varepsilon_{\mathrm{rel}} \|G|_{\hat{t}\times\hat{s}}\|_F}{\sqrt{\#\operatorname{sons}^*(t)}} \qquad \text{for all } u \in \operatorname{sons}^*(t)$$

and thus obtain

$$\|G|_{\hat{t}\times\hat{s}} - V_t V_t^* G|_{\hat{t}\times\hat{s}}\|_F^2 \le \sum_{u\in\operatorname{sons}^*(t)} \varepsilon_{u,s}^2 = \varepsilon_{\mathrm{rel}}^2 \|G|_{\hat{t}\times\hat{s}}\|_F^2 \frac{\#\operatorname{sons}^*(t)}{\#\operatorname{sons}^*(t)} = \varepsilon_{\mathrm{rel}}^2 \|G|_{\hat{t}\times\hat{s}}\|_F^2,$$

i.e., the desired estimate. It is also possible to refine this approach by assigning a more relaxed error tolerance to smaller clusters while demanding a higher accuracy for larger clusters: since the number of smaller clusters typically far outweighs the number of large clusters, the total storage complexity can be expected to improve significantly [8, 35].

4.3 *Vector- or Matrix-Valued Matrices*

We can extend the theory presented so far to vector- or matrix-valued matrices. We assume

$$g_{ij} \in \mathbb{R}^{p\times q} \qquad \text{for all } i \in \mathscr{I},\ j \in \mathscr{J},$$

where $p, q \in \mathbb{N}$. Depending on the nature of the matrices g_{ij}, different ways of turning G into a standard matrix H that can be treated by the presented algorithm have to be considered.

A very simple approach would be to introduce

$$\mathscr{I}_{p,q} := \mathscr{I} \times \{1,\dots,p\} \times \{1,\dots,q\}$$

and let $H \in \mathbb{R}^{\mathscr{I}_{p,q}\times\mathscr{J}}$ be defined by

$$h_{(i,\alpha,\beta),j} = (g_{ij})_{\alpha\beta} \qquad \text{for all } (i,\alpha,\beta) \in \mathscr{I}_{p,q},\ j \in \mathscr{J}.$$

We can extend the labels of a given cluster tree $\mathscr{T}_{\mathscr{I}}$ for $\mathscr{I}$ to the index set $\mathscr{I}_{p,q}$ by using

$$\hat{t}_{p,q} := \hat{t} \times \{1,\dots,p\} \times \{1,\dots,q\} \qquad \text{for all } t \in \mathscr{T}_{\mathscr{I}},$$

and this construction ensures that indices of $\mathscr{I}_{p,q}$ corresponding to the same row $i \in \mathscr{I}$ appear always together in a cluster.

If the matrices g_{ij} for indices $i \in \hat{t}$, $j \in \hat{s}$ in an admissible block $b = (t,s) \in \mathscr{L}^+_{\mathscr{I}\times\mathscr{J}}$ can be approximated by a low-dimensional subspace of $\mathbb{R}^{p\times q}$, the compression algorithm applied to H is able to take advantage of this property to reduce the storage requirements and improve the efficiency of the compression.

4.4 Three-Dimensional Data

The compression algorithm can be easily generalized to treat three-dimensional arrays $G \in \mathbb{R}^{\mathcal{I}\times\mathcal{J}\times\mathcal{K}}$ instead of matrices: we use an additional cluster tree $\mathcal{T}_{\mathcal{K}}$ for the third dimension and construct a block tree $\mathcal{T}_{\mathcal{I}\times\mathcal{J}\times\mathcal{K}}$ using triples $b = (t, s, r)$ of clusters $t \in \mathcal{T}_{\mathcal{I}}$, $s \in \mathcal{T}_{\mathcal{J}}$ and $r \in \mathcal{T}_{\mathcal{K}}$ instead of the pairs used before. The factorization (4) is replaced by the Tucker representation

$$G|_{\hat{t}\times\hat{s}\times\hat{r}} = (V_t \otimes W_s \otimes X_r)S_b, \qquad S_b \in \mathbb{R}^{K_t\times L_s\times M_r},$$

and the first phase of the compression algorithm looks for a cluster basis $(V_t)_{t\in\mathcal{T}_{\mathcal{I}}}$ that works uniformly well for all $s \in \mathcal{T}_{\mathcal{J}}$, $r \in \mathcal{T}_{\mathcal{K}}$ with $(t, s, r) \in \mathcal{L}^+_{\mathcal{I}\times\mathcal{J}\times\mathcal{K}}$: we require

$$\|G|_{\hat{t}\times(\hat{s}\times\hat{r})} - V_t V_t^* G|_{\hat{t}\times(\hat{s}\times\hat{r})}\|_F \le \varepsilon_t \qquad \text{for all } s \in \mathcal{T}_{\mathcal{J}},\ r \in \mathcal{T}_{\mathcal{K}} \text{ with } (t, s, r) \in \mathcal{L}^+_{\mathcal{I}\times\mathcal{J}\times\mathcal{K}}.$$

Here $G|_{\hat{t}\times(\hat{s}\times\hat{r})}$ denotes the matrix with row index set $\hat{t}$ and column index set $\hat{s} \times \hat{r}$, so that the multiplication by the projection matrix $V_t V_t^*$ can be performed as usual.

As before, the cluster basis $(W_s)_{s\in\mathcal{T}_{\mathcal{J}}}$ can be computed using the results of the first phase. A third phase is required to construct a cluster basis $(X_r)_{r\in\mathcal{T}_{\mathcal{K}}}$ for the third coordinate axis, but as before the number of operations required by the second and third phase can be expected to be significantly lower than for the first phase since the algorithm can take advantage of intermediate results of the compression. According to the error analysis in [12, Sect. 6.1], the order of the coordinate axes can, in the worst case, change the total error by a negligible factor of 3.

5 Summary

$\mathcal{H}^2$-matrices [12, 13] can be used to approximate dense $n \times n$ matrices in $\mathcal{O}(nk)$ units of storage, where k is a parameter controlling the accuracy. Compared to other approximation techniques, e.g., wavelet methods, the $\mathcal{H}^2$-matrix technique offers the advantage of using general bases that can be chosen adaptively by a relatively simple algorithm. Using intermediate results of this algorithm, an efficient error control strategy can be applied that can guarantee any desired accuracy.

The approach can be extended in several directions: three-dimensional tensors instead of matrix can be treated, tensor fields can also be covered.

In order to handle very large data sets, the algorithm would have to be extended to work on distributed memory computers. First promising results in this direction already exist [5].

An important focus of current research is the development of efficient matrix arithmetic operations like matrix multiplication, inversion or factorization. If the cluster bases are given, the best approximation of the product of two $\mathscr{H}^2$-matrices can be computed in $\mathscr{O}(nk^2)$ operations [7]. There are preliminary results indicating that the same order of complexity can also be reached for the inversion and factorization of matrices. The cluster bases can also be chosen adaptively during the course of the matrix multiplication in order to guarantee a given accuracy in $\mathscr{O}(nk^2 \log n)$ operations [12]. We are currently working on two approaches that may help to extend this result to the more challenging task of solving linear equations involving a system matrix in $\mathscr{H}^2$-representation.

Acknowledgements Most of the research results described in this chapter are the result of several years of work as a researcher at the Max Planck Institute for Mathematics in the Sciences, Leipzig, Germany, and I particularly owe a debt of gratitude to Wolfgang Hackbusch and Lars Grasedyck for many insightful discussions.

References

1. Bebendorf, M.: Approximation of boundary element matrices. Numer. Math. **86**(4), 565–589 (2000)
2. Bebendorf, M.: Why finite element discretizations can be factored by triangular hierarchical matrices. SIAM J. Numer. Anal. **45**(4), 1472–1494 (2007)
3. Bebendorf, M., Hackbusch, W.: Existence of $\mathscr{H}$-matrix approximants to the inverse FE-matrix of elliptic operators with L^∞-coefficients. Numer. Math. **95**, 1–28 (2003)
4. Bebendorf, M., Grzhibovskis, R.: Accelerating Galerkin BEM for linear elasticity using adaptive cross approximation. Math. Meth. Appl. Sci. **29**, 1721–1747 (2006)
5. Bendoraityte, J., Börm, S.: Distributed $\mathscr{H}^2$-matrices for non-local operators. Comput. Vis. Sci. **11**, 237–249 (2008)
6. Beylkin, G., Coifman, R., Rokhlin, V.: The fast wavelet transform and numerical algorithms. Commun. Pure Appl. Math. **44**, 141–183 (1991)
7. Börm, S.: $\mathscr{H}^2$-matrix arithmetics in linear complexity. Computing **77**(1), 1–28 (2006)
8. Börm, S.: Adaptive variable-rank approximation of dense matrices. SIAM J. Sci. Comput. **30**(1), 148–168 (2007)
9. Börm, S.: Data-sparse approximation of non-local operators by $\mathscr{H}^2$-matrices. Linear Algebra Appl. **422**, 380–403 (2007)
10. Börm, S.: Construction of data-sparse $\mathscr{H}^2$-matrices by hierarchical compression. SIAM J. Sci. Comput. **31**(3), 1820–1839 (2009)
11. Börm, S.: Approximation of solution operators of elliptic partial differential equations by $\mathscr{H}$- and $\mathscr{H}^2$-matrices. Numer. Math. **115**(2), 165–193 (2010)
12. Börm, S.: Efficient Numerical Methods for Non-local Operators: $\mathscr{H}^2$-Matrix Compression, Algorithms and Analysis. EMS Tracts in Mathematics, vol. 14. EMS, Zürich (2010)
13. Börm, S., Hackbusch, W.: Data-sparse approximation by adaptive $\mathscr{H}^2$-matrices. Computing **69**, 1–35 (2002)
14. Börm, S., Grasedyck, L.: Low-rank approximation of integral operators by interpolation. Computing **72**, 325–332 (2004)

15. Börm, S., Grasedyck, L.: Hybrid cross approximation of integral operators. Numer. Math. **101**, 221–249 (2005)
16. Börm, S., Grasedyck, L., Hackbusch, W.: Introduction to hierarchical matrices with applications. Eng. Anal. Bound. Elem. **27**, 405–422 (2003)
17. Daubechies, I.: Ten Lectures on Wavelets. SIAM, Philadelphia (1992)
18. Ewald, P.P.: Die Berechnung optischer und elektrostatischer Gitterpotentiale. Annalen der Physik **369**(3), 253–287 (1920)
19. Grasedyck, L.: Theorie und Anwendungen Hierarchischer Matrizen. Doctoral thesis, Universität Kiel (2001)
20. Grasedyck, L.: Adaptive recompression of $\mathcal{H}$-matrices for BEM. Computing **74**(3), 205–223 (2004)
21. Grasedyck, L.: Existence of a low-rank or $\mathcal{H}$-matrix approximant to the solution of a Sylvester equation. Numer. Linear Algebra Appl. **11**, 371–389 (2004)
22. Grasedyck, L., Hackbusch, W.: Construction and arithmetics of $\mathcal{H}$-matrices. Computing **70**, 295–334 (2003)
23. Grasedyck, L., LeBorne, S.: $\mathcal{H}$-matrix preconditioners in convection-dominated problems. SIAM J. Math. Anal. **27**(4), 1172–1183 (2006)
24. Grasedyck, L., Hackbusch, W., Khoromskij, B.N.: Solution of large scale algebraic matrix Riccati equations by use of hierarchical matrices. Computing **70**, 121–165 (2003)
25. Grasedyck, L., Hackbusch, W., LeBorne, S.: Adaptive geometrically balanced clustering of $\mathcal{H}$-matrices. Computing **73**, 1–23 (2004)
26. Grasedyck, L., Kriemann, R., LeBorne, S.: Domain decomposition based $\mathcal{H}$-LU preconditioning. Numer. Math. **112**(4), 565–600 (2009)
27. Greengard, L., Rokhlin, V.: A fast algorithm for particle simulations. J. Comput. Phys. **73**, 325–348 (1987)
28. Greengard, L., Rokhlin, V.: A new version of the fast multipole method for the Laplace equation in three dimensions. In: Acta Numerica 1997, pp. 229–269. Cambridge University Press, Cambridge/New York (1997)
29. Hackbusch, W.: A sparse matrix arithmetic based on $\mathcal{H}$-matrices. Part I: introduction to $\mathcal{H}$-matrices. Computing **62**, 89–108 (1999)
30. Hackbusch, W.: Hierarchische Matrizen: Algorithmen und Analysis. Springer, Berlin/Heidelberg (2009)
31. Hackbusch, W., Nowak, Z.P.: On the fast matrix multiplication in the boundary element method by panel clustering. Numer. Math. **54**, 463–491 (1989)
32. Hackbusch, W., Khoromskij, B.N.: A sparse matrix arithmetic based on $\mathcal{H}$-matrices. Part II: application to multi-dimensional problems. Computing **64**, 21–47 (2000)
33. Hackbusch, W., Khoromskij, B.N., Sauter, S.A.: On $\mathcal{H}^2$-matrices. In: Bungartz, H., Hoppe, R., Zenger, C. (eds.) Lectures on Applied Mathematics, pp. 9–29. Springer, Berlin (2000)
34. Rokhlin, V.: Rapid solution of integral equations of classical potential theory. J. Comput. Phys. **60**, 187–207 (1985)
35. Sauter, S.A.: Variable order panel clustering. Computing **64**, 223–261 (2000)
36. Tausch, J., White, J.: Multiscale bases for the sparse representation of boundary integral operators on complex geometries. SIAM J. Sci. Comput. **24**(5), 1610–1629 (2003)

Harmonic Field Analysis

Christian Wagner, Christoph Garth, and Hans Hagen

Abstract Harmonic analysis techniques are established and successful tools in a variety of application areas, with the Fourier decomposition as one well-known example. In this chapter, we describe recent work on possible approaches to use Harmonic Analysis on fields of arbitrary type to facilitate global feature extraction and visualization. We find that a global approach is hampered by significant computational costs, and thus describe a local framework for harmonic vector field analysis to address this concern. In addition to a description of our approach, we provide a high-level overview of mathematical concepts underlying it and address practical modeling and calculation issues. As a potential application, we demonstrate the definition of empirical features based on local harmonic analysis of vector fields that reduce field data to low dimensional feature sets and offers possibilities for visualization and analysis.

1 Introduction

Among a multitude of techniques for the visualization of scalar, vector, and tensor fields, feature analysis methods play a crucial role in enabling the visualization of large datasets. Here, application-oriented feature definitions are matched against a dataset to highlight interesting regions. Especially for highly complex fields arising from modern computer simulations, visualization efforts can be reduced or made feasible by depicting significant structural components and their interactions, allowing for an abstracted view. Existing methods are typically based on topological structures or application specific feature definitions. While the former leverage a deep mathematical framework to generate a topological skeleton of a field and are

C. Wagner (✉) · C. Garth · H. Hagen
University of Kaiserslautern, Kaiserslautern, Germany
e-mail: wagner@cs.uni-kl.de

D.H. Laidlaw and A. Vilanova (eds.), *New Developments in the Visualization and Processing of Tensor Fields*, Mathematics and Visualization,
DOI 10.1007/978-3-642-27343-8_19,

uniformly applicable to general fields, the latter requires intimate knowledge of the application domain. Here, we apply concepts of harmonic analysis towards this goal. In general, applications of harmonic analysis (often also called signal processing) can be manifold. Ranging from simple and generic processing such as smoothing over a variety of computer vision algorithms, it can also be applied to the extraction, detection and classification of features.

In this chapter, we provide an introduction to harmonic analysis of vector fields and describe an application to feature-based visualization, with a focus on practical aspects. As is required for a basic understanding of the concepts we discuss, we attempt to summarize harmonic analysis techniques for discrete field representation. Harmonic analysis as a general concept is neither specific to application domains, nor is it restricted to specific field types (scalar, vector, tensor) or domain geometry. We illustrate a global analysis approach for generic fields and discuss its computational implications, which leads to define local approaches that are much more feasible computationally.

To achieve the goal of feature-based visualization, we define a feature space over small neighborhoods of a given discretized field's domain that transforms it—using harmonic analysis—into a low dimensional feature vector. This transformation is achieved by formulating a discrete Laplacian over the discretization of the neighborhood and computing eigenvalue decomposition. This yields a basis of eigenfunctions over the neighborhood. The coefficients of the field representation in this basis form the feature vector. The latter then provides a means to define, locate and compare features in an empirical fashion. Our method is closely related to Fourier analysis that is used extensively e.g. in image processing and computer vision applications.

2 Related Work

Harmonic analysis is a concept with a rich and well-developed mathematical background, and has many applications. For example, in disciplines related to visualization, applications to geometry processing and mesh filtering have been discussed in depth, such as surface smoothing [25], matching [22], quadrangulation [4, 26] or the design of tangent vector fields over surfaces [11]. Vallet and Lévy [27] provide an overview of recent results. In the following, we concentrate on previous work that is immediately relevant to the practical presentation in this paper.

Eigenanalysis techniques are often applied in the context of Fourier transformations, convolution or pattern matching. Since classical Fourier techniques are not applicable to vector or tensor fields in a direct and meaningful manner, [6, 7, 23] employ complex invariant moments or Clifford algebra in order to define a suitable setting. A different approach is based on a discrete formulation of the Laplacian operator, which is the central concept behind harmonic analysis, such that its eigenfunctions can be directly formulated for vector fields.

However, discrete Laplacian formulations, such as the Discrete Exterior Calculus (DEC) approach introduced by Hirani [16], are not yet well understood regarding their application to vector-valued eigenvalue problems. Therefore, we compare numerical results with a classical Finite Element formulation of the Laplacian, for which many good textbooks are available like [17] as well as for eigenvalue problems itself [15] and the underlying functional calculus [19].

The implementations underlying our experiments (cf. Sect. 5) are based on the use of the ARPACK [20, 24] and SUPERLU [2] packages that facilitate eigenanalysis and decomposition of large matrices [21].

3 Harmonic Analysis

To provide a base understanding of our methods to readers not familiar with harmonic analysis, we will briefly touch on and illustrate a number of fundamental concepts. Beginning with the Fourier decomposition as a direct example, we will discuss the spectral theorem known from functional calculus. Then, a short discussion about arbitrary domains and field types is followed by describing low pass filtering as a global approach. Due to the limited space, we will necessarily limit ourselves to a high-level, phenomenological overview and refer the interested reader to [5, 10, 18] for an in-depth treatment of the topic.

3.1 Fourier Decomposition

The well-known technique of Fourier analysis is an example of harmonic analysis techniques, and we will discuss it here briefly to motivate the use of harmonic analysis on generic field types and geometries as a tool with manifold capabilities.

A periodic signal can be decomposed into a combination of sinusoidal functions with varying frequency and amplitude. Mathematically, sine and cosine functions are used as basis for the space of periodic functions, and the original signal is transformed to this basis. Specifically, a periodic function f with periodic length $T > 0$ can be described as a sum of sine and cosine functions

$$f(t) = \frac{a_0}{2} + \sum_{k=1}^{\infty}(a_k \cdot \cos(k\omega t) + b_k \cdot \sin(k\omega t)),$$

where the coefficients a_k and b_k are given by the projection of the function f onto the Fourier basis functions as

$$a_k = \frac{2}{T}\int_c^{c+T} f(t) \cdot \cos(k\omega t)\,\mathrm{d}t$$

and

$$b_k = \frac{2}{T} \int_c^{c+T} f(t) \cdot \sin(k\omega t)\, \mathrm{d}t.$$

Furthermore, these sinusoidal functions can also be interpreted as the real and imaginary parts of the complex valued functions e^{-ikt}. These are specifically the eigenfunctions to the second derivative operator, i.e.

$$-\frac{\partial^2 e^{-ikt}}{\partial^2 t} = -i^2 k^2 e^{-ikt} = \lambda e^{-ikt}.$$

In this simple case, the second derivative operator is the specific form of the Laplacian operator on the one-dimensional periodic domain $[0, T]$.

In summary, the Fourier decomposition is a Laplacian eigenvalue decomposition, where a periodic function is represented as a linear combination in a basis of eigenfunctions of the Laplacian operator.

3.2 Spectral Theorem

The mathematical background we are using has its origins in the spectral theorem of functional calculus. The theorem provides a strong and useful relationship between an operator $\mathscr{T} : V \to V$, e.g. $V = \mathbb{R}^n$, and its eigenfunctions with

$$V = \ker \mathscr{T} \oplus \overline{\mathrm{lin}}\{e_1, e_2, \ldots\}.$$

In a nutshell, for a compact operator $\mathscr{T}$ on a function space (such as the Laplacian), the spectral theorem states that the function space can be decomposed into a direct sum of the operator's kernel (which maps functions to zero) and the linear space spanned by its eigenfunctions. The non-zero functions e_i and non-zero values λ_i which fulfill

$$\mathscr{T} e_i = \lambda_i e_i$$

are called the *eigenfunctions* e_i to the corresponding *eigenvalues* λ_i of the operator $\mathscr{T}$. The set of λ_i is called the *spectrum* $\sigma(\mathscr{T})$ of the operator.

In the following, the Laplacian Δ will be the central operator we are concerned with. Since the Laplacian is a symmetric operator, its eigenfunctions are orthogonal, and the projection of a function onto the Laplacian's eigenfunctions is the simple inner product (typically in a L_2-sense) of two continuous functions f and g,

$$< f, g >:= \int f(x) \cdot g(x)\, \mathrm{d}x.$$

Consequently, the projection of the function f on the eigenfunctions e_i results in the basis coefficient a_i and has the continuous version

$$a_i =< f, e_i >:= \int f(x) \cdot e_i(x)\, \mathrm{d}x.$$

3.3 Discrete Setting

In practical applications, one has to consider a discrete setting, where a field under consideration is described over a computational grid. Here, the Laplacian Δ is represented by a matrix A that describes its action on a discrete representation of a given field. Its eigenfunctions are the eigenvectors e of A and the spectrum is given by

$$\sigma(A) = \{\lambda \in \mathbb{C} \,|\, \exists e \neq 0 : Ae = \lambda e\}.$$

The projection in the discrete case simplifies to the sum

$$a_i = x^T \cdot w \cdot e_i = \sum_{j=1}^{n} x_j\, w_j\, (e_i)_j,$$

where x^T is the function vector to be projected, e_i the i-th eigenvector of A and w is an area weighting function given by the size of the area around each vertex j.

3.4 Arbitrary Domain and Field Type

As motivated in Sect. 3.1, one aspect of harmonic analysis is the study of Laplacian eigenfunctions and eigenvalues. For the classical Fourier decomposition the functions are scalar-valued and periodic. Thus, they can be interpreted as scalar functions on a simple manifold: the unit circle S^1. However, the definition of the Laplacian is more general and can be stated in other circumstances. Especially, it can be defined for scalar, vector or tensor fields, and over different domain geometries including arbitrarily-shaped manifolds. Figure 1 shows an example of formulating the Laplacian for vector fields on an irregular shaped region, and the corresponding first eigenfunctions (in this case a vector field) are shown.

3.5 Global Harmonic Analysis

In similarity to the Fourier decomposition, harmonic analysis can be applied to an input field's global domain. We term this a *global approach* in the following.

Fig. 1 Example of vector-valued Laplacian eigenfunctions, which are vector fields themselves, illustrated by line integral convolution. Eigenfunctions of the vector-valued Laplacian are shown on an irregular-shaped region with cells on an unstructured grid. The eigenvalue multiplicity for the shown eigenvalues is two, therefore two corresponding eigenvector fields (*top* row vs. *bottom* row) are orthogonal

As an application example, we consider a turbulent vector field dataset arising from a CFD simulation and described on an unstructured mesh with about 50,000 vertices. The resolution of the mesh is slightly adaptive with vertex distance decreasing in the middle of the domain. As in this example, vector fields arising from e.g. computational fluid dynamics solvers, are often given on arbitrary shaped geometries discretized on unstructured grids. As outlined above, this yields the necessity of discretizing the Laplacian on the underlying discrete function space and solving for eigenvalues and eigenfunctions of the resulting matrix representation explicitly. We discuss details of the discretization process in Sects. 5.1 and 5.2, and the eigenvalue computation is discussed in Sect. 5.4.

After the eigenvalues of the matrix resulting from the discretization of the Laplacian over the domain of consideration are computed, any spectral method operating on the eigenfunctions can be applied and typically modifies the representation of the field in this basis. The resulting linear combination of eigenfunctions can then be evaluated on the original basis (e.g. per vertex) to allow interpretation of the result in the original context. Using this approach, Fig. 2 illustrates the effect of applying a low pass filter to the vector field that dampens the high frequency vector field oscillations by reducing the coefficients of eigenfunctions proportional to the corresponding eigenvalue magnitude.

Similarly, high-pass or band-bass filters can be applied to enhance different components of the vector fields. For example, increasing higher frequencies can amplify the vorticity to allow easier exploration of small scale structures.

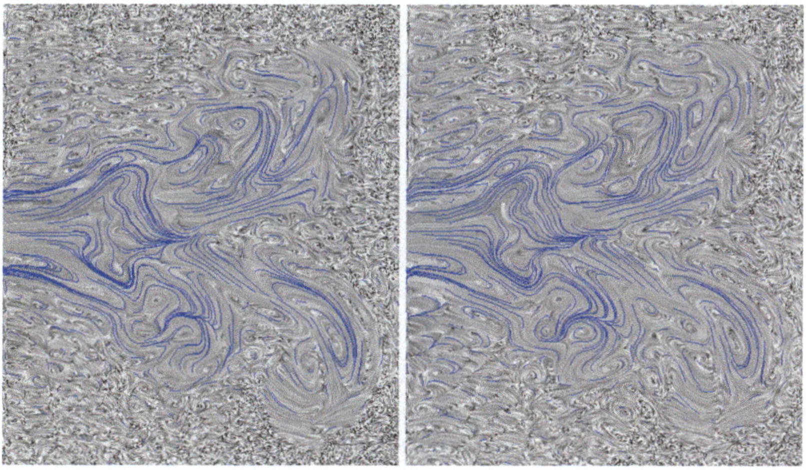

Fig. 2 The spectrum of the *left* vector field is dampened by removing high eigenvalue terms resulting in the *right* vector field. Some *streamlines* are drawn in the vector field to illustrate small turbulent structures

4 Local Harmonic Analysis

The dominating problem of global approaches is the large and significant computational cost arising from the computation of the Laplacian eigenvector basis. Since the storage space and calculation complexity grows quadratically in the number of vertices, global approaches are quickly limited in the feasible field size to be treated. In addition, many interesting features are local in nature and can be distributed spatially unequally, and are well represented only using a very large number of Laplacian eigenfunctions. For this reason, we introduce a local approach by defining small regions around each point in the domain and describe how to define, visualize, and analyze features in such a local neighborhood.

4.1 Locality and Local Feature Definition

We introduce a region $\epsilon(v_i)$ around each vertex v_i in the domain of interest. We then determine a local eigenbasis for this region and project only the local field inside the region on this local basis. From this, we obtain a local feature vector for every point in space consisting of the basis coefficients of the local field around each point, namely each vertex for a discretized field.

By using the same local region discretization for every point in space, only a single (small) eigenbasis must be computed and can be reused for every region. As long as the region's spatial resolution is high enough, scaling the region around each vertex can be done without recomputing the eigenvalue system.

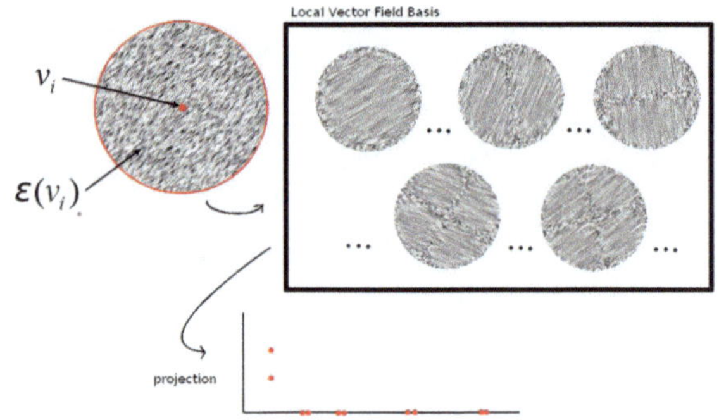

Fig. 3 Local feature definition using local eigenvector basis. For the region $\epsilon(v_i)$ around v_i the Laplacian eigenfunctions are determined. By projecting the original field onto these eigenfunctions the vector of basis coefficients is composed

Figure 3 illustrates these relationships. For a region of interest in a vector field, a local vector field basis is determined by the Laplacian eigenbasis fields corresponding to the region. Successively, the vector field is projected onto the elements of the eigenbasis, resulting in basis coefficients. These are then interpreted as a local feature descriptor or feature vector to define features and corresponding features strengths in the vector field as follows.

A feature f is defined by a feature strength function ρ_f, which is a mapping from the Laplacian eigenbasis coefficients a_i to the interval [0, 1]:

$$\rho_f : \mathbb{R}^n \rightarrow \mathbb{R} \tag{1}$$

$$(a_1, a_2, \ldots, a_n) \mapsto [0, 1] \tag{2}$$

Here, resulting feature strength of 0 implies a vanishing response to the feature type f in the selected region while a resulting value of 1 implies the definition matches exactly.

Consider the following example. The first eigenfield in the local eigenbasis for the example vector field introduced above is shown in Fig. 3. Together with an orthogonal eigenfield of the same eigenvalue, they form the first eigenspace. As depicted in the figure, the first eigenspace field contains exactly the linear component of the vector field. Therefore, we define the local vector field linearity ρ_L as a feature strength function by the ratio of the first two basis coefficients to the remainder of the basis coefficients on the spectrum as

$$\rho_L(a_1, \ldots, a_n) := \frac{\sum_{i=1}^{2} |a_i|}{\sum_{i=1}^{n} |a_i|}.$$

The purpose of this definition of ρ_L is to capture the relative importance of the basis functions that encode linear flow—as given by their coefficients a_1 and a_2—normalized by the overall size of the coefficients ($\sum_{i=1}^{n} |a_i|$). As ρ_L approaches 1, the contribution of non-linear eigenbasis elements represented by $a_3, \ldots, a_n$ necessarily tends to zero, signifying diminishing non-linear behavior.

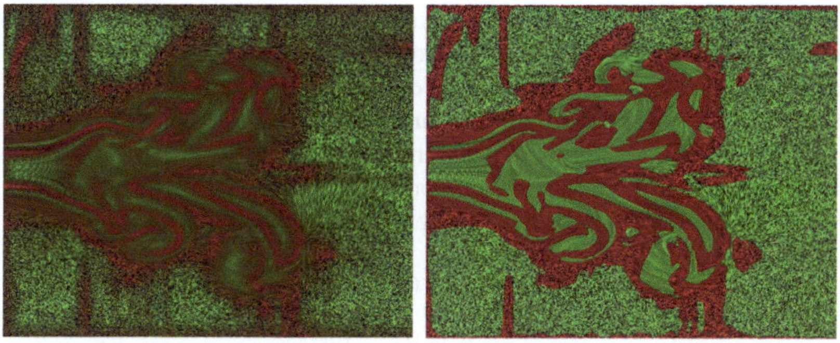

Fig. 4 Application of local feature definitions with local vector field linearity on the *left* and its segmentation on the *right*

In this case depicted in Fig. 3, only the first two basis coefficients of the local field basis are non-zero, and the other coefficients vanish, thus the field is perfectly linear in the considered neighborhood.

In other words, the feature strength functions measures the similarity of the feature vector to a desired feature type, where both are expressed in the Laplacian eigenbasis.

Naturally, local feature strength criteria can be applied globally for each point in the domain of the considered field. For example, Fig. 4 (left) illustrates the local linearity feature strength for the inflow dataset. High local linearity is colored in green while low local linearity is shown in red. The turbulent regions, where linear flow is not prevalent, are clearly distinguishable from the remainder of the field. In the right image, a simple threshold on ρ_L is used to perform a binary segmentation and separate mostly local linear regions from the turbulent parts of the vector field.

5 Discretizations and Computational Issues

The harmonic analysis technique can be used on arbitrary field and domain types. However, the choice of Laplacian discretization is crucial to guarantee good results. In the following, we summarize the most important smooth Laplacian versions, followed by a brief introduction to a finite element and a discrete exterior calculus discretization and their comparison. Finally, we will discuss computational costs.

5.1 Finite Element Discretization

To translate the concepts described above into practice, a discretization of the Laplacian operator acting over an unstructured mesh must be chosen. The simplest choice for this problem is finite element (FEM) formulation to discretize the Laplacian on the global domain or the local region. In this setting, the action of the Laplacian on a vector field is approximated by its action on interpolatory basis functions that form a basis of the vector field function space over the discrete domain. For example, for linear finite elements over triangles covering a two-dimensional domain, the interpolation functions are simple hat functions defined over the one-ring neighborhood of a vertex. For vector fields, there is one basis function per vertex and per vector component.

For a given domain Ω and the Laplace eigenvalue problem

$$\Delta u - \lambda u = 0 \qquad \text{in } \Omega \tag{3}$$

$$u = 0 \qquad \text{on } \partial\Omega \tag{4}$$

an approximation $\tilde{u}$ for the function u is sought according to Galerkin's method (cf. [13, 14]) by linearly independent functions $N_1, \ldots, N_m$ satisfying the homogeneous boundary condition

$$\tilde{u} = \sum_{k=1}^{m} u_k N_k. \tag{5}$$

In this formulation, the residual $||u - \tilde{u}||$, weighted with certain weight functions, is required to vanish. In Galerkin's method the basis functions $N_1, \ldots, N_m$ are reused as weight functions, and one obtains

$$\int_{\Omega} (\Delta\tilde{u} + \lambda\tilde{u}) N_j \, d\Omega = 0 \qquad \forall j = 1, \ldots, m. \tag{6}$$

Applying Green's formula and inserting (5) the Laplace eigenvalue problem is stated as the eigenvalue problem

$$\sum_{i=1}^{m} u_i \int_{\Omega} \nabla N_i \cdot \nabla N_j \, d\Omega = \lambda \sum_{i=1}^{m} u_i \int_{\Omega} N_i N_j \, d\Omega \qquad \forall j = 1, \ldots, m, \tag{7}$$

that is equivalent to the matrix eigenvalue problem

$$Au = \lambda M u \tag{8}$$

with the matrix coefficients

$$a_{ij} = \int_{\Omega} \nabla N_i \cdot \nabla N_j \, d\Omega \quad \text{and} \quad m_{ij} = \int_{\Omega} N_i N_j \, d\Omega.$$

5.2 Discrete Exterior Calculus (DEC) Discretization

A general framework for exterior calculus on discrete manifolds is the Discrete Exterior Calculus (DEC) framework introduced in [16]. It has been used for a variety of tasks in a variety of settings such as mesh processing [27], fluid simulation [9], and others, and is conceptually straightforward. Since it is less well known than the FEM approach, we provide a brief overview of the fundamental concepts in the following.

In the DEC framework, the geometry of discrete manifolds is described using simplicial complexes, which are constructed from simplices. k-simplices have $k+1$ vertices, that is 0-simplices are vertices, 1-simplices represent edges, 2-simplices are triangles, and 3-simplices are tetrahedra of a mesh. We make use of this framework to investigate a discrete version of the smooth Laplace-de Rham operator $\Delta = \mathrm{d}\delta + \delta\mathrm{d}$.

A central concept in exterior calculus is the concept of differentiable k-forms, differentiable forms that capture the notion of integrability over (sub-)manifolds of corresponding dimension (cf. [1, 12]). For example, a 1-form—which can be used to describe a vector field—can be integrated over a 1-dimensional submanifold or curve, thus describing a line integral.

The DEC framework is intimately tied to this notion and directly represents the integrals of k-forms for each k-simplex of the discrete manifold, i.e. one scalar value is assigned to each vertex, edge, triangle, and tetrahedron of a discretized manifold in 3-space. Therefore, scalar functions can be mapped to scalar values on vertices, vector valued functions as 1-forms are mapped to scalar values on edges and so on.

The implementation of the exterior derivative d uses Stokes' theorem, giving a unique relationship between the exterior derivative d and the boundary ∂ of a k-form ω on a region σ with

$$\int_\sigma d\omega = \int_{\partial\sigma} \omega .$$

In other words, the evaluation of the exterior derivative d can be interpreted as the evaluation on the boundary of the same simplex.

Thus, the exterior derivative d on k-forms can be computed as a $K^{k+1} \times K^k$-matrix D^k, where K^i is the number of i-simplices in the simplicial complex. This matrix D^k is the transposed incidence matrix of $k+1$-simplices and k-simplices.

The implementation of the DEC framework requires a second entity from exterior calculus, the Hodge star $\star$. In a simplicial complex, each cell has a dual cell. For example, in a tetrahedral mesh contained in a 3-dimensional embedding, each tetrahedron has a dual vertex. The Hodge star maps a k-form to the complementary $(n-k)$-form on the corresponding dual cell. Since it is natural to require an integral to be proportional to the volume of its domain of integration, we can define the Hodge star with the relation of dual and primal volumes,

$$\star = \frac{\mathrm{vol(dual)}}{\mathrm{vol(primal)}} .$$

In matrix representation, this gives a diagonal $K^k \times K^k$-matrix H^k with the fractions of dual and primal volumes as matrix entries.

The application of all possible exterior derivatives and Hodge stars to each k-form results in a discrete version of the de Rham complex:

$$\begin{array}{ccccccccccc} 0 \longrightarrow & \Omega^0 M & \xrightarrow{d} & \Omega^1 M & \xrightarrow{d} & \Omega^2 M & \xrightarrow{d} \dots \xrightarrow{d} & \Omega^n M & \longrightarrow 0 \\ & \downarrow \star & & \downarrow \star & & \downarrow \star & & \downarrow \star & \\ 0 \longleftarrow & \Omega^n M & \xleftarrow{d} & \Omega^{n-1} M & \xleftarrow{d} & \Omega^{n-2} M & \xleftarrow{d} \dots \xleftarrow{d} & \Omega^0 M & \longleftarrow 0. \end{array}$$

This finally allows the definition of the co-derivative δ for every $(k+1)$-form as the inverse mapping to the exterior derivative d with

$$\delta = (-1)^{n \cdot k + 1} \star d \star .$$

Finally, the discrete Laplace-de Rham operator L, approximating the smooth Laplace-de Rham operator $\Delta = \mathrm{d}\delta + \delta\mathrm{d}$, can be implemented by a concatenation of matrices D and H on the primal and dual simplicial complex. Extended descriptions about DEC and its implementations can be found in [3, 8].

5.3 Comparison of FEM and DEC Discretizations

It is not a priori obvious which discretization—FEM or DEC—approach gives the best results for harmonic analysis of vector fields as described above. Furthermore, it can be shown that there is no discretized version of the Laplacian operator that captures all the properties of the corresponding continuous one [28]. Hence, we evaluate different discretization with respect to their suitability towards harmonic analysis in a discrete setting.

To obtain a qualitative understanding of the approximation qualities of FEM and DEC that can inform the choice of discretization method used in local feature analysis, we define a simple test problem. We consider a rectangular region in the plane $\Omega = [0, a] \times [0, b]$ with edge lengths $a, b > 0$ and the scalar eigenvalue problem

$$\Delta u - \lambda u = 0 \qquad \text{in } \Omega \tag{9}$$

$$u = 0 \qquad \text{on } \partial\Omega \tag{10}$$

as a test case for which an analytic solution is known. This problem can be understood as a tensor product of the Fourier decomposition described in Sect. 3.1. Here, the eigenfunctions e_{ij} are products of sinusoidal functions and the eigenvalues λ_{ij} are sums of the corresponding one-dimensional eigenvalues, namely

$$e_{ij}(x, y) = \sin\left(\frac{i\pi x}{a}\right) \sin\left(\frac{j\pi y}{b}\right),$$

Table 1 The eigenvalues of the finite element and the discrete exterior calculus discretizations on increasing grid resolutions are compared to the ones known from calculus. The finite element method is found to deliver more accurate results

Analytic	FEM				DEC			
	11 × 11	25 × 25	51 × 51	101 × 101	11 × 11	25 × 25	51 × 51	101 × 101
2	20,441	20,076	20,017	20,004	21,234	21,514	21,560	21,570
5	52,408	50,412	50,095	50,023	51,808	53,559	53,847	53,912
8	87,147	81,223	80,281	80,070	81,002	85,349	86,076	86,240
10	108,610	101,456	100,334	100,083	99,416	106,356	107,518	107,781

$$\lambda_{i,j} = \pi^2 \left[\left(\frac{i}{a}\right)^2 + \left(\frac{j}{b}\right)^2 \right],$$
$$i, j = 1, 2, \ldots.$$

In the following, we compare the numerically approximated eigenvalues and eigenfunctions obtained by the FEM and DEC approaches to these analytical solutions.

Table 1 gives the eigenvalues of Eq. 9, as numerically determined from the FEM and DEC discretizations of a square region $(a, b = \pi)$, with increasing grid resolution. We observe that the convergence to the correct eigenvalue known from the analytic solution is better for the finite element method. Furthermore, even for relatively higher grid resolution the DEC approach seems to diverge to a slightly different value. This is an important observation for the situation in which eigenvalues are used for harmonic analysis (cf. Sect. 4.1).

Figure 5 illustrates the eigenfunctions of Eq. 9. The first eigenfunction as well as one of the two fourth eigenfunctions are shown using a color map on the left. The middle and right columns illustrate the pointwise error of the FEM and DEC approximations, respectively, using a greyscale mapping.

The error distribution over the squares is very similar for both approximation methods, and they appear qualitatively equivalent for the given test case. In general, we do not observe much difference between the FEM and DEC solution. However, for the first eigenfunction, the FEM solution is consistently better due to the property that FEM minimizes the integral error over the domain.

In summary, eigenfunctions computed by each of the two discretizations are quite similar. Therefore, the FEM approach as well as the DEC approach can be used in cases focusing on the usage of eigenfunctions. As an example, this is true for the local vector field linearity feature strength function rho_L we define in Sect. 4.1.

However, our comparison shows a clear advantage of the FEM discretization in approximating eigenvalues. For this reason, in cases relying on accurate eigenvalues the FEM approach should be preferred.

In a general setting, there are also other factors to consider when choosing a discretization for the Laplacian. While applying different boundary conditions is often possible in a finite element approach, the adaptation to arbitrary embeddings is not trivial. Conversely, this embedding is trivial in the discrete exterior calculus

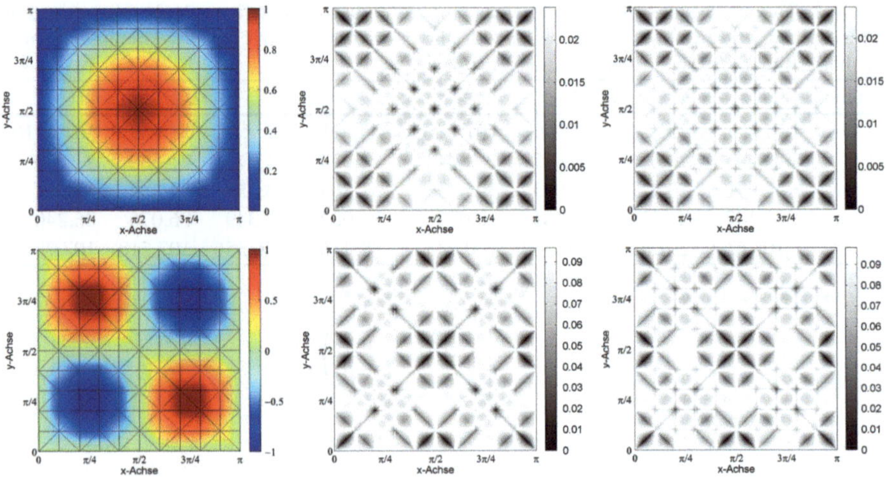

Fig. 5 The point-wise error of the finite element field solutions (*middle*) and the exterior calculus field solutions (*right*) compared to the analytic solution (*left*) on a 11 × 11-grid

setup, since its formulation only depends on simplex incidences and automatically handles cases like a two-dimensional surface in an arbitrary three-dimensional embedding.

5.4 Computation of Large Eigenvalue Sets

Solving a matrix eigenvalue system for its eigenvalues and eigenvectors is typically a time-consuming task. In addition, the amount of memory required to represent the corresponding matrices grows quadratic with the number of grid points or variables. This makes harmonic field analysis computationally expensive and must be considered a bottleneck—especially for global approaches where many eigenvectors must be computed.

Vallet and Lévy [27] combine the shift-and-invert method with band-by-band computation for manifold harmonics. This method can also be used for harmonic field analysis since it is a general method applicable to any matrix.

In a nutshell, the shift-and-invert method is a common spectral transformation. Instead of computing the eigenvalues and eigenvectors around an eigenvalue σ, the spectrum can be shifted such that only computing the eigenvalues around zero is needed by the examination of $A - \sigma I x$ instead of the initial matrix A. Since most algorithms for eigenvalue computation perform best for large eigenvalues modulus, the spectrum can additionally be inverted by inverting the matrix. Finally, not the eigenvalue λ is computed but the modified eigenvalue Θ with

$$A_\sigma^{-1} x = (A - \sigma I)^{-1} x = \Theta x, \qquad \Theta = \frac{1}{\lambda - \sigma}.$$

To solve this system, the shift-and-invert method, e.g. as implemented in ARPACK, is used. This interface is accelerated using SUPERLU for matrix vector operations and LU matrix decomposition. There is no actual need to compute and store the inverted matrix, because the computation only requires the evaluation of $x = A_\sigma^{-1}z$ for an arbitrary vector z. This can be computed very efficiently with the LU decomposition of A_σ by performing a back-substitution.

The entire spectrum corresponding to the Laplacian discretization matrix can then be computed in multiple parts. A small band of eigenvalues is computed at once; successively, this band is shifted with a small overlap until the entire spectrum is completed. With this technique, the sub-linear behavior of eigenvalue solvers can also be compensated.

In our experiments, the global eigenvalue problem for the inflow dataset with 36,000 edge values was solved in about 2 h on a commodity PC (Intel Q6600 quad-core machine at 2.4 GHz). Also applying classical eigenvalue solvers, we were only able to determine the first 3,000 eigenvectors in about 3 days of computation time. In comparison, for the two-dimensional local field analyses examples in Sect. 4.1 all the six to ten vector field basis functions were determined in less than 100 ms on the same hardware.

6 Conclusion

Harmonic analysis techniques are a fundamental tool in scalar field processing and analysis. Well-known applications are techniques based on the Fourier decomposition. In this chapter, we provide a number of ideas on applying harmonic analysis on scalar, vector and tensor fields over general domains, with a focus on visualization and analysis applications.

The typical global approach to harmonic analysis was illustrated on a small vector field example given an unstructured grid, and the problems for global approaches were identified and discussed. These issues led us to the introduction of a local approach using local regions for every point in space. We used this local approach to define a local feature strength measure, based on properties described in Laplacian eigenbases.

Since two direct possibilities for the choice of the discretization are apparent, we gave an overview of both methods and compared their behavior numerically. In addition, we discussed numerical aspects of the eigenvalue computation that has proven difficult to master in our experiments.

For the future, there are still many open avenues of investigation. Besides improving the computational system in accuracy, convergence and computational costs, for special cases there might be explicit algorithms not depending on a numerical eigensolver. Furthermore, other possibilities for feature measures going beyond the local vector field linearity criterion should be explored, and it is conceivable to apply the described technique to achieve pattern matching of empirically defined features. Finally, the adaptation of existing signal and image processing algorithms should also be considered.

References

1. Bishop, R., Goldberg, S.: Tensor Analysis on Manifolds. Dover Publications, New York (1968)
2. Demmel, J.W., Gilbert, J., Li, X.S.: SuperLU users' guide. Tech. rep. CSD-97-944, University of California (1997)
3. Desbrun, M., Kanso, E., Tong, Y.: Discrete differential forms for computational modeling. In: SIGGRAPH '06: ACM SIGGRAPH 2006 Courses, pp. 39–54. ACM, New York (2006)
4. Dong, S., timo Bremer, P., Garl, M.: Spectral surface quadrangulation. ACM Trans. Graph. **25**, 1057–1066 (2006)
5. Dragomir, S., Perrone, D.: Harmonic Vector Fields: Variational Principles and Differential Geometry. Elsevier Science Ltd, Oxford (2011)
6. Ebling, J., Scheuermann, G.: Clifford convolution and pattern matching on vector fields. In: Proceedings of the 14th IEEE Visualization 2003 (VIS'03), p. 26. IEEE Computer Society, Piscataway (2003)
7. Ebling, J., Scheuermann, G.: Clifford Fourier transform on vector fields. IEEE Trans. Vis. Comput. Graph. **11**(4), 469–479 (2005)
8. Elcott, S., Schröder, P.: Building your own DEC at home. In: SIGGRAPH '05: ACM SIGGRAPH 2005 Courses, p. 8. ACM, New York (2005)
9. Elcott, S., Tong, Y., Kanso, E., Schröder, P., Desbrun, M.: Stable, circulation-preserving, simplicial fluids. ACM Trans. Graph. **26**, 4-es (2007). doi:http://doi.acm.org/10.1145/1189762.1189766
10. Evans, L.C.: Partial Differential Equations. Graduate Studies in Mathematics. Oxford University Press, Oxford (1998)
11. Fisher, M., Schröder, P.: Design of tangent vector fields. ACM Trans. Graph. **26**, 56 (2007)
12. Flanders, H.: Differential forms with applications to the physical sciences. Dover Publications, Mineola (1989)
13. Fletcher, C.: Computational Galerkin Methods. Springer Series in Computational Physics. Springer, New York (1984)
14. Galerkin, B.G.: On electrical circuits for the approximate solution of the laplace equation. Vestnik Inzh. **19**, 897–908 (1915)
15. Heath, M.: Scientific Computing. McGraw-Hill, Boston (2002)
16. Hirani, A.: Discrete exterior calculus. Ph.D. thesis, California Institute of Technology (2003)
17. Huebner, K., Dewhirst, D., Smith, D., Byrom, T.: The Finite Element Method for Engineers. Wiley India Pvt. Ltd., New York (2008)
18. Katznelson, Y.: An Introduction to Harmonic Analysis. Cambridge Mathematical Library. Cambridge University Press, Cambridge/New York (2004)
19. Kreyszig, E.: Introductory Functional Analysis with Applications. Wiley, New York (1978)
20. Lehoucq, R., Sorensen, D.C., Yang, C.: Arpack users guide: solution of large-scale eigenvalue problems with implicitly restarted Arnoldi methods. Communication **6**(3), 147 (1998). Tech. Rep., SIAM, Philadelphia. URL citeseer.ist.psu.edu/article/lehoucq97arpack.html
21. Loan, G., Golub, G.: Matrix Computations. Johns Hopkins University Press, Baltimore (1996)
22. Reuter, M., Wolter, F., Peinecke, N.: Laplace-spectra as fingerprints for shape matching. In: Proceedings of the 2005 ACM Symposium on Solid and Physical Modeling, Cambridge, p. 106. ACM, New York (2005)
23. Schlemmer, M., Heringer, M., Morr, F., Hotz, I., Hering-Bertram, M., Garth, C., Kollmann, W., Hamann, B., Hagen, H.: Moment invariants for the analysis of 2D flow fields. IEEE Trans. Vis. Comput. Graph. **13**(6), 1743 (2007)
24. Sorensen, D.: Implicitly Restarted Arnoldi/Lanczos Methods for Large Scale Eigenvalue Calculations. Institute for Computer Applications in Science and Engineering, Hampton. In: Keyes, D.E., Sameh, A., Venkatakrishnan, V. (eds.) Contractor, pp. 1–34. Kluwer, New York (1996)

25. Taubin, G.: A signal processing approach to fair surface design. In: Proceedings of the 22nd Annual Conference on Computer Graphics and Interactive Techniques, pp. 351–358, ACM, New York (1995)
26. Tong, Y., Alliez, P., Cohen-Steiner, D., Desbrun, M.: Designing quadrangulations with discrete harmonic forms. In: Proceedings of the Fourth Eurographics Symposium on Geometry Processing, SGP '06, pp. 201–210. Eurographics Association, Aire-la-Ville (2006). URL http://dl.acm.org/citation.cfm?id=1281957.1281983
27. Vallet, B., Lévy, B.: Spectral geometry processing with manifold harmonics. Comput. Graph. Forum (Proceedings Eurographics) **27**, 251–260 (2008)
28. Wardetzky, M., Mathur, S., Kälberer, F., Grinspun, E.: Discrete laplace operators: no free lunch. In: SIGGRAPH Asia '08: ACM SIGGRAPH ASIA 2008 Courses, pp. 1–5. ACM, New York (2008)

Index

D.H. Laidlaw and A. Vilanova (eds.), *New Developments in the Visualization and Processing of Tensor Fields*, Mathematics and Visualization,
DOI 10.1007/978-3-642-27343-8, © Springer-Verlag Berlin Heidelberg 2012

MIX
Papier aus verantwortungsvollen Quellen
Paper from responsible sources
FSC® C105338

If you have any concerns about our products,
you can contact us on
ProductSafety@springernature.com

In case Publisher is established outside the EU,
the EU authorized representative is:
Springer Nature Customer Service Center GmbH
Europaplatz 3, 69115 Heidelberg, Germany

Printed by Libri Plureos GmbH
in Hamburg, Germany